Oceanography

Edited by **Theodore Roa**

R CALLISTO REFERENCE

New York

Published by Callisto Reference,
106 Park Avenue, Suite 200,
New York, NY 10016, USA
www.callistoreference.com

Oceanography
Edited by Theodore Roa

International Standard Book Number: 978-1-63239-488-0 (Hardback)

Printed in the United States of America.

Contents

Permissions

List of Contributors

Preface

This book is a single binding account covering all the major aspects of oceanography. Arthur C. Clarke once said, "How inappropriate to call this planet Earth when it is quite clearly Ocean". The origin of life stems from the oceans and the entire diversity of life across the world is modulated by marine and oceanic processes. All the processes ranging from micro-scale, like coastal processes, to macro-scale, like those in the seas, the oceans and the marine life, play a vital role in maintaining the equilibrium of this planet, both from chemical as well as physical viewpoints. Since the beginning of time, the exploration and discovery of world's oceans has brought advancements and immense knowledge to humankind. The metaphors of Jason and Ulysses represent the cultural progress achieved through these discoveries and explorations. Contemporary oceanographic research portrays one of the most crucial frontiers of knowledge about Earth, depending on the exploration of oceans. Therefore, it is closely related to the advancement of novel technologies. Additionally, other social and scientific disciplines can also contribute several basic inputs for a conclusive elucidation of the complete ocean ecosystem. Therefore, this book will greatly help the readers to learn better ways to preserve the Blue Planet: Our Earth.

Significant researches are present in this book. Intensive efforts have been employed by authors to make this book an outstanding discourse. This book contains the enlightening chapters which have been written on the basis of significant researches done by the experts.

Finally, I would also like to thank all the members involved in this book for being a team and meeting all the deadlines for the submission of their respective works. I would also like to thank my friends and family for being supportive in my efforts.

Editor

Part 1

Methods in Oceanography

New Technological Developments for Oceanographic Observations

Marco Marcelli, Andrea Pannocchi,
Viviana Piermattei and Umberto Mainardi
University of Tuscia, Laboratory of Exerimental Oceanology and Marine Ecology
Italy

1. Introduction

Measurement is the foundation of any branch of science, and no less so in oceanography (Thorpe, 2009).
Radical changes in early instruments that were largely mechanical, have been realized due to electronic, computing and data transmission advances.
These advances produced new ways in oceans observing and monitoring which improve operational and forecasting oceanography to better understand the oceans.
Operational forecasting of marine physical and biochemical state variables is now becoming an important tool for modern management and protection of the oceans and their living resources (Marcelli et al., 2007).
Despite all these rapid advances in ocean measuring capabilities, the number of variables necessary to solve oceanographic problems is still big and increasing, generating a continuous gap between the available instruments and what we want to measure.
In addition, the time and space scales of key processes span over ten orders of magnitude; to solve the risk of undersampling it is necessary to expand the rates of data acquisition, temporal coverage and spatial coverage.
The recent advances in ocean monitoring system and new sampling strategies will going to face all these problems. This modern approach is based on the use of in situ autonomous sampling together with satellite observations, integrating SOOP (Ship Of Opportunity) monitoring programs. Moreover, programs like the Pan European Seadatanet (www.seadatanet.org) are set up to promote the use of common vocabularies in datasets in order to allow interoperability and data exchange.

2. State of the art

Oceanographic investigations, traditionally, are both limited and strictly dependent by time and space scales and by sensors technological developments (Dickey & Bidigare, 2005).
To observe oceanographic phenomena and processes, several and different methods can be utilized providing different results.
Even if some methods still have remained almost unvaried, even if they have been improved and perfected (as for example the continuous plankton recorder, CPR), in many cases, technological developments lead us to a noteworthy advancing of the knowledge of processes and phenomena, formerly undetectable and just intuitable.

2.1 Platforms and sensors

Relating to a defined phenomena, observation has to be carried out with the appropriated method in order to detect processes and trends of such phenomena. To this aim, platforms and sensors have to be choose as in respect to the time and space scale of the phenomena, as taking into account the nature (e.g. chemical, physical or biological) of what we are looking to (Dickey & Bidigare, 2005). The environment (e.g. shallow water, deep sea or coastal areas) is also fundamental to the choice of the adequate sampling sensors and operative methods.

Many physical, chemical and biological aspects can be described by means of punctual observations, e.g. stand alone moored instruments or oceanography buoys and multi-purpose platforms. Also, dedicated oceanographic surveys can be set up in order to study a particular aspect of a defined marine area.

In order to reduce operative costs and to enhance spatial resolution of data for water column characterization, towed vehicles can be utilized to continuous measure of physical, chemical and biological variables, along water column and along horizontal trajectories. Also the use of expendable probes and the ship of opportunity, are fundamental to this aim.

From an operative point of view, observation platforms can be divided basically into the following categories:

Ships (Research Vessels RV);

Mooring;

Underwater vehicles and lowered devices (Remote Operated Vehicles ROV; Towed Vehicles; Crewed Deep Submersible Vehicles; Autonomous Underwater Vehicles; Gliders; Lagrangian buoys; Expendable probes; Drifters)

Voluntary Observing Ships (VOS);

Remote sensing platforms (Airplains and Satellites);

2.1.1 Ships

Oceanographic research vessels can be considered shipborne platform as they can be utilized to conduct many kind of scientific investigations, such as mapping and charting, marine biology, fishery, geology and geophysics, physical processes, marine meteorology, chemical oceanography, marine acoustics, underwater archaeology, ocean engineering, and related fields (Dinsmore, 2001 in Thorpe, 2009).

The United Nations International Maritime Organization (IMO) has established a category of Special Purpose Ship which includes ships engaged in research, expeditions, and survey. Scientific personnel are also defined as "persons who are not passengers or members of the crew, who are carried on board in connection with the special purpose" (www.imo.org).

The design of a modern Research Vessel can vary significantly in respect to the research activities. Typical caracteristics can be as follow described:

a.   RV have to be as general as possible in order to allow multidisciplinary studies and researches;

b.   the size is determined by the requirements, but the length-over-all should not exeed 100 m LOA;

c.   speed of 15 knots cruising should be sustainable through sea state 4 (1.25–2.5 m) and the seakeeping should be able to maintain science operations in the following speeds and seastates:

-   15 knots cruising through sea state 4 (1.25– .5 m);
-   13 knots cruising through sea state 5 (2.5–4 m);
-   8 knots cruising through sea state 6 (4–6 m);

- 6 knots cruising through sea state 7 (6–9 m).
d. Work environment: lab spaces and arrangements should be highly flexible to accommodate large, heavy, and portable equipments.
e. Suite of modern cranes, in order to reach all working deck areas and offload vans and heavy equipment, and in order to work close to deck and water surface.
f. Oceanographic winches permanently installed should provide a wire monitoring systems with inputs to laboratory panels allowing local and remote data and operational controls.

The above general characteristics, can be modulate relating to the particular purpose of the vessels itself.

Therefore it is more convenient and appropriate divides RV in the following categories:

General Pourpose Vessels

Multidiscipline Ships represent the classic oceanographic research vessels and are the dominant class in terms of numbers today. Current and future multidiscipline oceanographic ships are characterized by significant open deck area and laboratory space. Also accommodations for scientific personnel are greater than for single purpose vessels due to the larger science parties carried. Flexibility is an essential feature in a general purpose research vessel.

Mapping and Charting Vessels

These group of ships were probably the earliest oceanographic vessels, traditionally involved in exploration voyage. Surveys were carried out using wire sounding, drags, and launches. Survey vessels are also characterized by less deck working space than general purpose vessels. Modern survey vessels, however, are often expected to carry out other scientific disciplines. Winches, cranes and frames can be observed on these ships.

Fisheries Research Vessels

Fisheries research generally includes studies on environment, stock assessment, and gear testing and development.

The first of these are carried out by traditional surveys to collect biological, physical, and chemical parameters of sea surface and water column as well as geological informations. These surveys can be accomplished from a general purpose oceanographic research vessel.

Geophysical Research Vessels

The purpose of marine geophysical research vessels is to investigate the sea floor and sub-bottom, oceanic crust, margins, and lithosphere ranging from basic research of the Earth's crust to resources exploration. The highly specialized design requirements for a full-scale marine geophysics ship usually precludes work in other oceanographic disciplines.

Polar Research Vessels

Polar research vessels are defined by their area of operations. The special requirements defining a polar research vessel include increased endurance, usually set at 90 days, helicopter support, special provisions for cold weather work, such as enclosed winch rooms and heated decks, and icebreaking capability.

Support Vessels

These include vessels that support submersibles, ROVs, buoys, underwater habitats, and scientific diving.

Other Classes of Oceanographic Research Vessels

Into this category it can be include research ships which serve other purposes as ocean drilling and geotechnical ships, weather ships, underwater archaeology, and training and education vessels.

2.1.2 Moorings
Much of what we know about the oceans processes is the result of ship-based expeditionary science, dating back to the late 19th century.

It is clear that, to answer many important questions about oceans and Earth science, it is necessary a co-ordinated research effort based on long term investigations (Favali & Beranzoli, 1996).

One of the most important aspect of sea investigation, in fact, deals with the possibility to obtain continuous data from fixed stations or from a net of fixed stations, fundamental for forecasting systems (coupled with meteorological data and time data series).

Mooring platforms, both for upper and deep water, allow continuous observations of phenomena of very different disciplines like geophysical and biological once. World wide initiatives and programs, such as European Seafloor Observatory Net (ESONET NoE) and the European Multidisciplinary Seas Observatory (EMSO) have been developed in order to increase the capacity in the research, with the purpose of better understanding of physical, geological, chemical, ecological, biological and microbial processes, that take place in the oceans: from the surface down to the highest depth.

Much of our latest knowledge, infact, stems from studying the seafloor: its morphology, geophysical structure, and characteristics, and the chemical composition of rocks collected from the ocean floor.

Furthermore, in the late 1970s, at the midocean ridge (MOR) crest, deep sea biological observations led to the discoveries of deep sea 'black smoker' hydrothermal vents and to their chemosynthetic-based communities.

Such discovery changed the biological sciences, provided a quantitative context for understanding global ocean chemical balances. Deep sea observatory are also fundamental in the study of the physical oceanography of the global ocean water masses and their chemistry and dynamics (Moran, 2009 in Thorpe, 2009).

Traditionally, fixed platforms consist of one surface buoyant unit moored on sea bottom by means of an instrumented chain. Most often surface unit is equipped with meteorological sensors, while along the chain (from the surface till the bottom) are mounted many different sensors in order to collect water column variables.

A mooring design can vary relating both to the kind of investigation and to physical characteristics of the environment of destination (coastal or open ocean).

Fixed measurement platforms have the following features and play the following roles:

1) Stand alone data collecting system allowing real time or quasi-real time data acquisitions and transmission

Surface platforms are traditionally able to store sampled data into an internal storage devices, and most often they are equipped with a communication system that allows data transmission to data centers. Typical systems involve satellite (open ocean), GSM or H3G phone systems but also radio, ethernet or LAN communication can be utilized.

2) Operational forecasting system

Continuous data coming from a buoy or a buoy network provide fundamental input for ocean and weather forecasting models.

3) Remote sensing calibration system

In situ data can represent the sea-truth for remote sensing as they are provided from standard and calibrated instruments. Because of the high maintenance possibility and the continuos data control, fixed platforms and mooring buoys can serve successfully to this aim. The calibration of a moored buoy, for example, can be ongoing, with retrieval at certain intervals to check sensor degradation and biofouling.

4) Data network

The importance of realizing a network of buoys is related to the possibility to put all the information of each platform together, in order to provide an even more rich data base. It is necessary not only for the oceanographic research and monitoring, but also for climate and global change investigations.

Following there are reported some of important buoy networks.

a) An important example is represented by the United States National Data Buoy Center (NDBC). In the 1960's, there were approximately 50 individual United States programs conducted by a variety of ocean-oriented agencies. These programs were consolidated into the National Data Buoy Center under the control of the US Coast Guard. Nowadays the United States NDBC manages the development, operations, and maintenance of the national data buoy network. It serves as the NOAA focal point for data buoy and associated meteorological and environmental monitoring technology. It provides high quality meteorological/environmental data in real time from automated observing systems (www.ndbc.noaa.gov).

It also manages the Volunteer Observing Ship (VOS) program to acquire additional meteorological and oceanographic observations supporting the NOAA National Water Service (NWS) mission requirements. It operates the NWS test center for all surface sensor systems. It maintains the capability to support operational and research programs of NOAA and other national and international organizations.

b) The Australian National Moorings Network of the Integrated marine Observing System (www.imos.org.au) comprises a series of national reference stations and regional moorings designed to monitor particular oceanographic phenomena in Australian coastal ocean waters.

Nine national reference stations are located around the Australian coasts in order to provide a baseline informations, decadal time series of the physical and biogeochemical properties of Australia's coastal seas, to inform research into ocean change, climate variability, ocean circulation and ecosystem responses.

Each National Reference Station consists of a mooring with sensors for conductivity, temperature, depth, fluorescence, dissolved oxygen, photosynthetically available radiation (PAR), fluorescence and measurement of turbidity at three depths: the surface, seabed and an intermediate depth. On the seafloor, acoustic doppler current profilers (ADCPs) are also deployed. All reference stations telemeter a reduced data set via Iridium satellite for real time monitoring. Boat-based water sampling is also undertaken at each of the reference stations on a monthly basis. These samples are analysed for nutrients, plankton species, both visibly and genetically, and pCO2.

A number of the National Reference Station moorings are also equipped with passive acoustic listening arrays, containing sea noise loggers to record sounds in the ocean. Furthermore, three National Reference Stations are equipped with three instruments determining surface CO2, dissolved oxygen, temperature and salinity.

Regional moorings can measure physical, chemical and biological parameter of sea water. These moorings can hold a range of instrumentation including acoustic doppler current profilers (ADCPs), Water Quality Meters (WQMs), fluorometers, instruments to measure turbidity, dissolved oxygen, photosynthetically active radiation (PAR), nutrients, pCO2, dissolved inorganic carbon, total alkalinity, as well as imaging flow cytometry, spectroradiometry, profiling conductivity-temperature-depth instrumentation, laser In- situ scattering and transmissometry.

The regional moorings monitor the interaction between boundary currents and shelf water masses and their consequent impact upon ocean productivity and ecosystem distribution and resilience. Operation of the moorings network facility is coordinated nationally and distributed between several sub-facilities.

d) The Italian National Mareographic Net (RMN) was set up by the National Mareographic System. It consists of 33 measuring stations uniformly positioned throughout the entire national seas.

The main feature of Italian RMN is the measure of sea level. Each buoy is equipped with a sea level microwave sensor coupled with another one level sensor with back-up function. In addiction there is the traditional ultrasound hydrometric sensor working since 1998. Each buoy also carries a meteorological station and a sea water temperature sensor. Ten buoys are equipped with a multiparametric probe in order to measure temperature, pH, conductivity and redox parameters.

All data collected are available for historical series updates, real time observations, astronomical tides forecast and for scientific investigations. (www.mareografico.it).

Other important examples are represented by the deep sea observatories.

a) The Circulation Obviation Retrofit Kit (CORK) is a seafloor observatory that measures pressure, temperature, and fluid composition, important parameters for the study of the dynamics of deep-sea hydrologic systems. CORKs are installed by the International Ocean Drilling Program (IODP) for measurements over long periods of time (months to years) (Moran, 2009in Thorpe, 2009)

b) The MARS Observatory of Monterey Bay Aquarium Research Institute (MBARI) is the first deep-sea ocean observatory offshore of the continental United States and cosists of a metal pyramid on the seafloor off the coast of Central California at 900m depth. Working since 2008, it is involved in the FOCE experiment (Free-ocean carbon dioxide enrichment) to study the effects of increased carbon dioxide concentrations in seawater on marine animals (www.mbari.com). The heart of observatory consists of two titanium pressure cylinders packed with computer networking and power distribution equipment. These cylinders are nested within a protective metal pyramid on the deep seafloor. This central hub is connected to shore by a 52-kilometer-long cable that can carry up to 10,000 watts of power and two gigabits per second of data. Most of the cable is buried a meter below the seafloor.

c) The Ocean Observatory Initiative (OOI), a project funded by the National Science Foundation, is planned as a networked infrastructure of science-driven sensor systems to measure the physical, chemical, geological and biological variables in the ocean and seafloor (www.oceanobservatories.org). The OOI will be one fully integrated system collecting data on coastal, regional and global scales. Three major Implementing Organizations are responsible for construction and development of the overall program. Woods Hole Oceanographic Institution and its partners, Oregon State University and Scripps Institution of Oceanography are responsible for the coastal and global moorings and their autonomous vehicles. The University of Washington is responsible for cabled seafloor systems and moorings. The University of California, San Diego, is implementing the cyberinfrastructure component. Rutgers, The State University of New Jersey, with its partners University of Maine and Raytheon Mission Operations and Services, is responsible for the education and public engagement software infrastructure.

The OOI will consist of six arrays with 56 total moorings and 763 instruments. Moored platforms (Fig.1) provide oceanographers the means to deploy sensors at fixed depths between the seafloor and the sea surface and to deploy packages that profile vertically at one location by moving up and down along the mooring line or by winching themselves up and down from their point of attachment to the mooring.

An oceanographic mooring is anchored to the sea floor by a mooring line extending upward from the anchor to one or more buoyant floats, which can be located in the water column or at the sea surface.

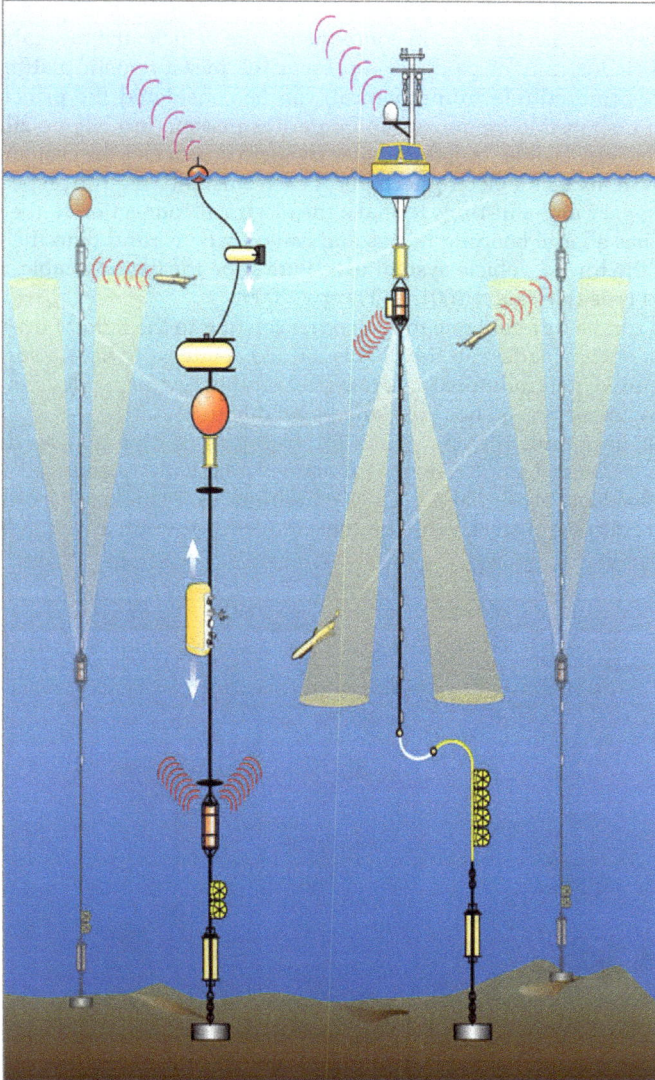

Fig. 1. Ocean Observatory Initiative moored platforms. (Credit: Jack Cook, Woods Hole Oceanographic Institution)

2.1.3 Underwater vehicles and lowered devices
Underwater vehicles can be considered as platforms that combines the advantages of a ship-mounted instruments collecting data while moving, and a lowered device that profiles the water column.

Underwater Vehicles can be divided into three categories related with their moving capabilities: towed, remoted operated and autonomous. Traditionally they are called Towfish or Towed body, ROV and AUV respectively.

Towed vehicles

A towed vehicle system has three main components: the vehicle, the tow cable and a winch. Vehicles, with their instrument payload, represent the measurement platforms. The cable, whenever double-armored (electromechanical) can be considered the principal part of the towfish. Cable is responsible for power and data trasmission, but, above all, because of its total cross sectional area, cable drag dominates the performance of the system.

Towed-vehicle systems using electromechanical cables usually require a special winch with accurate spooling gear and slip-rings to make the electrical connection to the rotating drum. The faired cable has a large bending radius and can only be wound onto the winch drum in a single layer. If the towed vehicle system uses wire rope for the tow cable, then the winch can be a standard type (Helmond 2001, in Thorpe, 2009).

Towfish payload can be different one to each other relating to the field of investigation.

Typical instrument payloads can be CTDs added with specific sensor (i.e. pH and fluorometer) required for traditional hydrological studies. Acoustic devices and biological sampler can be mounted on "ad hoc" designed towed body.

One of the most important towed vehicle for biological sampling was designed by Sir Alister Hardy in 1925, and until now it is remained relatively unchanged in its sampling mechanism (Richardson et al., 2006). Called Continuous Plankton Recorder, (CPR) this towfish is able to collect and stock zooplankton samples continuously while it is towed from a ship (Fig. 1a.).

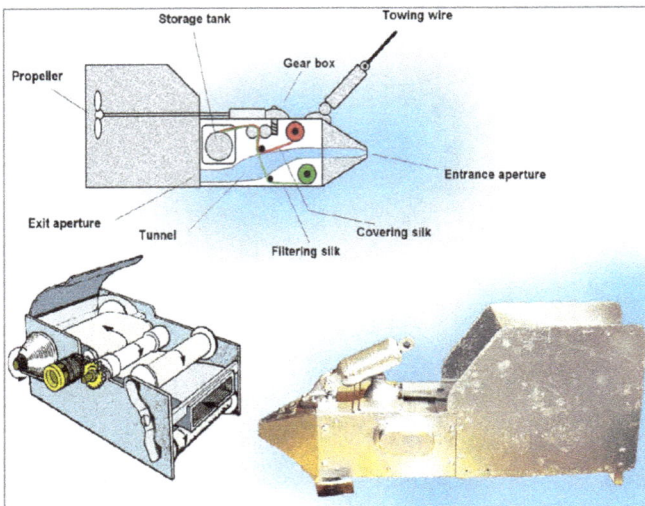

Fig. 1a. Continuous Plankton Recorder (Richardson et al., 2006).

The continuous plankton recorder (CPR) survey is the largest multi-decadal plankton monitoring programme in the world (Richardson et al., 2006). Since the prototype was deployed, until 2004, over 4 000 000 miles of towing have resulted in the analysis of nearly 200 000 samples and the routine identification of over 400 species/groups of plankton, used to study biogeography, biodiversity, seasonal and interannual variation, long-term trends, and exceptional events (John 2001, in Thorpe, 2009). Another and notably CPR most important aspect is the ability to collect hundreds of samples throughout an ocean basin because it is can be towed behind ships of opportunity (SOOPs) at their conventional operating speeds of 15–20 knots.

Recently, at the Woods Hole Oceanographic Institution (WHOI), was designed and realized a Bio-acoustic system, the BIOMAPER II (Wiebe et al. 2002). In its original conception BIOMAPER II (Fig.1b) was designed primarily for acoustic monitoring of plankton and includes both up- and down-looking acoustic transducers of different frequencies, as well as a suite of conventional environmental sensors (including conductivity, temperature, pressure, chlorophyll fluorescence and beam transmission). The upgraded vehicle has been integrated with a pair of dual path absorption and attenuation meters (AC-9, Wet Labs, Inc.), one for whole water and the other for a filtered fraction (0.2 Pm), and two spectral radiometers (OCI/OCR-200 series, Satlantic, Inc.) for measuring downwelling irradiance and upwelling radiance. The BIOMAPER II is particularly well suited to assessment of apparent optical properties during towed operation because the vehicle is designed to maintain a horizontal attitude regardless of flight pattern.

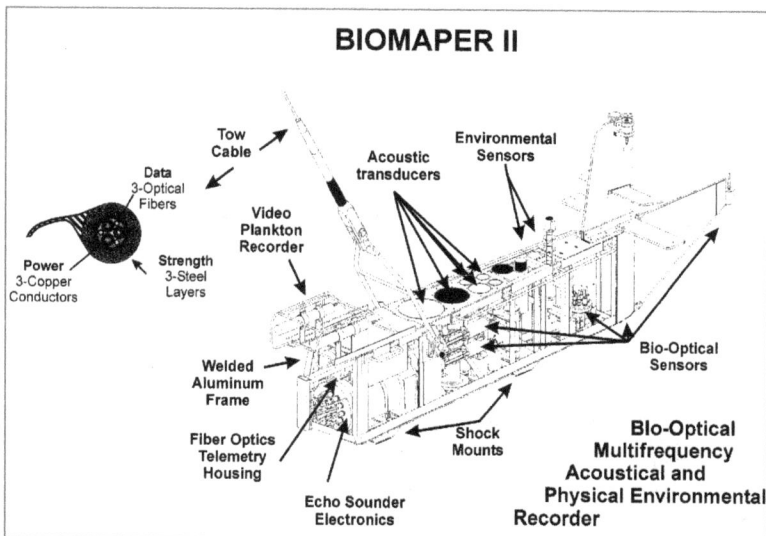

Fig. 1b. BIOMAPER II (Wiebe et al. 2002)

During the 1970s the firsts undulating vehicles were realized at the Bedford Institute of Oceanography (BIO), Canada and the Institute of Oceanographic Sciences (IOS of National Oceanography Centre, Southampton, UK). The BIO Batfish was a variable-depth towed body equipped with a CTD and various other sensors used for rapid underway profiling of the upper layers of ocean.

Its evolution is the SeaSoar (Fig.1c) developed by Chelsea Technologies Group from the original design of IOS. It is a versatile towed undulating vehicle used to deploy a wide range of oceanographic monitoring equipment. Its typical instrumentation payload is a combination of: CTD, Fluorimeter, Transmissometer, Turbidity, Bioluminescence, Irradiance meter, Nitrate/Nitrate sensor, Plankton Sampler, and SeaWifs bands sensors.

Fig. 1c. SeaSoar (www.chelsa.co.uk)

In 1990 the italian underwater undulating vehicle SARAGO (Bruzzi & Marcelli, 1990) (Fig. 2) was planned at ISMES Laboratories in collaboration with University of Tor Vergata, Rome, Italy. Its main objective was to provide quasi-synoptic measurements of ecological and physic-chemical variables in marine environment. Its main application were the basic scientific research, the environmental monitoring and the assessment of biological resources. Therefore it was created a modular system capable to support different instruments and to expand its own working operating capabilities.

SARAGO characteristic payload consists of a SBE 19 CTD for physical variables and a pressure transducer to control depth separately from CTD sensors and a double impulse fluorometer called Primprod 1.11, created in collaboration with the Biophysics Institute of Moscow University. With this configuration, the SARAGO was used in the project PRISMA2 CNR from 1995 to 1998, to continuous measure photosynthetic efficiency and PAR, and to compute high resolution estimations of phytoplankton primary production (Piermattei et al., 2006).

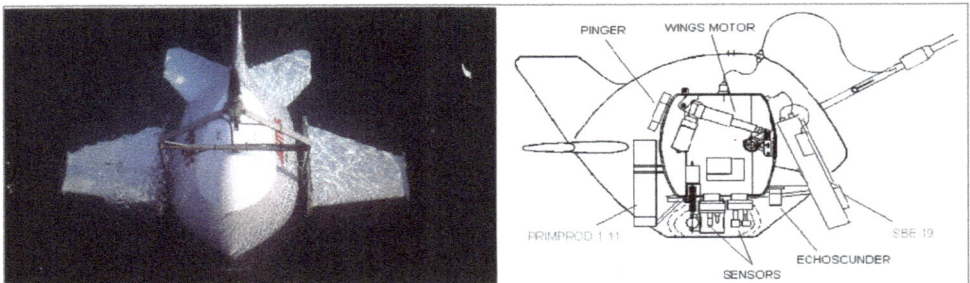

Fig. 2. SARAGO

SARAGO paths are checked by means of a pressure transducer for the wings and a Sonar. The pressure transducer allows to control depth separately from CTD sensors. The sonar works both as a path controller (in case SARAGO is programmed to navigate at constant distance from the bottom) and as a bottom alarm (Marcelli & Fresi, 1997)

A different approach was taken by the researcher of Institute for Marine Environmental Research (IMER), Plymouth, UK. They targeted ships of opportunity with the robust and compact Undulating Oceanographic Recorder. Unlike its predecessors the UOR did not require a multi core tow cable all power being generated by an onboard impeller/alternator thus removing the need for a dedicated research vessel.

The Undulating Oceanographic Recorder UOR Mark 2 is a self-contained, multirole, oceanographic sampler. It is independent of the vessel for any service and can undulate from research vessels and merchant ships at speeds up to 13.5 m. s−1 (26 knots) between the surface to depths of 55 m (at 4 m.s−1) and 36 m (at 10 m.s−1) with a preset undulation length between 800 and 4000 m.

It can record (internally on a miniature digital tape recorder, resolution 0.1% full scale) measurements by sensors for temperature, salinity, depth, chlorophyll concentration, radiant energy, or other variables (Aiken J., 1985).

Since emerging oceanographic programmes demanded large synoptic data sets over extended periods of time with inherent reliability and high data recovery, so new towed vehicles are required to accommodate larger payloads, be highly versatile yet meet or exceed the performance of existing vehicles. Improved production techniques should also reduce the manufacturing cost.

One example is represented by the ScanFish MK II (Fig. 3) of Graduate School of Oceanography, University of Rhode Island. ScanFish is a towed undulating vehicle system designed for collecting profile data of the water column in oceanographic, bathymetric and environmental monitoring applications at either fixed depth or to pre-programmed undulating flight path. The ScanFish MK II is an active unit, which generates an up or down force in order to position the tow-fish in the water column. This instrument may carry on a number of sensors measuring parameters like conductivity (salinity), temperature, pressure, fluorescence, photosynthetically active radiation (PAR) and other parameters.

Fig. 3. ScanFish (www.eiva.dk)

ROV, AUV and GLIDERS

The technology of underwater vehicles focus on seafloor research or/and deep sea observation using human occupied submersibles (i.e. Alvin) or remotely operated vehicles

(ROVs), and most recently, autonomous underwater vehicles (AUV). Despite AUVs represent the ultimate advances of oceanographic platforms, they are very expensive due to involved technologies and deployment systems.

GLIDERS, are the most recently developed class of underwater vehicles. Due to their autonomous capacity to perform saw-tooth trajectories driven by changes of buoyancy, they can be consider as low cost AUVs (Eriksen 2009 in Thorpe, 2009)

The first modern deep *Human occupied vehicles*, was the William Beebe bathysphere able to dive down to 923 m depth.

Realized in 1934, the bathysphere was a small spherical shell made of cast iron, 135 cm in inside diameter designed for two observers.

The second generation of deep submersibles began in 1964 with Alvin (Fig 4a). It was funded by the US Navy under the guidance of the Woods Hole Oceanographic Institution (WHOI). Alvin was a three-person, self-propelling, capsule-like submarine about 8 meters long that could dive up to 3.600 meters. It was equipped with underwater lights, cameras, a television system, and a mechanical manipulator.

In 1975, scientists of Project FAMOUS (French-American Mid-Ocean Undersea Study) used Alvin to dive on the Mid-Atlantic Ridge for the first direct observation of seafloor spreading. Alvin has allowed the discovering of the hot springs and associated sea life that occur along the East Pacific Rise. Moreover, it provided early glimpses of the Titanic wreck site in 1985. Since Alvin, other manned submersibles have been used successfully to explore the ocean floor.

Remotely operated vehicles (ROVs) (Fig 4b) are underwater platforms remotely controlled from the surface by a cable that provides power and control communication to the vehicle. ROVs can be divided into three main types of vehicle:

a.   free-swimming tethered vehicles have thrusters that allow manoeuvring along the three axes, and provides visual feedback through onboard video cameras. Generally their use is for mid-water or bottom observation or intervention;

b.   bottom-crawling tethered vehicles move with wheels and can only manoeuvre on the bottom. Visual feedback is provided by onboard video cameras. Their mainly use is for cable or pipeline works;

c.   towed vehicles are carried forward by the surface ship's motion, and are manoeuvred up and down by the surface-mounted winch. Towed vehicles usually carry sonar, cameras, and sometimes sample equipment.

ROVs have been developed in the 1950s and have been used worldwide since the 1970s and 1980s to explore previously inaccessible underwater supporting scientific missions.

Autonomous underwater vehicles (AUVs) (Fig. 4c) are untethered mobile platforms, computer controlled, used for survey operations by ocean scientists, marine industry, and the military. Their main feature is the completely free/or very little human operator interaction while operating. This characteristic allow AUVs to access otherwise-inaccessible regions as for example under Arctic and Antarctic ice.

Despite the off-shore oil industry represents the main field of use of AUV platforms, since 1990s the oceanographic scientific community has adopted these autonomous vehicles for oceanographic investigation as they can provide, for example, the rapid acquisition of very distributed data sets (Bellinghman, 2009 in Thorpe, 2009).

Today a wide range of AUV are available for commercial use as well as for oceanographic surveys, and a large number of companies develop subsystems and sensors to install onboard AUVs.

AUVs are highly integrated devices, containing a variety of mechanical, electrical, and software subsystems that can be summarized as follow:

1. hardware and software components in order to program specific tasks;
2. energy storage to provide power;
3. propulsion system;
4. stability control system;
5. measurement instruments
6. communication devices;
7. locating devices;
8. emergency recovery system.

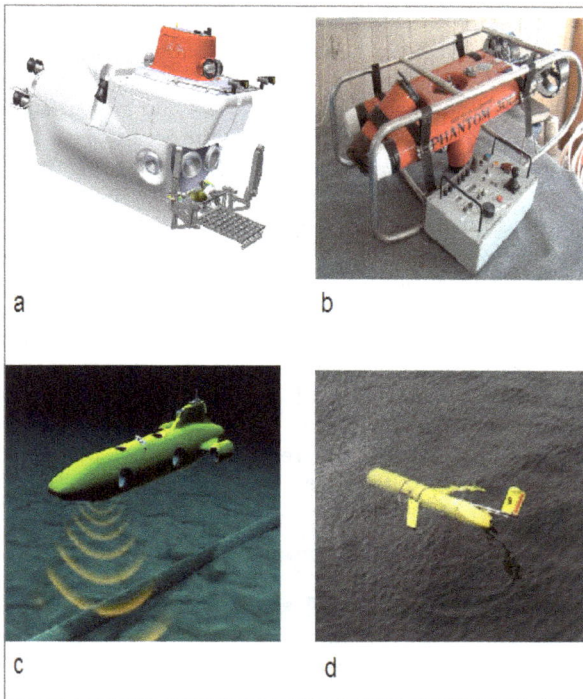

Fig. 4. a Alvin; b ROV; c AUV, d GLIDERS

Gliders (Fig. 4d) have been developed specifically to better investigate the main characteristics of mesoscale eddies, fronts, and boundary currents. The salient characteristic of gliders can be found in the realization cost that correspond to about few days of research vessel operation while they can be operated for a few months (Eriksen, 2009 in Thorpe, 2009).

Three operational gliders, Slocum (Webb Research Corp.), Spray (Scripps Institution of Oceanography and Woods Hole Oceanographic Institution) and Seaglider (University of Washington) were realized within the Autonomous Oceanographic Sampling Network program set up by the US Office of Naval Research (ONR) in 1995.

The four key technical elements of gliders are :

1. small reliable buoyancy engines;
2. low-power computer microprocessors;
3. GPS navigation system;
4. low-power duplex satellite communication.

The three operational gliders, are descendent floats and differ from drifter and lagrangian profilers essentially because of their characteristics: the wings that allow forward motion mainly while diving. During this gliding cycle gliders are fully autonomous, and at the surface they use GPS to determine next target direction.

Lagrangian devices, expendable probes and drifter

Traditional methods for oceanographic measures and observations are carried out through ad hoc surveys in order to collect data of specific areas, with the consequently limitation of time and costs.

To overcome this limitation, world oceanographic community, has developed different programs and methodologies.

Starting in the 1970s, a great number of surface *drifters* and subsurface neutrally buoyant floats have been developed, improved, and tracked in the ocean.

Drifters are able to provide worldwide maps of the surface and subsurface water velocity at a few depths. *Lagrangian profiling buoys* like WOCE floats and their successors Argo (Fig. 5a and 5b), are measuring and reporting in real time the temperature and salinity structure of the upper 2km of the ocean (Richardson 2009, in Thorpe, 2009). Drifters and floats provide fundamental surface and sub surface data allowing the description of world ocean's temperature and salinity structure, ocean circulation and its time variability.

The basic elements of a lagrangian drifter designed, for example, to track mean currents at a fixed depth beneath the ocean surface are: the drogue, the surface float and the connecting tether. The drogue is the draging element that locks the drifter to a water parcel.

The surface float contains the telemetry system, antenna, batteries, and sensors. The drifter position can be calculated by an installed GPS receiver. Data and position can be transmitted by Argos or Iridium systems to the data server.

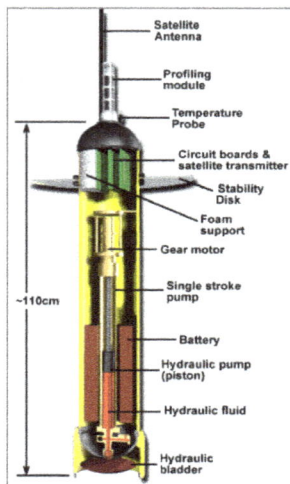

Fig. 5a. Argo (www.argo.net)

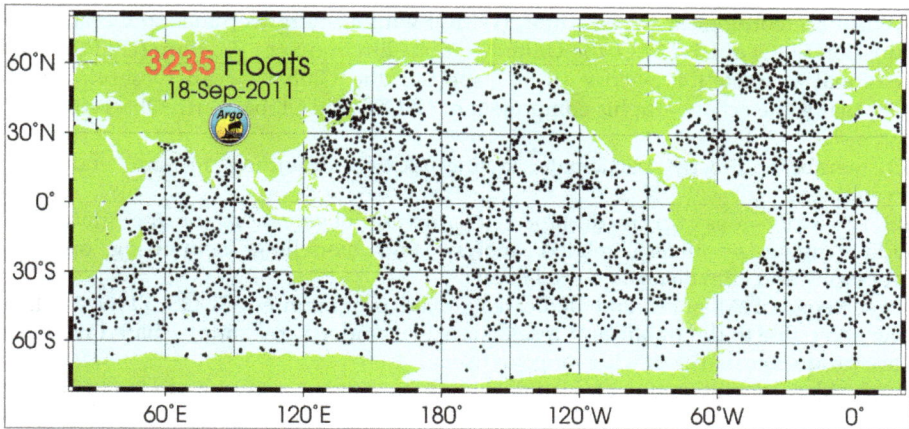

Fig. 5b. Argo distribution map (www.argo.net)

Autonomous WOCE and Argo float, usually drifts submerged for a few weeks. Periodically, when it rise up to the sea surface, transmit data and position by the Argos satellite system. After around a day drifting on the surface, the float re-submerges to its operative depth, in the upper kilometre of the ocean, and continues to drift for another few weeks. Around 100 round trips are possible over a lifetime up to 6 years.

Despite Argo floats can autonomously measure physical, chemical and some biological parameters, in practice this potential capability is limited by cost and power supply.

Recently, expendable probes like *eXpendable Bathy Termographs* (XBT) (Fig. 5c), have assumed particular relevance for the measure of physical variables. Their main future is the possibility to be launched directly from a moving ship, saving time and costs.

Despite all technological developments, the challenge of oceanography still remains the synopsis of data acquisition in order to observe high spatial and temporal variability phenomena such as oceans primary production. This variable is fundamental as it is strictly related with oceans water characteristics (physical, chemical and biological) and it is regulated by ocean dynamics (current and water mass properties).

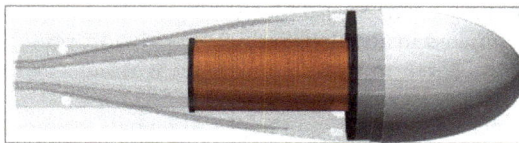

Fig. 5c. XBT (www.earthobservatory.nasa.gov)

2.1.4 The JCOMM Ship Observations Team (SOT) programmes: VOS and SOOP

From the origin of the climate studies, oceanographic measures has been acquired by ships, the only available platforms to survey the oceans.

Although there are now several other platforms (i.e. satellites, drifting buoys, floatings and radar) ships still play a very important role providing sea truth for the calibration of satellite observations and because of they allow measurements not yet obtainable by other means, such as air temperature and dew point (www.bom.gov.au/jcomm/vos/).

The World Meteorological Organization (WMO) within Intergovernmental Oceanographic Commission (IOC) of UNESCO, the United Nations Environment Program (UNEP), and the International Council for Science (ICSU) sponsor the Global Ocean Observing System (GOOS) that is the oceanographic component of GEOSS, the Global Earth Observing System of Systems.

WMO members directly participate in the GOOS, providing in situ and satellite observations.

GOOS (www.ioc-goos.org) is designed to provide observations of the global ocean (including living resources), related analysis and modelling, supporting operational oceanography and climate change predictions.

Through Regional Alliances and dedicated linked programmes, GOOS allow all data to become accessible to the public and researchers by standard and simple shared methods to produce products useful to a wide range of users.

It is not solely operational, but includes work to convert research understanding into operational tools. In this way GOOS should sustain the management of marine and coastal ecosystems and resources, supporting human activities like mitigate damage from natural hazards and pollution, protect life and property on coasts and at sea and suggest scientific research activities.

Since 1999, the marine activities of WMO, as well as those of IOC, have been coordinated by the Joint Technical Commission for Oceanography and Marine Meteorology (JCOMM - www.jcomm.info).

The creation of this Joint Technical Commission results from a general recognition that worldwide improvements in coordination and efficiency may be achieved by combining the expertise and technological capabilities of WMO and IOC.

The Observations Program Area of JCOMM is primarily responsible for the development, coordination and maintenance of moored buoy, drifting buoy, ship-based and space-based observational networks and related telecommunications facilities.

In this framework it was created a Ship Observations Team (SOT) which coordinates two programs: the Voluntary Observing Ship Scheme (VOS) and the Ship of Opportunity Program (SOOP).

The WMO Voluntary Observing Ships (VOS) program

Developed about 150 years ago, VOS is the international scheme for taking and transmitting meteorological observations by ships recruited by National Meteorological Services (NMSs), comprising member countries of the World Meteorological Organization (WMO).

With a fleet strength, estimated in about 4000 ships worldwide, presently the VOS supports measures useful to comprehend and predict extreme weather events, climate variability, and long-term climate changes (www.vos.noaa.gov).

In this way, providing increasingly data to the global climate studies, VOS makes a highly important contribution to the Global Observing System (GOS) of the World Weather Watch (WWW).

The Ship-of-Opportunity Program (SOOP)

The SOOP is based on the possibility to use merchant ships that join to the programme, in routinely strategic shipping routes. At predetermined sampling intervals, Expendable Bathythermographs (XBTs) are launched to acquire temperature profiles in the open ocean in order to be assimilated into operational ocean models.

SOOP plays a relevant role as it serves as a platform for other observational programmes, communicating closely with the scientific community. In this way, SOOP assumes a big relevance to seasonal and interannual climate prediction.

The programme is managed by the SOOP Implementation Panel (SOOPIP). Along the strategic shipping routes, SOOPIP identifies and coordinates the measuring necessities in terms of type of instruments, their use modality and deployment.

So, thanks to the use of different instruments such as XBTs, TSGs, CPR, it is possible to obtain long measuring series of physical, chemical and biological parameters.

Moreover SOOP looks at new instrumental technological development and coordinates the exchange of technical information; in particular: functionality, reliability and accuracy, and recommended practices about relevant oceanographic equipment and expendables.

The main important aspect of VOS and SOOP programs deals with the possibility of collect data in a cost effective way.

VOS operates without ship cost, whilst the communication charges, the observing equipments and the consumables are furnished by the National Meteorological Services.

Within these programs (VOS and SOOP), many international initiatives are presently carried out in order to develop new observing devices and systems, suitable to be utilized from as many vessels as possible.

New expendable probes and devices (low cost and user friendly) represent the ultimate effort of oceanographic technological development in order to collect even more accurate measurements and to enhance the range of variables to be measured in automatic way, particularly for the biological ones.

2.1.5 Remote sensing platforms

Data from satellites allow us understanding and monitoring ocean processes at large spatial and temporal scale. Also if the same sensors payload equips airborne remote sensing, this platform is otherwise dedicated to the surface analysis of meso-scale processes, particularly to that of shallow waters and coastal environment.

Even if remote sensing can provide a unique synoptic view of phenomena such as chlorophyll concentration and sea surface temperature, however the measurement validation requires constant scrutiny and much sea-truth data as possible (Marcelli et al. 2007).

For this reason (and also to develop ecosystem forecasting models) the integration between field sampling and remote sensing technique should be improved. Actually this is a primary necessity also because of still a lack both of basin wide and of coastal ones, operational, multidisciplinary, in situ observing system.

In fact today, thanks to new satellites capabilities, there is the possibility of sampling on smaller scales to describe relevant phenomena, especially in the coastal zone where scales are considerably smaller than those in the open ocean.

To comprehend the complexity of physical and biological processes, the human activities impacts, the climate changes, and every process involving seas and oceans, it is not possible to use a single approach.

To face this problem, the best strategy seems to be the integration of data derived from different platforms.

For all these reasons, the oceanographic investigation needs more and more new instruments and sensors with well defined measurement characteristics also in order to calibrate the synoptic data providing by remote sensing observations.

Whilst the monitoring of physical variables is continuously improving, the biological ones still lack and furthermore they have to be observed in situ.

There are several empirical algorithms commonly used to assess primary production in the euphotic zone, based on data provided by satellites Coastal Zone Colour Scanner.

These models are parameterized both on water visible radiance (obtained by means of satellites), and other simple variables, and also on the knowledge of photoadaptive parameters, such as maximum photosynthesis which can be obtained only by in situ sampling (Balch & Byrne, 1994).

3. Requirements of oceanographic observations

One of the most important problem of operative and operational oceanography is related to the availability, quality and accessibility of the data.

To meet these needs it is necessary to acquire more data than possible, but, as already mentioned, data collection is often limited by survey costs. Moreover data distribution is often discontinuous and it does not allow a synoptic understanding of many dynamic processes.

As already mentioned, the ships of opportunity are fundamental, since they integrate the ocean surface temperature (from satellites) with water column data. So they contribute to the Marine Forecasting System capability of providing near real time analysis of the oceans.

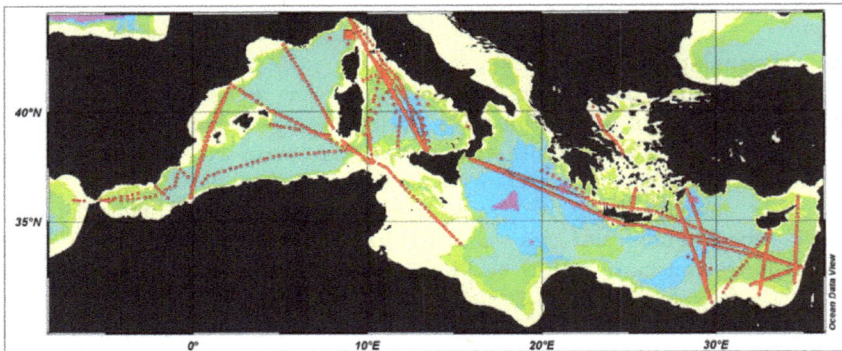

Fig. 6. Mean Mediterranean VOS acquisitions (MFSTEP Project)

As described in detail in par. 2.1.4, within the international projects knowledge necessities, it is fundamental the development of innovative technologies according to requirements of high performances, data quality and low-cost.

Even if data provided by models, by high resolution remote sensing and by the new autonomous sampling platforms are increasing temporal and spatial sampling capabilities (Dickey & Bidigare, 2005), however, there is still a lack of operational, multidisciplinary, in situ observing systems.

This fact is particularly related to the biological variables that need to be observed in situ more than the physical once. Especially in the mid-high latitudes, a complete upper layer observation of the water column is needed, because of the typical distribution of phytoplankton's biomass (Mann &Lazier, 1991).

Traditional methods (water sampling, storage and laboratory analysis) are too expensive and do not allow to have enough measures to describe variability of marine natural phenomena with sufficient temporal and spatial detail.

Therefore, bio-optical measures, used to study the main environmental characteristics, such as phytoplankton biomass, CDOM (Coloured Dissolved Organic matter), turbidity, can be integrated in new technological developments to give continuous measures.

In such a context our aim was to answer to all the above issues, developing new user friendly technologies, based on low cost materials and suitable to different configurations (Moored, stand alone, expendable, continuous and towed).

In order to face the necessity of the world oceanographic observations, we identified two main development necessities:

- - a new low cost flexible modular instrument, to be used like profiler, expendable, stand alone and instrumental payload;
- - a towed vehicle able to be used both in coastal and open oceans, capable to host the new low cost instrument

Low cost modular instrument

Expendable probes (see before XBT), represent an approach to ocean measurements in which the high accuracy of measurements may be sacrificed considering lower costs and operational expediency (Thorpe, 2009). Relating to operational oceanography, this kind of technology derives from the needs of adequate spatial sampling on timescales commensurate with temporal variability.

These probes can provide measures of temperature quickly as they can be used by a ship moving up to 20-30 knots. This technique also provides standard results and it became the central component of programmes such as the Global Ocean Observing System (GOOS).

Current expendable probe capabilities include also the measurements of sound speed, conductivity, ocean current and (most recently) optical irradiance and suspended particle concentration (Thorpe, 2009).

Such expendable measurements could be fundamental for satellite data calibrations and for programmes such as Marine Forecasting System (MFS).

With support from GOOS, GEO, POGO, IOCCG and Plymouth Marine Laboratory, the Chlorophyll Globally Integrated Network (ChloroGIN) was set up in order to provide a network of researchers with the aim of integrating in situ time series of chlorophyll measurements with satellite ocean colour-based observations. From these integrated measurements it is possible to provide maps of ocean chlorophyll and sea surface temperature, as indicator of the state of the ecosystem, and measures of light penetration to calculate primary production (Hardman-Mountford, 2008).

In addition to the prototypes made by Marcelli et al. (2007) during the MFS-TEP project, at now there are no expendable probes for the measurement of chlorophyll and other bio-optical variables.

This last feature is very important as it may concern in situ measurements of water optical property dealing directly with phytoplankton biomass and Coloured Dissolved Organic Matter (CDOM).

The measure methods currently available consist infact in expensive instruments, or they require expensive methods (sampling and laboratory analysis).

In this field, the maximum effort should be made to develop innovative tools looking to: economicity, flexibility and with a sufficient level of precision and accuracy.

They must be modular to be adapted to measure the widest number of variables and they must be flexible to be used by the largest possible number of measurement platforms, and also as expendable.

Towed vehicle
In order to increase our understanding of ocean processes it is necessary to have the most comprehensive and synoptic data sets.
Significant advances have been made in remote sensing from both satellite and aircraft, which nevertheless increase the requirement of in situ data for sea truth.
Methods of traditional oceanographic research and marine environment monitoring are limited to the time scale of the observer and therefore they cannot carry out synoptic observations (Marcelli & Fresi 1997).
The large coverage collection, of long term data sets or high resolution field studies, should require high costs.
The problem of monitoring of physical, chemical and biological parameters of the oceans, both for operational and forecasting, remains one of the most important issue that ocean science community is facing today.
It is commonly accepted that the most cost effective method for spatially and temporally monitoring is represented by the use of both towed instrumentation and expendable low-cost probes (Reseghetti et al., 2006) from ships of opportunity.
Towed vehicles, in use since the 1930s, in recent years have been advanced significantly becoming extremely robust and high reliable. Typical payloads are CTDs, that can be be interfaced with other sensors such as fluorometers, transmissometers, pH, dissolved oxygen, and other optical sensors dedicated to sea truth.
Undulating vehicle main feature is the capability to make continuous measurements along water column by moving ship, with a considerable operative time saving. Furthermore continuous measurements can resolve fine structures of high variability phenomena such as phytoplankton patchiness.
Distribution of phytoplankton biomass is characterized by a great variability. Its assessment is a fundamental issue since it modulates the carbon cycle. While primary production measurements are expensive for time and costs, phytoplankton biomass can be detected by in-vivo chlorophyll a measurements.
The possibility of using a continuous profiling probe, with an active fluorescence measurement, is very important in real time phytoplankton's study; it is the best way to follow the variability of sea productivity. In fact, because of the high time and space variability of phytoplankton, due to its capability to answer in a relatively short time to ecological variations in its environment and because of its characteristic patchiness, there isn't a precise quantitative estimation of the biomass present in the ocean (Piermattei et al., 2006).
Despite this fundamental aspect, the use of such towed vehicle often needs dedicated surveys or ships with ad hoc deploying and towing systems. However, towing vehicle will continue to have a key role if deployed from ships of opportunity looking to oceanic processes and it will become fundamental for the study of coastal high variability processes.

4. Material and methods

The first innovative probe that we present in this work is an expendable probe for chlorophyll a fluorescence and temperature measurements: the T-FLAP (Marcelli et al.2007). It represents an evolution of the XBT (eXpendable BathyThermograph), from which differs mainly for its electronic system and the and the addition of a fluorescence sensor.

The simple use of this probe and its low cost were the principles of its development, in order to realize an industrial production which will diffuse its employment by any ship who participates to the VOS (Voluntary Observing Ship) program.

The second is the Sliding Advanced VEhicle (SAVE) which allows a rapid and detailed physical and biological characterization of the water column. The system is composed by a depressor unit, towed at fixed depth, and an underwater instrumented vehicle able to slide along the towing cable.

The main features of the system are the cable drag reduction, that allows a higher velocity and the absence of the fairings (fundamental for the application of the system on board of ship of opportunity), and the electromagnetic induction data transmission system.

This towed vehicle consists in a "sliding unit" able to be used in shallow water and from small boats. The sliding system was chosen to reduce the "data anisotropy" deriving from the undulating and sinusoidal motion of the traditional towed vehicle.

4.1 Temperature Fluorometric Launchable Probe

In the framework of MFSTEP project (Manzella et al., 2003; Pinardi et al., 2003), and in the follow ADRICOSM-STAR Project we started to develop and realize a new low cost expendable probe able to measure fluorescence of Chl a, with sufficient accuracy to detect the main ecological structures in the water column (DCM).

The Temperature Fluorescence LAunchable Probe T-FLAP (Fig. 7), was charted in order to answer to the claim of a cost effective temperature and fluorescence expendable profiler, to be used on ships of opportunity. The development of the expendable fluorometer has followed similar concepts of the XBT, but differently the T-FLAP was developed with an electronic system which can be improved and adapted to several measure channels. T-FLAP (Marcelli et al. 2007) was tested during several oceanographic surveys (Fig.8).

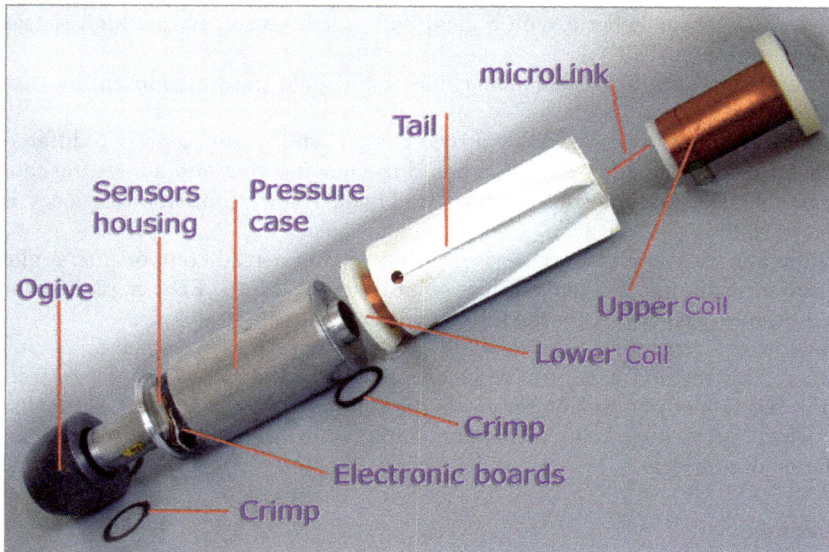

Fig. 7. T-Flap outline

Fig. 8. T-Flap Launched from a ship

The probe measurement cell is made of a tube where the water flows into, getting directly in touch with the sensors. The fluorescence sensor is composed by a light source (blue LEDs between 430nm and 470nm) and a photo-diode positioned orthogonally with respect to the light source.

The temperature sensor is build with a glass bulb micro sensor, with a high sensitivity to temperature variations (0.01°C).

Electronic and firmware for sensor management and data transmission are located inside the pressure case.

Digital data transmission is assured by a twin copper wire wrapped on two different reels: one in the probe tail, the other one is on board the moving ship and allows the connection with the pc till the signal interruption for wire break. The transmission frequency is about 5.6Hz and the falling velocity 4.5m/s.

To reach the aim of a low-cost probe, were utilized commercial components: a glass bulb temperature resistor for the temperature measurement, blue LEDs, a photo-diode and available selective glass filters, for the fluorescence measurement.

The measurement principle employed to detect phytoplankton biomass, is the active fluorescence. This method is an in vivo chlorophyll estimation, that can get the immediate biophysical reaction of phytoplankton inside the aquatic environment; it is a non-disruptive method which gives real time estimation and avoids the implicit errors due to the manipulation of samples.

4.1.1 Sensors
Fluorometer
The fluorometer is composed by:

- a source of light composed by 3 different wavelengths (blue LEDs);
- an optical filter which selects the blue wavelength from 430 to 480 nm positioned over the light source;
- an amplified semiconductor element (with electronics integrated) as sensitive receptor;
- an optical filter which selects the red light up to 600 nm over the sensitive receptor.

Temperature measurement

The temperature measurement is performed through a glass bulb micro sensor which comes out from the measure cell for 10 mm, the sensitive part is composed by resistive sensor inside a spherical glass bulb with the diameter of 1.5 mm. This sensor has a high sensitivity to temperature variations (0.01°C) and to the dynamic variations (0.05 ms).4.1.2.

4.1.2 Calibration

The calibration procedure is a very important step in the development of a fluorometer, especially if it is an expendable one. In vivo fluorescence (IVF) is commonly used to estimate Chl a (i.e., phytoplankton biomass) in natural waters (A. E. Alpine and J. E. Cloern 1985). The main problem of an expendable probe is that it is not possible to calibrate the instrument through in situ samples, because ships of opportunity does not stop to collect water samples like in a traditional survey.

Fluorometer calibration

Different solutions of Chlorella sp. measured with a calibrated fluorometer (PrimpProd 1.11) were used for the calibration of T-FLAP. Figure 9a shows the relation between fluorescence measurements of PrimProd and T-FLAP.

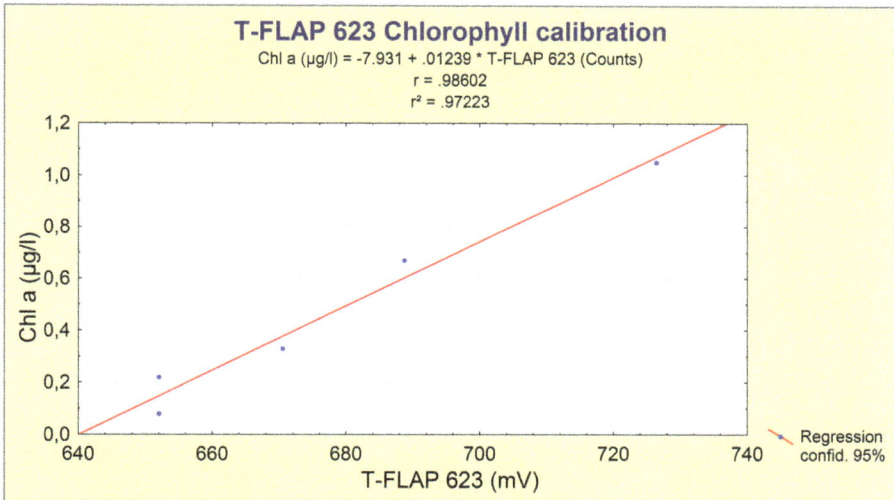

Fig. 9a. On X axis there is the Chl "a" measured by T-FLAP (millivolt) while on Y there is the Chl "a" from PrimProd 1.08 (µg/m3)

The fluorometer (PrimpProd 1.11) was calibrated by spectrophotometric analysis of the samples (Lazzara et al. 1990).

In order to acquire data simultaneously it was built a water circuit where the T-FLAP and the Primprod 1.11 were together in line.

The water flowed through the reference probe and through the T-FLAP thanks to a bulk insulated tube system connected to a circulation/feeder pump (0.7 bar, 5.7 l/min) which kept a constant flow.

In this way it has been possible to perform a dynamic calibration, which is necessary to test an instrument like T-FLAP, that descends at constant speed in the water column.

The probes were connected to the pc by a serial interface so that the values were available in real time.

Temperature calibration

The temperature sensor was calibrated using the same flow system employed for the fluorescence calibration.

The response of the temperature resistor is converted in a tension signal value. To find out the conversion law, which allows to transform the mV signal value in °C, the T-FLAP was immersed in temperature controlled water and the degree temperature value has been obtained by a reference temperature sensor.

The temperature measurement inside the calibration circuit was done by means of a OTM Falmouth temperature sensor with Platinum Resistance Thermometer which has got an accuracy of 0.003°C.

The probe is connected to the pc by the serial interface so that the values were available in real time.

When the water volume reached a stable temperature the T-FLAP acquisition got started and the output values were saved on the pc.

The calibration was realized in different correspondent temperatures, in a range of 0 to 25°C. For each of the temperature values, correspondent T-FLAP values were extracted from sensors and a regression analysis carried out. In Fig. 9b is represented the fit curve for the calibration of the temperature sensor of T-FLAP.

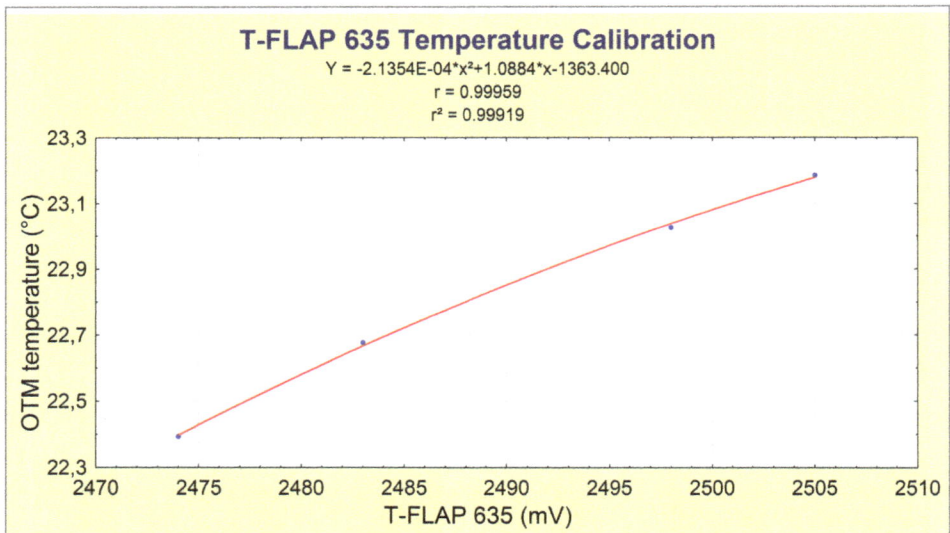

T-FLAP 635 Temperature Calibration
$Y = -2.1354E-04*x^2+1.0884*x-1363.400$
$r = 0.99959$
$r^2 = 0.99919$

Fig. 9b. On X, temperature from T-Flap (mV); on Y temperature from OTM (°C).

4.1.3 Digital transmission system

The digital data transmission system is allowed by a twin copper wire (µlink) wrapped on two separate coils; one of these is mobile and positioned in the tail, while the other is fixed in the canister launcher, through which the probe is connected to the pc on board.

Data acquisition doesn't need specific software, it is enough to have a pc terminal as Windows ® HyperTerminal. The T-FLAP firmware provides an interactive menu, which allows different work activities including programming and controlling.

Coil's capacity

The mobile coil, inside the probe, has to contain a wire with a length as the maximum depth it could reach plus a tolerance length: the pre-serie has 700 meters wire.

Limits

Being subjected to a progressive hydrostatic pressure, the functional limits of T-FLAP, concerning to the implosion of the materials, has reached 500 dBar.

A second limit is the capability of the coils to host a longer or thicker couple of conductors: for each one it was prevented at the most a double length with the same diameter, or a bigger diameter with the present length.

4.2 Sliding Advanced Vehicle SAVE

SAVE was designed to carry out detailed and rapid profiles, in order to have quasi-synoptic measures of physical and biological variables.

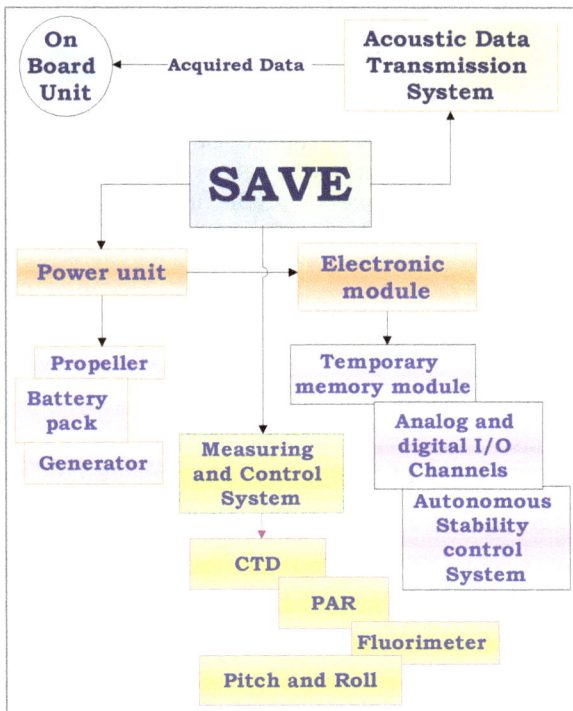

Fig. 10. Flux diagram of SAVE Architecture System

The system permits to perform a detailed characterization of the water column, following the phenomena dynamic evolution.

The system is composed by a depressor towed by a surface unit at fixed depth and an instrumented vehicle, that slides along a cable from the surface to the depressor. This approach starts from a previous development that was built in the mid-80s (Nomoto et al. 1986).

Numerical Model

In order to test all the system (Depressor and SAVE vehicle) a numerical analysis program, able to simulate the SAVE navigation, was developed.

The system is subdivided in:
- depressor
- SAVE vehicle
- cable

In this way the system can be improved with different operative conditions. A special attention was given to the definition of the geometry of the depressor and of the hydrodynamic control surfaces of the SAVE vehicle and of their moving, in order to obtain the expected performances at the different operative conditions.

For every solution the navigation limits were verified, in order to reach a compromise between system efficacy and subsystems measuring. We optimized the executive design of the single subsystems making it more effective.

4.2.1 Depressor unit

The depressor was designed to work between surface and 200m depth and between 2 and 8 knots speed. The model allowed to define the balance configuration of the depressor/cable system at different speeds.

The wing geometry was selected after a comparison between different NACA profiles efficiency. The results conduced to select NACA 4412 profile.

Extremity plates was applied to improve the profile, in order to reduce the fluid motions between the overpressure margin and the depression margin.

In order to control the behaviour during the navigation, the depressor lodges a miniature pressure transducer (Depth range 0-500 dBar, Depth resolution 0.1 dBar), a biaxial inclinometer (Roll angle ± 90 degrees, Pitch angle ± 90 degrees, Angles resolution 1 degree) and a data acquisition board (Rate 1 sample/second, Storage up 4000 samples, Serial com 0.5 sample/second).

4.2.2 Underwater vehicle

The SAVE vehicle is constituted by an instrumented pressure hull; the guide system is located on the front side of the vehicle and is composed by pulleys connected to the cable.

The vehicle was designed to work up to the maximum depressor depth and between 2 and 12 knots. The model provided the vehicle movement with different towing speed. In figure 11 is represented the result at 4 Kn speed (Fig.11).

4.2.3 Sensors

The temperature and the conductivity measurement are effectuated through a CT double sensor, that includes a ceramic type inductive conductivity cell and a high stability platinum resistance temperature device (RTD) temperature sensor.

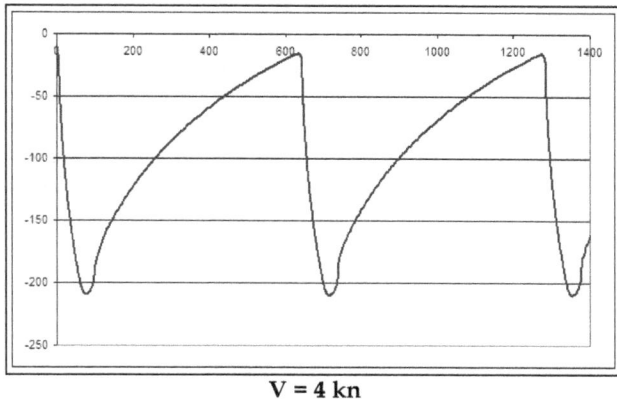

V = 4 kn

Fig. 11. Model results of vehicle movement at 4 kn speed. On X is represented the distance in m; on y is represented the depth in m.

Specification of the Conductivity sensor:
-    range 0->64 mmho/cm (0->6.5 S/m)
-    accuracy: ± .025 mmho/cm (.0025 S/m)
-    stability: ± .005 mmho/cm/month (.0005 S/m)
Specification of the Temperature sensor:
-    range -5° ->35° Celsius (ITS-90)
-    accuracy: ± .050° Celsius
-    stability: ± .002° Celsius/month
The CT sensor electronics is contained on a single printed circuit board, that provides two high level analog output signals 0-5 VDC.
The pressure measurement is effectuated through a 50 Bar pressure transducer supplied by a stabilized reference voltage (5V). A type Butterword (12dB/octave) low-pass filter is present In cascade to the amplifier, in order to eliminate the noise component induced by the water flux on the membrane.

4.2.4 Cable
The cable is one of the main elements of the system, because it sustains the weight and the hydrodynamic drag of the depressor. Beside to be a support, the cable is the sliding guide of the SAVE vehicle. The selected cable has the following characteristics: anti torsion zink-plated steel cable, 133 leads covered with RILSAN, external diameter 9mm.
The characteristics of the cable allow the electromagnetic induction transmission of the data between the underwater vehicle and the ship.
The power supply of the electronic components and of the electromechanical actuators comes from 12 rechargeable NiMH batteries (1.2 V, 7Ah).

4.2.5 Data acquisition/transmission system
Depressor and underwater vehicle transmit data to the on board unit during the navigation. The data transmission realizes both in descent and ascent in continuous way, thanks to a magnetic ring located between the pulleys. The cable passes through the ring, that realizes a coupling model, utilizing the electromagnetic induction to transfer the digital information.

5. Field tests: Results and discussions

Both systems have been developed and tested in the Laboratory of Experimental Oceanography and Marine Ecology in Civitavecchia.
32 T-FLAP prototypes were built and were launched by R/V Urania in several oceanographic surveys and from the university boat to perform specific tests.
One SAVE prototype was built and it has been tested to CNR IAMC of Messina from the R/V Luigi Sanzo.

5.1 T-FLAP field tests
In this paper we show the results of the T-FLAP field tests, carried out during two CNR oceanographic surveys both aboard the R/V Urania:
- MEDGOOS13, held from October 28 to November 8, 2006;
- ADR0208 Oceanographic Cruise, held from 17th of October 2008 to 28th of the same month.

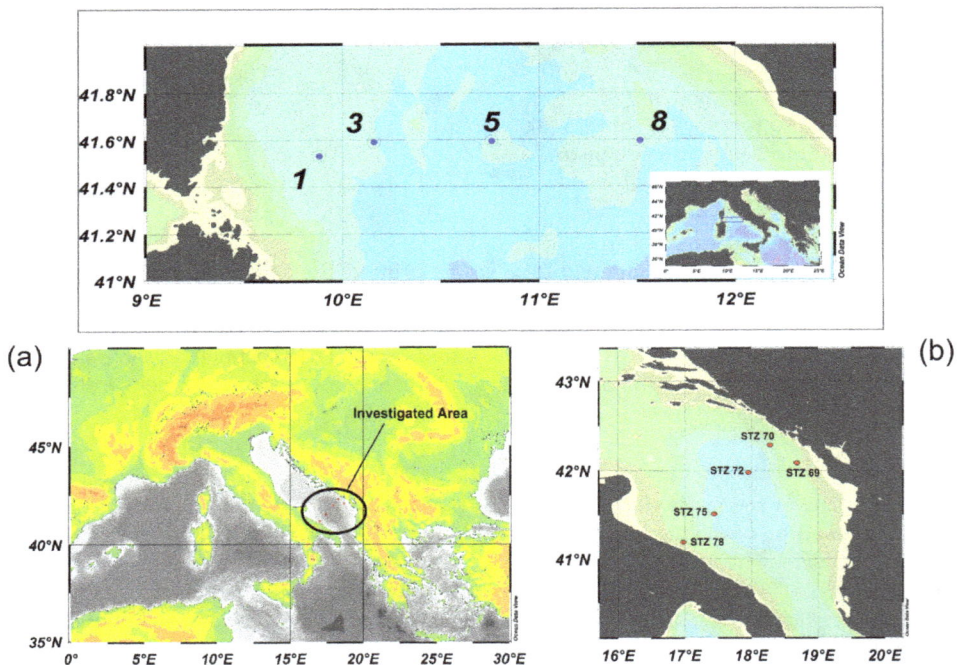

Fig. 12. On the left, the map of the Tyrrhenian sea stations (MEDGOOS13 cruise); on the right the map of the ADR0208 Oceanographic Cruise (a) and the location of the T-FLAP stations (b) (data analysed with ODV - Schlitzer, 2011)

In the Tyrrhenian Sea, the MEDGOOS13 cruise comprehended an oceanographic transect between Lazio and the Strait of Bonifacio, where several measurement stations were positioned in order to characterize the water column by multiparametric profiles and sampling water.

In each measuring station of this transect CTD casts were made by a Sea-Bird probe SBE 911 Plus equipped with an Aquatraka Chelsea MkIII chlorophyll a fluorometer.

In four station of the transect (Figure 12), T-FLAP prototypes were successfully launched, and profiles of chlorophyll a fluorescence, photosyntetic efficiency, PAR and temperature were measured by a high quality fluorometer probe: the PrimProd 1.08 (Antal et al 1999, 2001).

Data processing allowed the analysis of the upper oceanographic features along the transect as below described.

Because of this experiment looks to demonstrate the possibility to utilize T-FLAP from Ship Of Opportunity, it was decided to compare directly the results of data analysis obtained by means of T-FLAP profiles with the same analysis obtained by means of CTD and reference fluorometer casts.

This kind of approach derives from the impossibility for a Ship of opportunity to collect water sample for the post calibration.

Temperature and fluorescence profiles obtained from T-FLAP data were compared with the profiles obtained from Sea-Bird probe SBE 911 plus and from Aquatraka Chelsea MkIII chlorophyll a fluorometer.

As can be seen in fig. 13a, the T-FLAP has given good results both for temperature and fluorescence measures.

The sections infact represent the comparison between temperature and fluorescence distribution along the transect obtained by means of the T-FLAP and the reference sensors respectively.

CTD and T-FLAP stations have been represented in the same way in order to understand if T-FLAP is able to detect the water mass characteristics as the reference probe.

As can be seen in the left side of the figure, temperature sections of both T-FLAP and SBE CTD evidence the thermocline structure between 20 and 40 m depth with a mean value of about 14 °C.

Chlorophyll a sections obtained by means of both T-FLAP and Chelsea Aquatraka fluorometer, describe the trend of fluorescence of chl a indicating an higher biomass concentration in the central region of the transect decreasing in the western area. As it can be seen, the T-Flap can successfully describe the Chlorophyll distribution in the water column.

Results obtained by the analysis of depth-distance sections show that T-FLAP allows to well describe water mass main feature existing in the investigated area.

The T-FLAP probe has still some operative limits such as the launch method and the surface lightning. In some cruises, at the moment of the launch, outside the water, the high sensitive photodiode was saturated by the environmental light radiation, generating a measure error in the first water layer. This error was higher with a higher day lightning.

A similar approach was followed for the analysis of the results of the ADR0208 cruise. The CTD and T-FLAP data were represented in depth - distance sections, along the Dubrovnik-Bari transect. The objective was to test the possibility to detect mesoscale structures, in particular the Southern Adriatic gyre.

This gyre influences and modulates the DCM distribution, which has the main concentration, in the period of the ADR0208 cruise, near the Balkan coast. The temperature sections show a similar trend for both the probes, as shown in Fig. 13b.

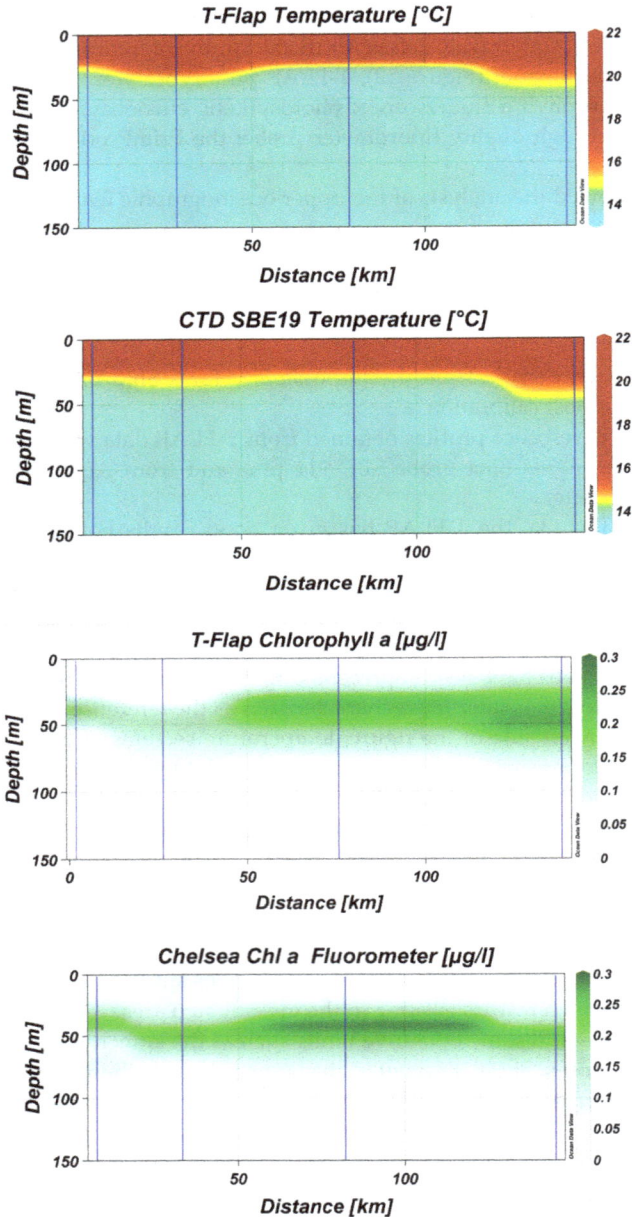

Fig. 13a. MEDGOOS13 cruise: depth-distance sections (W-E oriented) obtained by means of T-FLAP and SBE 911 plus CTD-Chelsea Aquatraka fluorometer: temperature upside, fluorescence downside. The mesoscale features are well idetified both from CTD and from T-FLAP (data analysed with ODV - Schlitzer, 2011)

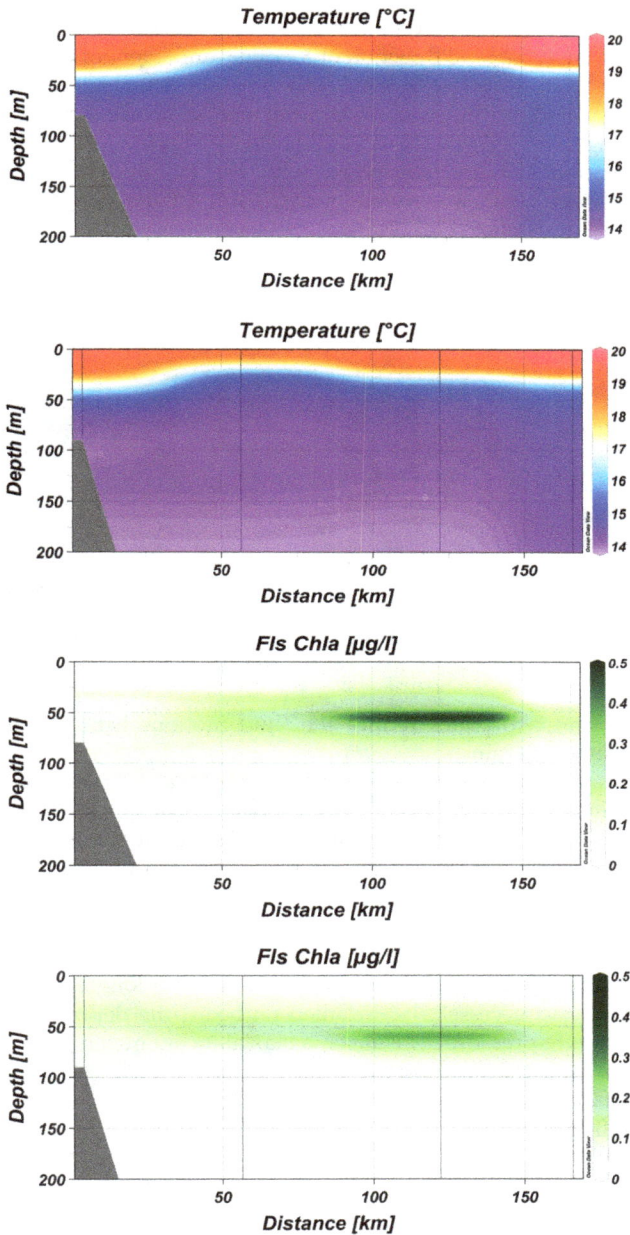

Fig. 13b. ADR0208 cruise: depth-distance sections (W-E oriented) obtained by means of T-FLAP and SBE 911 plus CTD-Chelsea Aquatraka fluorometer: temperature upside; fluorescence downside (data analysed with ODV - Schlitzer, 2011)

5.2 SAVE field tests

After the realization of the first prototypes, two different cruises were performed on board of the Luigi Sanzo, of the CNR IAMC Section of Messina, and a longer cruise on board of the Thetis, of the CNR IAMC.

The depressor was realized in PVC, with an internal steel structure to join together the ribs, that compose the wing.

To increase the rotation stability, compared with the longitudinal axis (rolling), a Λ configuration was chosen, with a consequent increasing of the wings half-opening, that allows a higher lift and resistance (Fig. 14).

Fig. 14. SAVE wing

At the end of the wing, two plates reduce the fluid motions between the overpressure margin and the depressor margin.

The first depressor was realized to test his behaviour in the water, at different depths.

The first tests were inherent in the signal of the stability control sensors (Fig. 15 left); the chosen final sensor is very little (65mm, but it can be reduced to 50 mm) so that it will not take up a lot of the pressure hull both of the vehicle and the depressor. It is a SUNDSTRAND sensor, with range of linear measure of $\pm 45°$ and non-linear up to $\pm 90°$. A test has been effected on the depth sensor too (Fig 15 rigth). It is a GE DRUCK miniature pressure transducer with thermal compensation, high signal outputs, good linearity, negligible hysteresis and good repeatability performance with long term stability. For this reason a cylindrical pressure vessel was installed on board the depressor. The vessel was equipped with a miniature pressure transducer, a biaxial inclinometer and a data acquisition board.

The characteristics of the instrumental payload are:
- depth range: 0-500 dBar
- depth resolution: 0.1 dBar
- roll angle: ± 100 degrees
- pitch angle: ± 100 degrees

The data acquisition performances are:
- rate: 1 sample/second
- storage: up 4000 samples
- serial com: 0.5 sample/second

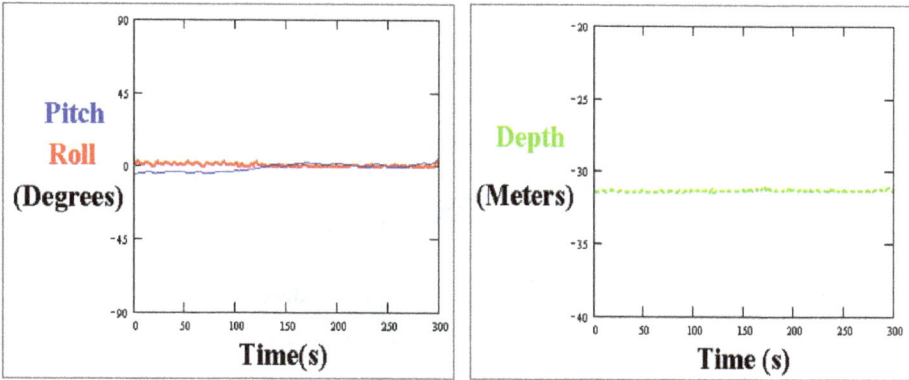

Fig. 15. On the left it is represented the stability of roll and the pitch movement; on the right the towing of the depressor at a fixed depth (of about 31m)

The transmission data method is a Frequency Modulation Encoded, with a spectrum: 500 Hz to 50 kHz.
The power supply was internal 7.2V Lithium battery, with a capacity of 30 days in continuous.
The basis payload has been completed with Falmouth OEM Conductivity and Temperature sensors. These analogic sensors do not need pumps, have high accuracy and stability and require low power.
A pressure cylinder equipped with these sensors was installed on the depressor. The cylinder was casted to test the response of the sensors in a towed application. In figure 16 it is represented a Temperature acquisition at a fixed depth.

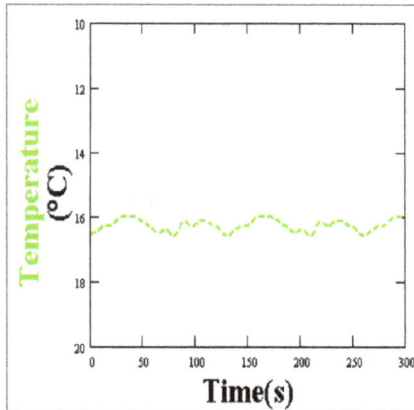

Fig. 16. Temperature distribution recorded at a fixed depth

The SAVE vehicle was a simulacre containing an Idronaut probe for CTD, fluorimetric and PAR measures. It was connected to the cable through a sliding system of Teflon, so that it was possible to acquire physical-chemical and biological parameters of the water column, from the surface till the depressor depth (Fig.15).

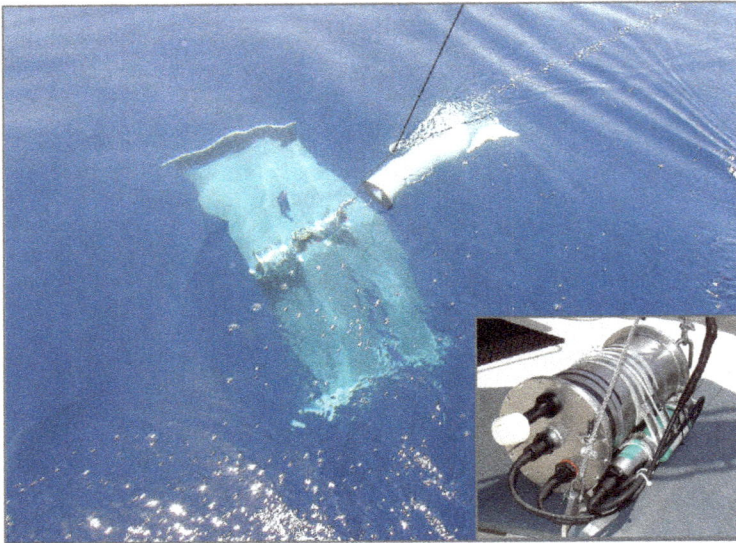

Fig. 17. CT sensor , data acquisition and data transmission unit on board the cylindrical vessel. SAVE was simulated by a IDRONAUT CTD probe.

6. Conclusions

Though many advances have been made in the technological innovation for the study of the seas and oceans, at present there is an increasingly need for technological development, both as regards the development of measurement platforms and so far with respect to the innovative sensors.

In fact, in most cases, either the measurement platforms are still too costly or they are limited in the use by the operational capabilities (eg, autonomy, depth, etc..).

To the development of new measurement platforms must also be joined efforts for the development of new sensors, especially for the physical measurement of biological variables.

Basically, on the one hand, the platforms to be developed must take into account the spatial and temporal scales of oceanographic phenomena, on the other hand, they must meet the need to be the sea truth of the satellites and mathematical models, that require large amounts of data for their calibration.

Even more it is to do to biological measures.

In fact, many solutions are still to be explored both for the measurement of nutrients and for the measurement of bio-optical variables, able to measure the optical absorption of light by phytoplankton.

The study of biological samples takes a lot of time and the use of highly specialized personnel, something that involves high costs.

Also commonly measured variables, such as oxygen and pH, require the development of new sensors: more stable, less fragile and less subject to attention from researchers (eg, calibration).

Other innovations are required to support marine technology: supporting innovations but fundamental.

For example, oceanographic research requires new materials (such as for the polar seas), new power systems, new computational capabilities.

For example:

-    the AUVs are limited because of the weight, the lack of autonomy, the volume of electronics, and the inability to accommodate complex measuring systems (eg. for nutrient analysis);
-    the gliders and floats profiling are limited by the ability to host complex and big sensors;
-    the moorings are limited by the power support and from foulings on sensors;
-    moreover satellites and models, for their calibration, require a very large number of measures that can be made available only by ad hoc developments of low-cost reliable and user friendly tools to be used extensively;
-    the coastal shallow water represents a sea area where the develop of monitoring networks, with high spatial and temporal resolution, become increasingly crucial. This will only be possible as affordable technologies will became available.

Our choices of technological development were born in this scenario, with the aim to realize:

-    a modular measurement technology, low cost and user-friendly, usable by different measurement platforms as more as possible: the T-FLAP;
-    a measurement platform for the study of shallow water phenomena with high spatial and temporal resolution: the SAVE.

The T-FLAP was designed in 2003-2006 during UE MFS-TEP project, and after a long period of prototype development it was tested in different operative conditions.

The in situ tests analysed in this work, were performed in November 2006 along a transect in the central Tyrrhenian sea. The different profiles obtained show in this case the full satisfaction of the project requirements.

Comparing the T-FLAP measurements with the reference tool, the temperature profiles of the two instruments overlap almost perfectly.

The resolution of temperature measurement required by the project, at least 0.1 °C, was achieved and exceeded by an order of magnitude, infact T-FLAP temperature sensor has a resolutiony of 0.01 °C and a dynamic response of 0.5 ms.

With regard to the measurement of chlorophyll a, the comparison between the T-FLAP and the Aquatraka profile, shows that the fluorescence peak was well identified also by T-FLAP.

Since then, it is possible to identify correctly the depth of the DCM, satisfying also in this case the project requirement (i.e. the localization of the Deep Chlorophyll Maximum).

It is necessary to point out that the Tyrrhenian Sea is a really oligotrophic area of the western Mediterranean basin, especially during the summer season, where the mean values of Chlorophyll a is 0.06 mg/m3 , for the top layer of the water column (Bosc et at. 2004) , while the DCM reaches 0.6 mg/m3 in late summer (Marcelli et al. 2005). The identification of DCM in the Tyrrhenian Sea, during the month of November, grants a much better response of the instrument if the same is used in most eutrophic areas, such as the Adriatic Sea.

Because of the estimate of chlorophyll a concentration expresses a fundamental biological characteristic of the seas, the T-FLAP, once commercially produced, should be targeted not only for oceanographic research, but also for fishery, sea water quality, and coastal management.

The T-FLAP temperature measurements, will instead integrate the now ten-year XBT measurements, contributing, with high resolution temperature measurements, to SOOP programs and operational ocean forecasting.

Regarding to the SAVE it is a modular, flexible and versatile monitoring system to provide continuous profiles of physical and optical measure, along the water column, from a moving ship.

The system was designed for the most possible manageability and ship of opportunity use and was composed by a depressor, towed at fixed depth, and a vehicle, that slides autonomously along the cable, performing continuous profiles in the water column, for a fine physical and biological characterization.

The main objective of this research was to create a system that would provide quasi-synoptic measurements of ecological and physic-chemical variables in marine environment. Its main application are the basic scientific research, the environmental monitoring and the assessment of biological resources. Therefore it was created a module system capable to support different instruments and to expand its own working operating capabilities (Marcelli & Fresi, 1997).

This sliding platform allows to ride over the methods of traditional oceanography research and of marine environment monitoring, which are limited to the time scale of the observer and therefore they cannot carry out synoptic observations.

7. References

Aiken, J. (1985). The Undulating Oceanographic Recorder Mark 2 - Mapping Strategies in Chemical Oceanography. Advances in Chemistry, American Chemical Society, Vol. 209 ISBN 13:9780841208629, pp.315-332

Alpineà A.E. & Cloern, J.E. (1992). Trophic interactions and direct physical effects control phytoplankton biomass and production in an estuary. *Limnol. Oceanogr.*, 37(S), pp. 946-955

Antal, T. K., Venediktov, P. S., Konev, Yu. N., Matorin, D. N., Hapter, R., Rubin, A. B. (1999). Assessment of vertical profiles of photosynthesis of phytoplankton by fluorescentic method. Oceanologia (Russia), 39(2), pp: 314-320

Antal, T.K., Venediktov, P.S., Matorin, D.N., Ostrowska, M., Wozniak, B., Rubin, A.B. (2001). Measurement of phytoplankton photosynthesis rate using a pump-and probe fluorometer. Oceanologia (Poland), 43(3), pp: 291-313

Balch, W. & Byrne, C.F. (1994). Factors affecting the estimate of primary production from space. *Geophys. Res. J.*, 99, pp: 7555-7570

Bosc, E., Bricaud, A., Antoine, D.; (2004). Seasonal and interannual variability in algal biomass and primary production in the Mediterranean Sea as derived from 4 years of SeaWiFS observations, Global Biogeochemical Cycles, 18 GB1005, 10.1029/2003 GB 002034

Bruzzi, D. & Marcelli, M. (1990). Sviluppo di una metodologia strumentale automatica per il monitoraggio dei fondali marini, *Proceedings of VI Colloquio AIOM*, pp. 79-84

Dickey, T. D. & Bidigare, R.R. (2005). Interdisciplinary oceanographic observations: the wave of the future, *SCI. MAR.*, 69 (Suppl. 1), pp. 23-42

Favali, P. & Beranzoli, L. (2006). Seafloor Observatory Science: a review, Annals of Geophysics, vol. 49, N. 2/3, pp: 515-567

Hardman-Mountford, N. (2008). The chlorophyll global integrated network (chlorogin), *Changing Times: An International Ocean Biogeochemical Time-Series Workshop*, La Jolla, California 5-7 November 2008 IOC Workshop Report No 217, pp. 11-12

Lazzara, L., Bianchi, F., Falcucci, M., Modigh, M., & Ribera D'Alcala`, M. (1990). Pigmenti Clorofilliani, *Nova Thalassia*, 2, pp. 207–223

Mann, K. H. & Lazier, J. R. (1991). Dynamics of Marine Ecosystem, Blakwell Sci., 23–90, ISBN 0-86542-082-3, Michigan, United States of America

Manzella, G. M. R., Scoccimarro, E., Pinardi, N., & Tonani, M. (2003). Improved near real time data management procedures for the Mediterranean ocean Forecasting System – Voluntary Observing Ship Program, *Ann. Geophys.*, 21, pp. 49–62

Marcelli, M. & Fresi, E. (1997): "The SARAGO" project: new undulating towed vehicle for developing, testing new technologies for marine research and environmental monitoring. Sea Technology pp 62-67

Marcelli, M., Di Maio, A., Donis, D., Mainardi, U. & Manzella, G.M.R. (2007). Development of a new expendable probe for the study of pelagic ecosystems from voluntary observing ships, *Ocean Sci.*, 3 pp. 1-10

Nomoto, M., Tsuji, Y., Misumi, A. & Emura, T. (1986) An advanced underwater towed vehicle for oceanographic measurements. Advances in underwater technology, *Ocean science and offshore engineering*, vol. 6, pp. 70-88

Piermattei, V., Bortoluzzi, G., Cozzi, S., Di Maio, A., & Marcelli, M. (2006). Analysis of mesoscale productivity processes in the Adriatic Sea: Comparison between data acquired by Sarago, a towed undulating vehicle, and CTD casts. Chemistry and Ecology , Vol. 22, Supplement 1/ pp: 275-292

Pinardi, N., Allen, I., Demirov, E., De Mey, P., Korres, G., Lascaratos, A., La Traon, P-Y., Millard, C, Manzella, G. & Tziavros, C. (2003) The Mediterranean ocean Forecasting System, first phase of implementation (1998-2001), *Ann.Geophysicae*, 21, pp. 3-20

Reseghetti, F., Borghini, M., & Manzella, G. M. R. (2006) Improved quality check procedures of XBT profiles in MFS-VOS. Ocean Sci. Discuss., 3, 1441–1480.

Richardson, A.J., Walne, A.W., John, A.W.G. , Jonas, T.D., Lindley, J.A., Sims, D.W., Stevens, D., Witt, M. (2006) Using continuous plankton recorder data, Progress in Oceanography, 68, pp: 27–74

Schlitzer, R., Ocean Data View, http://odv.awi.de, 2011.

Thorpe, S. A. (2009). *Encyclopaedia of Ocean Science: Measurement Techniques, Sensors and Platforms*, Elsevier Academic Press, ISBN: 978-0-08-096487-4, Italy

Wiebe, Peter, H., Stanton, Timothy, K., Greene, Charles, H., Benfield, Mark, C., Sosik, Heidi, M., Austin, Thomas, C., Warren, Joseph D., and Hammar Terry (2002) BIOMAPER-II: An Integrated Instrument Platform for Coupled Biological and Physical Measurements in Coastal and Oceanic Regimes, ieee journal of oceanic engineering, 27, 3, pp: 700-716

2

Application of the Long-Term Delayed Luminescence for Study of Natural Water Environments

Zdzisław Prokowski and Lilla Mielnik
West Pomeranian University of Technology in Szczecin,
Poland

1. Introduction

A part of photon energy absorbed by a photosynthetic apparatus of photosynthetic organisms and some organic compounds in water or bottom sediments is wasted for luminescence. Depending on the period elapsing from the exposure of the material until registration of its luminescence we may distinguish fluorescence (Fl) registered during exposure of the material to light and delayed luminescence (DL) registered in darkness some time after the exposure to light is terminated.

Despite the fact that the quantity of energy lost due to luminescence is not major, it provides important information, since it reflects processes related to the absorption of light, intramolecular energy transfer and structural changes in a given material.

For technical reasons, fluorescence and constituent parts of long-term DL registered t > 0.1 s after excitation are the most suitable for registration. Findings obtained so far suggest that using DL for testing water environments creates opportunities for developing fast and accurate luminescence assessment methods for photosynthetic properties of active algae cells and properties of bottom sediments.

2. Basic methods for registering long-term luminescence in suspensions

At the moment, suspension is excited for luminescence primarily using radiation of a visible light spectrum whereas luminescence is usually tested by registering its kinetic decay or changes in intensity during exposure of a sample (induction kinetics).

Determining the DL decay kinetics involves exposing a sample to light for a specified time and then registering intensity of decaying light in darkness in a continuous manner or at predefined intervals.

In luminometers designed to determine decay kinetics of long-term DL, a sample is placed in a light proof camera and exposed to light of specific properties with photodetector covered for a predetermined period. Then, the source of light is blocked and photodetector opened and connected to a recording system to register intensity of decaying light. The need to cover the photodetector during exposure of a sample to light of high density of photons and uncovering it when measuring DL signals of low density of photons stream results from limited dynamics of light-sensitive detectors and systems cooperating with them. It is necessary to separate streams of photons exciting luminescence from those emitted in the form of luminescence. When we use photomultipliers to detect luminescence instead of using shutters, we block their operation electronically during excitation luminescence of a sample (Lavorel et al., 1986).

In the case of solutions and suspensions, the intensity of delayed luminescence in a given time after exposure as well as kinetics of its induction can be registered continuously using flow methods (Strehler & Arnold, 1951; Krause et al., 1982; Krause & Gerhardt, 1984; Prokowski, 1991; Wiltshire et al., 1998). The methods involve exciting luminescence in a tested suspension beyond the place of detection and then pumping illuminated suspension to the photodetector which registers DL. Streams of exciting and emitted photons are separated by light traps. The kinetics of DL decay can be determined by halting the flow of the suspension.

To determine DL intensity in a given time span after exposure the suspension is pumped through an exposure tray to a light proof camera which contains measuring (emission) cuvette and photodetector combined with a system registering intensity of luminescence (Fig. 1) (Krause et al., 1982; Krause & Gerhardt, 1984; Wiltshire et al., 1998). By using monochromatic light for exciting of luminescence and placing an additional photodetector covered with a filter before the excitation cuvette, it is possible to measure Fl and DL at the same time (Yacobi et al., 1998).

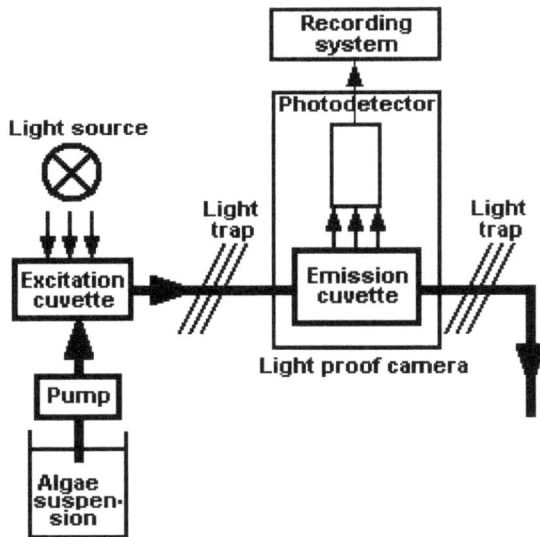

Fig. 1. Schematic diagram flow luminometer for registering intensity of long-term delayed luminescence in suspensions.

Exposure and DL registration times for suspension of algae depend on capacity of cuvette and flow. In the case of a fixed flow time between the end of exposure and DL registration depends on the capacity of pipes linking the two cuvette. Using monochromatic light of controlled wavelength to excite luminescence enables us to test the excitation spectrum of suspension delayed luminescence (Krause et al., 1982). The application of a special exciting cuvette liked with a measuring cuvette in the system simplifies the construction of such luminometers (Prokowski, 2001).

In the case of suspensions for which DL intensity changes during their excitation it is possible to register the induction kinetics for this type of luminescence. In order to determine induction kinetics for long-term DL using a flow method, the suspension is placed in a transparent container – excitation cuvette (Fig. 2) (Strehler & Arnold, 1951; Prokowski, 1991) – Fig. 2.

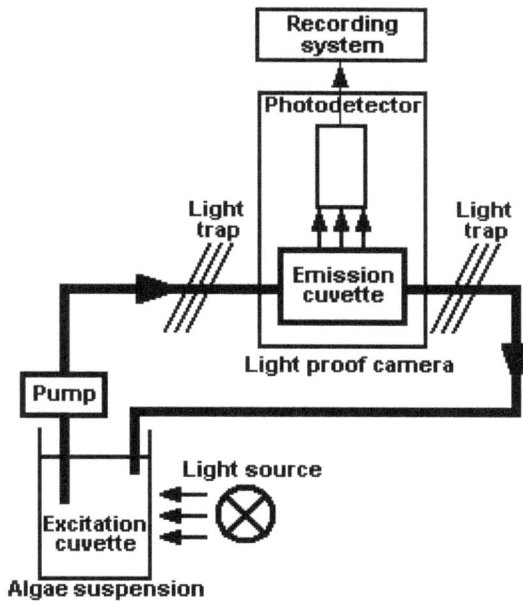

Fig. 2. Schematic diagram of flow luminometer used to determine induction kinetics for long-term delayed luminescence of suspensions.

The suspension is pumped from the container through non-transparent pipes to a light proof camera which contains measuring cuvette and photodetector. The suspension after leaving the measuring cuvette returns to the container. After the exposure of the suspension the photodetector registers changes in DL caused by the exposure.

Certain modifications of the flow method enabled registering a number of induction kinetics, both fluorescence and delayed luminescence using a single sample of algae suspension after different intervals after excitation (Prokowski, 1991, 1999). For this purpose, algae suspension after exposure in a transparent container is pumped through a series of connected light proof cameras with measuring cuvettes and detectors.

The intensity of luminescence registered is usually expressed in relative units. This is related to the fact that the registered value of emitted light from a given sample depends on a number of factors, such as geometry of the optic part of the luminometer and sample, emission intensity and spectrum of the excitation light source, optical properties of filters, mirrors and monochromators, as well as sensitivity spectrum of the photodetector.

3. Luminescence of chlorophyll *a* of the algal photosynthetic apparatus

In light-harvesting complexes, pigments in the photosynthetic apparatus of photoautotrophic organisms absorb light and while transforming absorbed photons into chemical energy some of the energy is reemitted in the form of luminescence. The excitation spectrum of the luminescence depends on the kind of pigments in light defusing complexes. In the case of land plants and algae, particles of chlorophyll *a* (Chl*a*) in photosystem II of their photosynthetic apparatus are responsible for the emission with its maximum at wavelength of $\lambda \approx 685$ nm. Luminescence Chl*a* accompanies light absorption by pigments in light-harvesting complexes and stabilisation of excitation energy in the photosystem. The luminescence excitation spectrum depends on the composition of pigments which accompany Chl*a* in the photosynthetic apparatus. The Chl*a* luminescence signal reflects particular stages of primary photosynthesis as well as influence of various biotic and abiotic factors on those reactions. Moreover, when exposure of photosynthetic organisms changes, their photosynthetic apparatus adjusts to new conditions. The processes are reflected in characteristic changes of the intensity of Chl*a* luminescence. The changes are described as luminescence induction and are observed during the first several minutes after new exposure to light. After that, the level of luminescence intensity stabilises (Prokowski, 1991).

Fluorescence (Fl_{Chla}) which accompanies photosynthesis is mainly the effect of the loss of photon energy absorbed by light-harvesting complexes during the photophysical stage of photosynthesis, whereas the delayed luminescence (DL_{Chla}) is related to the imperfection of photon energy transformation into chemical energy in the photosynthetic apparatus. Contrary to fluorescence, the emission of DL_{Chla} may occur from Chl*a* contained in the photosynthetic apparatus which has the capacity to promote primary photosynthesis. It occurs at some stages of primary photosynthesis and therefore DL_{Chla} has several components produced by various mechanisms (Jursinic, 1986). Particular DL_{Chla} constituent parts can be distinguished by registering the drop of luminescence after excitation light is turned off. The decay curve can then be divided into a number of exponential parts which last from several microseconds to several minutes (Jursinic, 1986).

In the case of long-term DL_{Chla}, its parts are registered within a range between hundreds of milliseconds to several dozens of seconds. This results from a recombination of negative charges accumulated at electron carriers Q_A and Q_B with positive charges in the water decomposition system (Jursinic, 1986). During that period we may distinguish at least three DL_{Chla} constituent parts: one lasting about 0.5 s (fast component DL_F) related to deactivation of reduced Q_A acceptor and two other of 2.5 s (middle component DL_M) and 14.5 s (slow component DL_S), both related to the deactivation of reduced Q_B acceptor (Prokowski, 1991, 1999). Those parts of delayed luminescence react to physical and chemical factors differently (Prokowski, 1991, 1993, 1999). In comparison with Fl_{Chla}, intensity of the DL_{Chla} component is more than two orders of magnitude smaller.

Figure 3 presents sample changes of luminescence intensity registered simultaneously for Fl_{Chla} (IFI) and three long-term component of DL_{Chla} (IDL_F, IDL_M and IDL_S) for test algae *Scenedesmus quadricauda* Brèb. in arbitral units – different for particular iterations, whereas figure 4 shows normalised values of those iterations in a semi-logarithmic scale.

Those iterations were registered for algae of 150 mgChl*a*·m^{-3} adapted in darkness for 2 h and then exposed to monochromatic light of wavelength λ_{max} = 660 nm and photons flux of 100 µmol(quanta)·m^{-2}·s^{-1}. Fluorescence was registered at wavelength λ_{max} = 690 nm during the exposure of algae suspension, whereas delayed luminescence intensity of particular components was registered at λ > 660 nm at the following intervals after excitation: IDL_F – from 0.3 to 0.7 s, IDL_M – from 3.0 to 3.6 s and IDL_S – from 12.0 to 12.6 s. F_0, F_L, F_D, F_P, F_M, F_S, F_T are used to mark intensity of fluorescence in specific points of the induction curve; DL_{Fm} – transient maximum of changes intensity in DL_F.

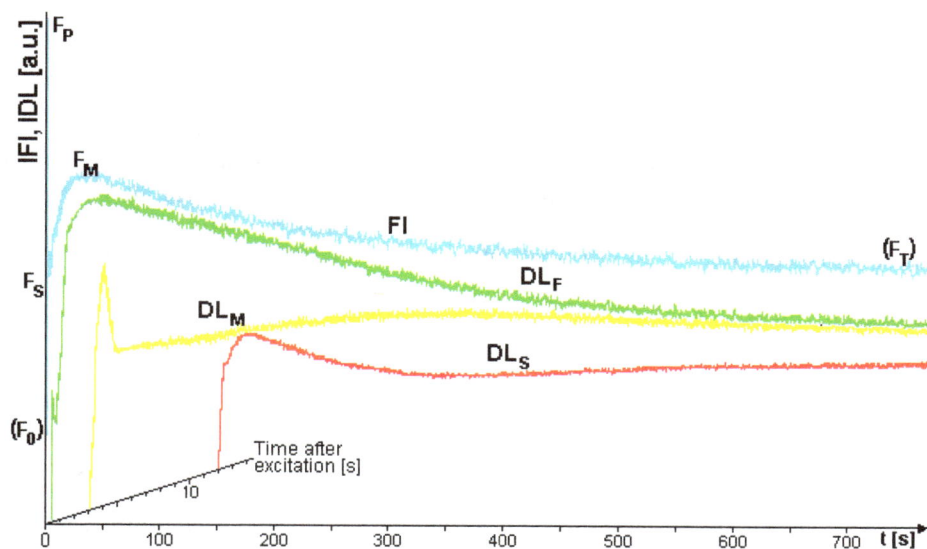

Fig. 3. Sample changes of fluorescence and three components of long-term delayed luminescence of test algae suspension.

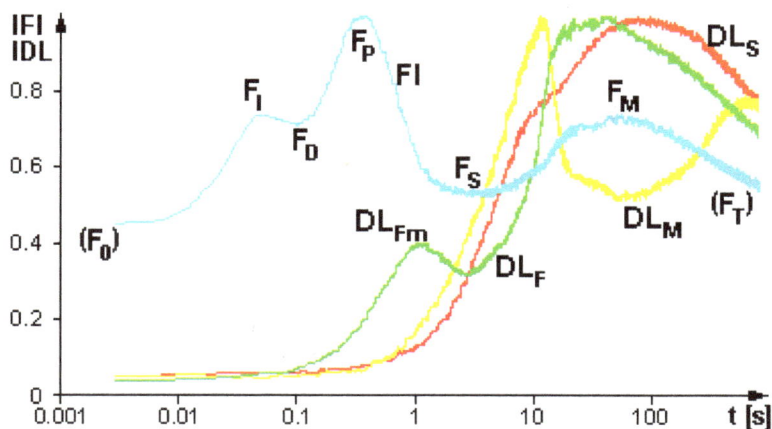

Fig. 4. Comparison of normalised changes of fluorescence and three components of long-term delayed luminescence of test algae suspension.

While analysing changes of luminescence Chl$a$, we notice that in the case of Fl_{Chla} typical changes of intensity occur during induction, namely abrupt increase of intensity to F_0 immediately after the excitation of the algae suspension. We also record fast increase to F_I after which there is a temporary slowdown to F_D and another increase to maximum F_P. During further exposure of the suspension, Fl_{Chla} intensity drops to F_S, then it increases again to F_M to drop later to the level of stationary intensity F_T.

In the case of DL_{Chla}, no temporary changes of intensity typical for Fl_{Chla} in the initial period of induction are observed. In the case of DL_F only, similarly to Strehler & Arnold (1951) we also observed a temporary maximum. The maximum was registered when intensity of Fl_{Chla} dropped from F_P to F_S, whereas during further exposure of the algae suspension changes of particular components of DL_{Chla} were different. Intensity of DL_F and DL_S reached maximum values and then decreased to a stable level after long time exposure of algae suspension whereas changes of DL_M before reaching that level showed two temporary increases in luminescence intensity. Usually during free excitation, several temporary changes of intensity occurred in delayed luminescence components. These changes were preceded by minor changes of fluorescence intensity. Moreover, both Fl_{Chla} and DL_{Chla} components react differently to physical and chemical factors (Prokowski, 1991, 1993, 1999).

Some research register Fl intensity changes during exposure of a sample to constant light with additional strong light impulses used to determine fluorescence parameters (for reviews see Seppälä, 2009). The technique used previously did not allow for testing reaction to such excitation in case of long-term delayed luminescence components. Figure 5 presents the induction kinetics of test algae suspension luminescence (C) registered for three levels of photon flux density (I) of exposure to continuous light and effect of the influence of a temporary increase of intensity of monochromatic light on luminescence of Chl$a$.

The research presented in the figure used photon stream impulses of 1000 µmol (quanta) $m^{-2} \cdot s^{-1}$ and duration of 0.6. The first impulse occurred 55 s before continuous excitation light was switched on. The next impulse occurred at the drop of Fl_{Chla} intensity to F_S, and further after every 60 s of exposure.

In the case of test algae luminescence, their Fl_{Chla} increased with the increase of photon flux density of light illuminating the algae suspension, whereas no major changes were registered in intensity of DL_{Chla} components and changes of conditions of illuminating algae (Fig. 5).

Additional light impulses cause different reactions as regards intensity of fluorescence and delayed luminescence components. In the case of fluorescence Chl$a$ we observe temporary increase in intensity when an impulse occurs. The increase depends on the induction moment in which an impulse occurred. The largest increase was recorded for an impulse which occurred in darkness, whereas the smallest for an impulse which occurred when Fl_{Chla} amplitude for test algae changed from F_P to F_S. Impulses occurring every 60 s did not cause any visible changes of the fluorescence induction curve comparing a flash and the reference value.

In the case of DL_{Chla}, the increase in intensity of all components after an impulse in darkness was much lower than its intensity registered immediately after excitation light was switched on. Continuous excitation light causes various reactions of particular delayed luminescence components depending on the excitation light intensity.

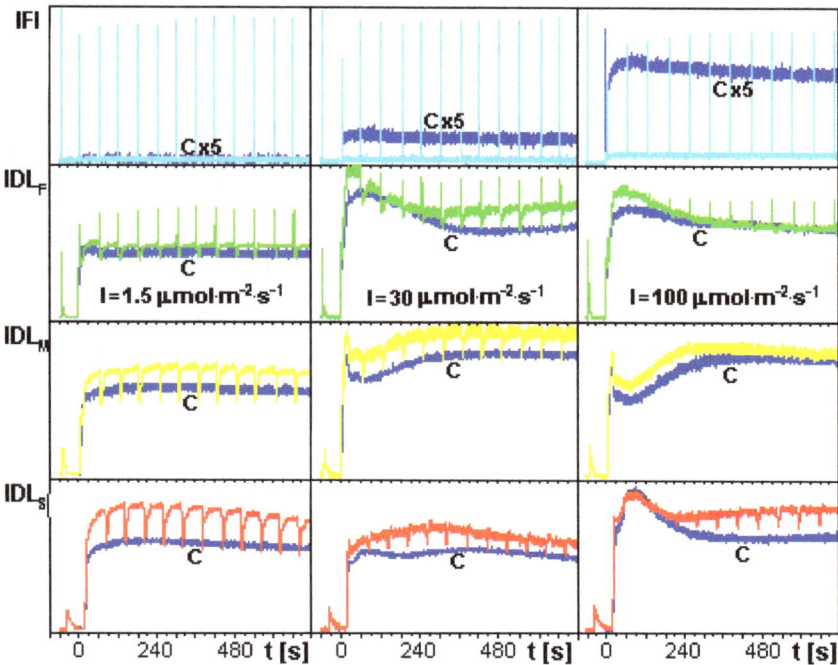

Fig. 5. Influence of additional light impulses on changes of test algae luminescence.

The first impulse at the beginning of the induction period caused small increase in intensity of all components. In the initial period of induction, we registered the smallest changes of intensity in components of delayed luminescence for all continuous light intensity levels tested. The nature of those changes depended on the excitation light intensity.

At the stream of continuous light photons exciting luminescence of 1.5 μmol(quanta)$\cdot$m$^{-2}\cdot$s^{-1}, consecutive light impulses caused temporary increase in DL$_F$ intensity in the algae suspension. After that we usually recorded a small temporary drop followed by another increase to the actual stationary level. The changes of intensity of the delayed luminescence component were correlated with temporary drops of component intensity registered after a longer period from the exposure of the suspension.

After increasing continuous light photons exciting luminescence to 30 μmol(quanta)$\cdot$m$^{-2}\cdot$s^{-1} the amplitude of DL$_F$ decreased and the drop amplitude increased. The intensity relaxation time also increased for DL$_{Chla}$ to the stationary level. At the same time, the intensity of continuous light exciting third, fourth and fifth impulse caused larger temporary drop of intensity for DL$_M$ and DL$_S$ after the impulse. Then impulses caused a drop of amplitude and stabilised for those components of the delayed luminescence. However, in the case of DL$_S$, after a temporary drop of intensity, its temporary increase could be observed (decaying slowly).

Due to the increase in the stream of continuous light photons which excited luminescence to 100 μmol(quanta) m$^{-2}\cdot$s^{-1}, the increase of DL$_F$ was smaller after the impulse and decay of intensity changes DL$_M$. DL$_S$ at the third, fourth and fifth impulse reacted with a small increase in intensity above the stationary level. However, after further consecutive impulses was could observe a temporary drop of intensity DL$_S$ followed by a small temporary increase.

4. The use of long-term delayed luminescence of chlorophyll *a* of the algal photosynthetic apparatus for testing phytoplankton

The interdependence of Chl*a* luminescence parameters and different biotic and abiotic factors is a basis for the luminescent methods for assessing properties of land plants and algae. However, the dependence of the phenomenon on conditions in which recording takes place creates problems in terms of measuring, as well as interpreting changes registered. We should remember, however, that luminescence methods provide information about luminescent properties of phytoplankton and we need to analyse a luminescence signals obtained in predefined conditions to determine other parameters. An additional difficulty while using long-term DL_{Chla} for phytoplankton testing is small intensity, which is particularly important when an *in situ* population of algae has small concentration.

Contrary to Fl_{Chla}, long-term DL_{Chla} intensity reflects concentration of reactive centres of photosystem II with separated charges. Findings of previous research show that in specific measurement conditions we may determine an important static relationship between DL_{Clha} intensity and primary productivity of phytoplankton (Krause & Gerhardt, 1984; Yacobi et al., 1998, Wilhelm et al., 2004).

Moreover, the use of monochromatic light of several wavelengths for exciting of Chl*a* luminescence enables us separating algae belonging to specific taxonomic from the population of phytoplankton (Krause et al., 1982; Gerhardt & Boedemer, 2005). However, from a practitioner point of view, the most desired is the use of luminescence for precise assessment of on line content of biomass of phytoplankton.

4.1 The use of long-term delayed luminescence for assessing the content of chlorophyll *a in situ*

The main difficulty while using luminescence for assessing the content of Chl*a* is the dependence of measurement results on several factors, such as specie composition of phytoplankton, temperature and growth conditions. This requires a frequent calibration of devices and adjustment of results. Some of those factors can be limited by incubating a sample of phytoplankton before luminescence measurement in the environment of constant temperature and light (Wiltshire et al., 1998). An additional way of limiting the impact of some of those factors is blocking electron transport in the photosynthetic apparatus of algae by adding a specific electron transport inhibitor to photosystem II of the sample, e.g. atrazine or 3-[3,4-dichlorophenyl]-1,1-dimethylurea (DCMU) (Prokowski, 1991, 2009). The inhibitors increase intensity of the DL_{Chla} fast components and reduce changes during excitation and eliminate DL_{Chla} slow components (Prokowski, 1991, 1999).

Previously, luminometers registering intensity of the DL_{Chla} fast component after a predefined exposure of algae suspension incubated in darkness were used for assessing Chl*a* content in phytoplankton (Wiltshire et al., 1998). Alternatively white light of small photons flux was used for that purpose (Prokowski, 2009). The exposure time is adjusted to register intensity of the component during temporary maximum at the induction curve (Fig. 4). White or monochromatic light is used for DL_{Chla} excitation. The signal registered is DL intensity (Prokowski, 2009; Wiltshire et al., 1998) or, in the case of luminometers capable of determining the excitation spectrum of phytoplankton, total intensity for particular wavelength of the excitation light (Yacobi et al., 1998). It was proved that in the case of a defined algae culture its DL_{Chla} intensity is a linear function of Chl*a* concentration through 5 order magnitudes (Krause et al., 1982). It was also proved that in the case of phytoplankton incubated in darkness at predefined temperature there is a linear relationship between DL_{Chla} intensity and Chl$a$ content in samples collected in a given measurement point at the same time but at different depths (Witshire et al., 1998). However, the relationship between DL_{Chla} intensity and Chl$a$ concentration depends on the date of sampling (Wiltshire et al., 1998).

Since the sensitivity of existing luminometers designated for determining the relationship between DL_{Chla} intensity and Chl$a$ concentration is limited, they are used to test the population of freshwater phytoplankton in which the concentration of phytoplankton is considerably high (Wiltshire et al., 1998; Istvánovics et al., 2005). Sensitivity can be enhanced by changing the construction of the optical part of the luminometer (Prokowski, 2001). Figure 6 includes a flow chart of a luminometer of improved sensitivity designed for continuous *on line* registration of delayed luminescence of phytoplankton incubated in predefined temperature and light with and without electron transport inhibitor.

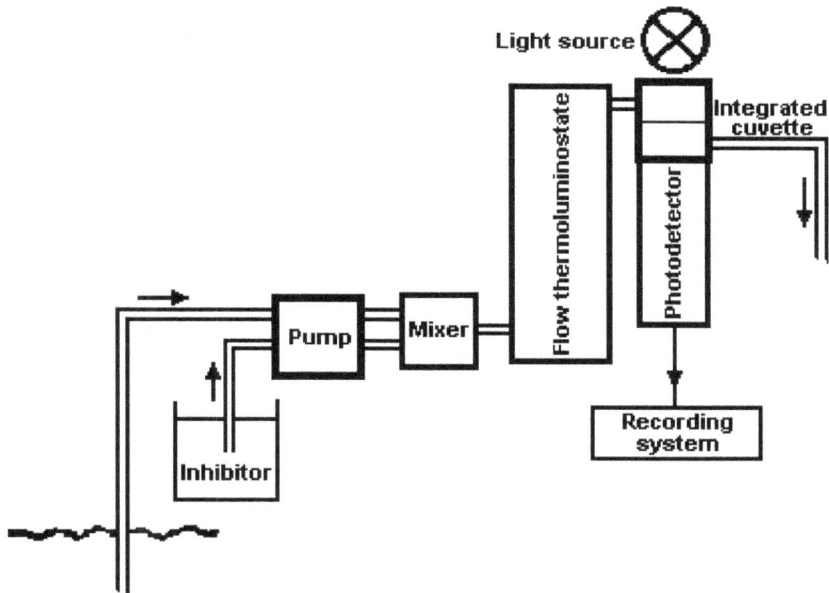

Fig. 6. Schematic diagram of flow luminometer for *on line* registration of delayed luminescence intensity of phytoplankton incubated in predefined temperature and light with and without electron transport inhibitor.

Figure 7 presents findings of research focusing on the relationship between DL_{Chla} intensity of sea phytoplankton registered *in situ* and extracted Chl$a$ content ([Chl$a$]). Long-term delayed Chl$a$ luminescence of phytoplankton was registered in time period of 0.3 s to 0.9 s for the suspension of algae incubated continuously for 3 min at 20 °C and in white light of 1 μmol(quanta)PAR·m^{-2}·s^{-1} of photon flux. The suspension was exposed to white light for 0.6 s with photons flux of 75 μmol(quanta)PAR·m^{-2}·s^{-1}. DL_{Chla} was detected using sounder with photomultiplier, which enabled registering light within the red part of the spectrum.

The study was conducted for phytoplankton from the depth of 1 m to 15 m sampled in 6 measurement points at the Gulf of Gdańsk. Chl*a* content was determined using a spectrophotometric method (Lorenzen, 1967). The ratio of chlorophyll *b* (Chl*b*) and Chl*a* determined using the spectrophotometric method by Jeffrey & Humphrey (1967) for the phytoplankton population tested was 0.263 ± 0.027, whereas the ratio of chlorophyll *c* (Chl*c*) and Chl*a* ranged 0.175 ± 0.018.

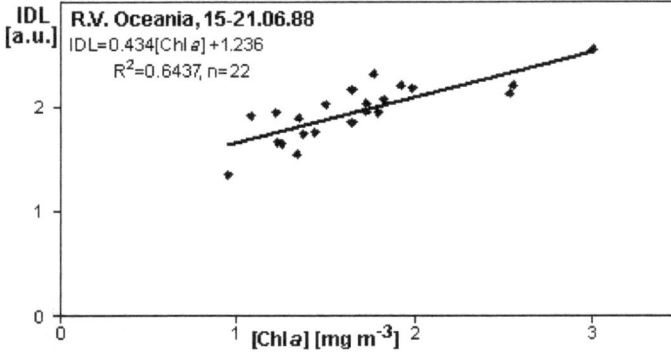

Fig. 7. Relationship between long-term delayed luminescence and concentration of chlorophyll *a* for sea phytoplankton samples.

While analysing data in figure 7, we may determine a statically significant linear relationship between delayed luminescence of phytoplankton from the Gulf of Gdańsk, which was incubated at constant temperature and lights, and concentration of Chl*a*. However, despite using incubation of samples at predefined temperature and light the value of determination coefficient R^2 for the tested population of phytoplankton was relatively low. Moreover, due to considerably small number of samples collected, it was not possible to determine whether there were any major differences of the relationship found between measurement points.

Also carried out studies to determine the relationship between DL_{Chla} intensity of sea phytoplankton registered *in situ* in the above mentioned conditions for the suspension of algae incubated continuously with the inhibitor of electron transport and the content of Chl*a* extracted using the spectrophotometric method. Karmex, urea herbicide, with DCMU as an active substance was used as inhibitor of electron transport. A saturated solution of herbicide was added to the suspension of phytoplankton to reach the concentration of the suspension 5 $\mu mol \cdot dm^{-3}$. Tests were carried out in 4 measurement points at the South Baltic Sea: R6, G2, 92A and B13. Additionally, Fl_{Chla} intensity was determined for the analogical incubated suspension of phytoplankton. The Fl_{Chla} level was recorded using a spectrophotometer with an attachment for measuring fluorescence with photomultiplier as a photodetector. Chl*a* fluorescence was excited using light of wavelength $\lambda_{max} = 436$ nm and its intensity registered at wavelength $\lambda = 690$ nm. Table 1 shows the location of measurement points, range of the sampling depth and ratios Chl*b*/Chl*a* and Chl*c*/Chl*a*.

Test points	Geographic coordinates	h [m]	Chl$b$/Chl$a$	Chl$c$/Chl$a$
R6	λ = 54° 57′ N; φ = 18° 24′ E	1-27	0.202 ± 0.043	0.269 ± 0.023
92A	λ = 54° 35′ N; φ = 18° 40′ E	1-30	0.188 ± 0.025	0.269 ± 0.022
G2	λ = 54° 50′ N; φ = 19° 20′ E	1-40	0.119 ± 0.017	0.237 ± 0.020
B13	λ = 54° 04′ N; φ = 14° 15′ E	1-10	0.204 ± 0.012	0.190 ± 0.015

Table 1. Location of measurement points, sampling depths and of chlorophyll ratios.

Figure 8 presents relationship between the intensity of long-term DL$_{Chla}$ (IDL) and Fl$_{Chla}$ (IFI) for samples of sea phytoplankton with inhibitor of photosynthesis and concentration of Chl$a$ extracted in particular measurement points. Inclination coefficients marked with different letters differed significantly from others (P < 0.01).

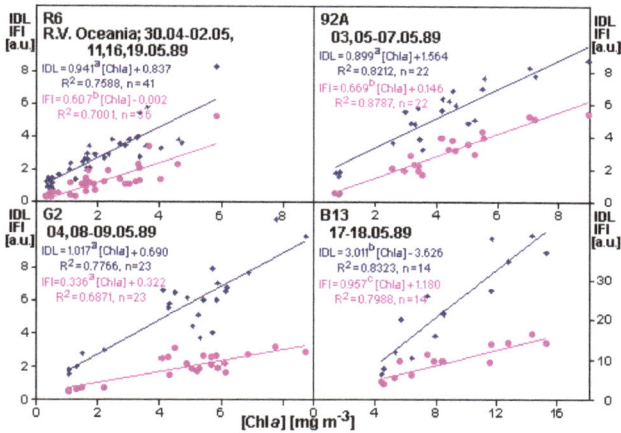

Fig. 8. Relationship between intensity of long-term delayed luminescence and fluorescence and concentration of chlorophyll a for samples of sea phytoplankton with inhibitor of electron transport in particular measurement points.

Comparing data in figure 8 we may notice major statistical differences between certain values of inclination coefficients determined for relations between luminescence from [Chl$a$] in particular sampling points. However, no major statistically significant differences were found for particular dates of sampling in a given measurement point.

In the case of linear relationship IDL = f([Chl$a$]) a major statistical difference in inclination values was found in the case of phytoplankton sampled at the estuary of the Oder to the Baltic Sea (point B13) if compared with corresponding relations registered in the case of phytoplankton collected in other points. Moreover, at the same point, the highest value of the coefficient and the lowest level of Chl$c$/Chl$a$ (Tab. 1). In the case of linear relationship IFl = f([Chl$a$]), statistically significant differences in the inclination coefficient were recorded between phytoplankton sampled at 92A and R6 and that sampled at other stations. The lowest value of the coefficient was recorded at G2 for which Chl$b$/Chl$a$ was also the lowest, whereas the highest coefficient was recorded at B13, for which Chl$c$/Chl$a$ was the lowest.

Figure 9 shows relationship between Fl_{Chla} intensity (IFI) and long-term DL_{Chla} intensity (IDL) for samples of sea phytoplankton with photosynthesis inhibitor at particular measurement points. The coefficients marked with different letters show significant differences ($p < 0.01$).

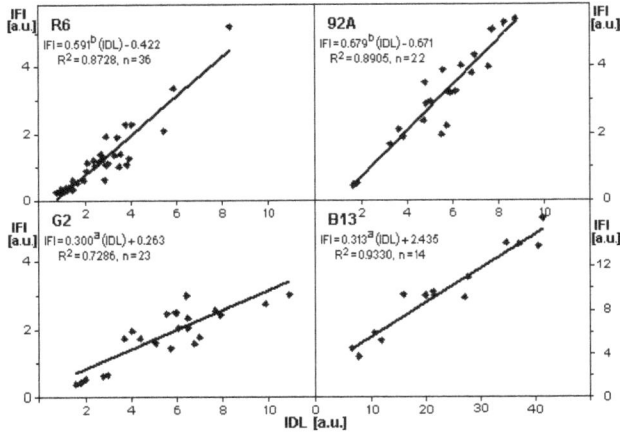

Fig. 9. Relationship between long-term DL_{Chla} intensity and Fl_{Chla} intensity for samples of sea phytoplankton with photosynthesis inhibitor.

In the case of linear relationship IDL = f(IFI), no statistically significant differences were recorded between directional coefficients of simple regression for phytoplankton sampled at R6 and 92A. Such differences were not found also for phytoplankton sampled at G2 and B13, whereas statistically significant differences were recorded between phytoplankton sampled at R6 and 92A and phytoplankton sampled at G2 and B13.

The findings show that despite incubating samples phytoplankton at identical temperature and light, with Karmex, both taxonomic composition of the phytoplankton suspension sampled at different stations and differences in conditions may have different impact on Fl_{Chla} and DL_{Chla} intensity of phytoplankton suspension samples.

In the case of IDL = f([Chla]), statistically significant differences between directional coefficients of simple regression were only recorded for phytoplankton sampled at B13 and other stations. It may indicate that in this particular case the taxonomic composition of phytoplankton suspension sampled at R6, 92A and G2 situated at the Gulf of Gdańsk as well as differences of conditions there did not influence DL_{Chla} of the phytoplankton suspension set at measurement conditions. However, the delayed luminescence of phytoplankton sampled at the estuary of the Oder River (B13) influenced factors which did not differentiate the population of phytoplankton tested at the Gulf of Gdańsk.

5. Long-term delayed luminescence of sediments

Sediments are important elements of aquatic ecosystems. They comprise a natural filter and indicate the degradation of the natural environment. Comprehensive analysis of sediments is crucial while deciding about measures aimed at protecting lakes and their reclaiming; it may also be used for more detailed chemical analysis of the aquatic environment.

An important component of such sediments is organic matter, mainly humic substances. Such substances are produced by complex chemical and microbiological processes of decomposition and secondary synthesis of plant and animal residues in water. Properties of humic substances are determined by the type of organic matter humified in a specific environment as well as various habitat related and anthropogenic factors which determine the direction of their transformation.

In the case of certain types of multiple molecular organic matter, including substances in sediments, we also observe long-term delayed luminescence (DL_{om}) (Prokowski, 2001; Istvánovics et al., 2005; Mielnik, 2009). The luminescence is emitted within the whole spectrum. Excitation spectrum shows its maximum at short wavelength (Istvánovics et al., 2005). Mechanisms leading to generating delayed luminescence in organic matter in sediments have not been sufficiently examined. Findings of some research show that in specific excitation conditions, conditions of decomposition of primary organic matter may influence DL_{om} parameters of humic substances contained in sediments (Prokowski, 2001; Mielnik 2009).

The study was conducted for sediments from ten Lobelia lakes of different trophy: 5 - oligotrophic and 5 - dystrophic. Bottom sediments were sampled during summer stagnation from the surface layer (~15 cm) at the maximum depth in a lakes. Sediment samples were lyophilised, grinded and sieved using a sieve of 1 mm mesh size. Next humic substances were extracted using two solutions: (i): alkaline solution of 0.1 mol·dm^{-3} NaOH and (ii): bi-distilled water. The weight ratio of sediment to solution was 1:10.

Research on excitation and recording of intensity of the DL_{om} of humic substances solutions were carried out with the use of a device for continuous recording of photo-induced luminescence of liquids and suspension (Prokowski, 2001). DL_{om} was excited for 0.6 s by halogen light of density of a photon flux 3000 μmol(quanta)PAR·m^{-2}·s^{-1}. DL_{om} intensity was registered 0.1 s to 0.7 s after the excitation at wavelength range from 400 nm to 600 nm. All measurements were done at 20°C.

The statistically significant differences in DL_{om} intensity between the examined humic substances were observed. The relatively lower values of the DL_{om} intensity for humic substances extracted from oligotrophic lakes sediments were registered, compared to humic substances isolated from sediments deposited at the bottom of dystrophic lakes. It may be evidence of the differences in structure of studied humic substances. In addition obtained differences may be evidence of the variable quantitative and qualitative contribution of photoluminophors in the structure of the examined humic substances molecules as well as their different photochemical reactivity. The quality of the organic matter is an important feature in respect of the physico-chemical conditions at the lake bottom. It is generally accepted that the quality or the conditions for decomposition are determined by the ratio of organic carbon content to total nitrogen – C/N (Meyers, 1997).

Figure 10 presents the relationship between ratios of organic carbon and total nitrogen in bottom sediments (Co/No), water extract (C/N H$_2$O) and alkaline extract (C/N NaOH) and the value of DL_{om} intensity registered for water extract to DL_{om} intensity for alkaline extract (IDL H$_2$O/IDL NaOH).

While analysing data in figure 10, we may notice highly statistically significant linear relationship between C/N in bottom sediments and alkaline extract and IDL H$_2$O/IDL NaOH. Moreover, in the case of sediments sampled from oligotrophic lakes, IDL H$_2$O/IDL NaOH was the lowest. The findings may indicate the impact of the primary composition of organic matter and conditions of its decomposition on DL_{om} registered.

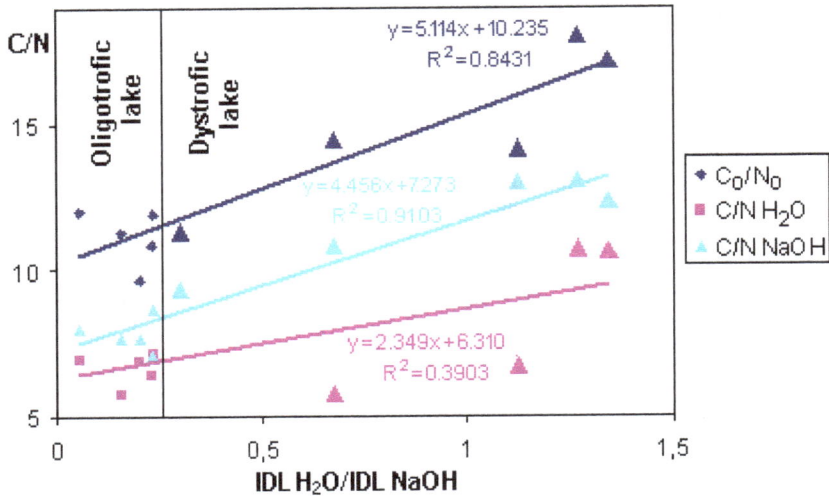

Fig. 10. Relationship between ratios of organic carbon and total nitrogen and ratios of intensity of delayed luminescence of humic substances in the water and alkaline extracts.

6. Conclusion

While studying the aquatic environment, luminescence can be a useful tool to determine certain properties of phytoplankton and assess threats to aquatic ecosystems, including the use of algae tests for assessing water quality. According to the research, in comparison with the delayed luminescence, mechanisms generating excited Chl$a$ particles in the case of fluorescence are different and therefore the kinetics of their induction and reaction for a given factor is different as well.

Changes of fluorescence during exposure of the phytoplankton suspension and changes of intensity of long-term components of delayed luminescence provide information respectively about processes taking place in the photosynthetic apparatus immediately after light absorption and electron transport in the photosynthetic apparatus.

Registering changes in fluorescence and several component of delayed luminescence during exposure of a given sample provides for selective determination of primary mechanisms of photosynthesis. Relationship between changes of luminescence emission also enables testing the influence of various stress factors on photosynthesis. Moreover, the method can be used to examine *in situ* the suspension of phytoplankton and provides much more information about a given object. At the same time, the method enables defining several luminescence indicators and can be particularly useful in algae tests for diagnosing the status of the environment.

Findings indicate that humic substance exposed to light undergo photochemical processes accompanied by creation of long-term fluorophores capable of producing electromagnetic radiation in the process of luminescence. The use of luminescence in testing organic substances of various origin may provide information about the structure and nature of organic links, as well as photoreactivity of those substances. Additionally, based on findings from luminescence tests it is possible to determine changes which occur in those substance in time, in particular in the process of humification.

The method of using long-term delayed luminescence can play a major role in testing humic substances from various ecosystems, especially that the method is exceptionally sensitive, easy and quick. Eventually, this should save time and cost of such tests. The use of the method may significantly reduce the number of analyses and loss of information necessary to draw conclusions.

Therefore, while researching water environments, luminescence may provide important information concerning specific conditions in the environment. It is necessary, however, to develop new methods to register on line luminescence properties of phytoplankton and sediments found in water bodies. It seems that on line methods enabling simultaneous registration of fluorescence and delayed luminescence have the largest potential in terms of information provided.

7. References

Gerhardt, V. & Bodemer, U. (2005). Delayed Fluoresence Excitation Spectroscopy – an in vivo Method to Determine Photosynthetically Active Pigments and Phytoplankton Composition. *Algas*, No. 33, pp.4-18, ISSN 1695-8160

Istvánovics, V.; Honti, M.; Osztoics, A.; Shafik, H.M.; Panisk, J.; Yacobi, Y. & Eckert, W. (2005). Continuous Monitoring of Phytoplankton Dynamics in Lake Balaton (Hungary) Using on-line Delayed Fluorescence Excitation Spectroscopy. *Freshwater Biology*, Vol.50, pp. 1950-1970, ISSN 1365-2427

Jeffrey, S.W. & Humphrey, G.F. (1975). New Spectrophotometric Equations for Determining Chlorophylls a, b, c1 and c2 in Higher Plants, Algae and Natural Phytoplankton, *Biochemie und Physiologie der Pflanzen,* Vol.167, pp. 191-194

Jursinic, P.A. (1986). Delayed Fluorescence: Current Concepts and Status, In: *Light Emission by Plants and Bacteria*, Govindjee; J. Amesz & D.C. Fork, (Ed.), pp. 291-328, *Academic Press*, ISBN 0-12-294310-4, Orlando, USA

Krause, H.; Helml, M.; Gerhardt, V. & Gebhardt, W. (1982). In vivo Measurements of Photosynthetically Active Pigment Systems in Fresh Waters Using Delayed Luminescence, *Archiv für Hydrobiologie Beiheft Ergebnisse der Limnologie*, Vol.16, pp. 47-54

Krause, H. & Gerhardt, V. (1984). Application of Delayed Luminescence in Limnology and Oceanography, *Journal of Luminescence*, Vol.31/32, pp. 888-891

Lavorel, J.; Breton, J. & Lutz, M. (1986). Methodological Principles of Measurement of Light Emitted by Photosynthetic Systems, In: *Light Emission by Plants and Bacteria*, Govindjee; J. Amesz & D.C. Fork, (Ed.), pp. 57-98, *Academic Press*, ISBN 0-12-294310-4, Orlando, USA

Lorenzen, C.J. (1967). Determination of Chlorophyll and Pheopigments: Spectrophotometric Equations, *Limnology and Oceanography.* Vol.12, pp. 343-346

Meyers, P.A. (1997). Organic Geochemical Proxies of Paleoceanographic, Paleolimnologic, and Paleoclimatic Processes. *Organic Geochemistry* Vol.27, No.5/6, pp. 213-250

Mielnik, L. (2009). The Application of Photoinduced Luminescence in Research on Humic Substances of Various Origins. *Oceanological and Hydrobiological Studies*, Vol.38, No.3, pp. 61-67, ISSN 1730-413X

Prokowski, Z. (1991). Effect of DCMU on Induction Kinetics of Long-lived Delayed Luminescence in Green Algae. *Photosynthetica*. Vol.25, pp. 223-226

Prokowski, Z. (1993). Effect of HgCl$_2$ on Long-lived Delayed Luminescence in Scenedesmus quadricauda. *Photosynthetica* Vol.28, pp. 563-566

Prokowski, Z. (1999). Luminescence of a Chlorophyll of Green Alga Scenedesmus quadricauda. *Advences of Agricultural Sciences Problem Issues*, Vol.469, pp. 643-650, ISSN 0084-5477

Prokowski, Z. (2001). Measurement Set for Continuous Determination of Photoinduced Luminescence of Liquids and Suspensions. *Acta Agrophysica*, Vol.48, pp. 157-166, ISSN 1234-4125 (in polish)

Prokowski, Z. (2009). The Use of the Delayed Luminescence Method for Determinations of Chlorophyll *a* Concentrations in Phytoplankton. *Oceanological and Hydrobiological Studies*, Vol.38, No.3, pp. 43-49, ISSN 1730-413X

Seppälä, J. (2009). Fluorescence Properties of Baltic Sea Phytoplankton. *Edita Prima Ltd*, pp. 1-80, Helsinki, ISBN 978-952-10-5589-8

Strehler, B.L. & Arnold, W. (1951). Light Production by Green Plants. *Journal of General Physiology*, Vol.34, No.3, pp. 809-820

Wilhelm, C; Becker, A.; Toepel J.; Vieler A. & Rautenberger R. (2004). Photophysiology and Primary Production of Phytoplankton in Freshwater. *Physiologia Plantarum*, Vol.120, pp.347-357

Wiltshire, K.H.; Harsdorf, S.; Smidt, B.; Blöcker, G.; Reuter, R. & Schroeder, F. (1998). The Determination of Algal Biomass (as Chlorophyll) in Suspended Matter from the Elbe Estuary and the German Bight: A Comparison of High-Performance Liquid Chromatography, Delayed Fluorescence and Prompt Fluorescence Methods. *Journal of Experimental Marine Biology and Ecology*, Vol.222, pp. 113-131

Yacobi, Y.Z.; Gerhardt, V.; Gonen-Zurgil, Y. & Sukenik, A. (1998). Delayed Fluorescence Excitation Spectroscopy: a Rapid Method for Qualitative and Quantitative Assessment of Natural Population of Phytoplankton. *Water Research*, Vol.32, No.9, pp.2577-2582

3

Open-Sea Observatories: A New Technology to Bring the Pulse of the Sea to Human Awareness

I. Puillat et al.*
*IFREMER Centre de Brest, REM Department, BP 70,
Plouzané,
France*

1. Introduction

Historically, observation in Marine Science was mainly based on *in situ* measurements made mainly over ship surveys and shore measurements. Unfortunately, ship surveys can only be episodic, and are constrained by weather and by the constant rise of ship-time cost. As the data provided by non-communicating moorings are stored in the measurement system, a ship intervention is needed to recover both the mooring and the data after several acquisition months. Further to the rather successful medium- and short-term deployment of these traditional devices, scientists have expected the development of long-term observations and permanent marine system-monitoring tools so as to gain more insight into the observed processes. By providing additional information, satellite technology can partly solve this gap between the reality and expectations. However, even though satellite images provide information over a large time frame (from minutes to years) and a wide range of spatial resolutions (from metres to thousands of kilometres), they only cover the upper layer of the sea. An Open-Sea Observatory is a complementary tool that allows one to make, in the water column and on the seafloor, long-term measurements of many environmental parameters and to acquire them in real-time, or near real-time. In addition to this real-time data transmission, these systems permit remote intervention by humans when needed, and thus can be considered as 2-way communicating devices. Because of these two characteristics, observatories are innovative systems that bring internet to the ocean and make the ocean reality visible to the human eye. According to our definition of an Open-Sea observatory, other very useful observation tools such as gliders, floats, repeated profiler transects, etc. will not be considered in this chapter to only focus on such ocean observatories.

Observatory initiatives have been spreading worldwide since the 1990s. In Europe, several initiatives started twenty years ago so as to upgrade free-fall systems from the sea surface (the so-called "landers") to make them 2-way communicating and to develop bottom

*N. Lanteri, J.F. Drogou, J. Blandin, L. Géli, J. Sarrazin, P.M. Sarradin, Y. Auffret, J.F. Rolin and P. Léon
IFREMER Centre de Brest, REM Department, BP 70, Plouzané, France

stations linked to a ground station by acoustic or cable connection. The development of observatory expertise in different ways and places has led scientists to federate their efforts around common projects, amongst which the most recent ones in Europe are EUROSITES, ESONET and EMSO (see next section). Worldwide, the most advanced network of cabled multidisciplinary observatories is the Neptune Canada infrastructure led by the University of Victoria; its 800-km-long cable was installed by Alcatel Submarine Networks in autumn 2007. Five nodes with their associated instruments were deployed in 2009. The infrastructure is currently running and the data are available on the main website http://www.neptunecanada.ca/. In the USA, the Ocean Observatory Initiative (OOI), driven by the National Science Foundation (NSF), is currently developing observatories in three steps: coastal observatories, cabled region observatories (the Neptune US observatory) and global observatories. In 2006 Taiwan decided to develop and implement a submarine observatory off the western part of the island (MACHO project). Japan chose to install a 20-node submarine network, devoted to seismic monitoring and tsunami alerts, DONET (http://www.jamstec.go.jp/jamstec-e/maritec/donet/). China recently implemented an observatory network off its eastern coasts (http://www.bulletins-electroniques.com/actualites/66736.htm); it is dedicated to research in oceanography and geosciences by scientists from the partner universities (Tongji university Jiaotong Shanghai University, Huadong Normal University and Zhejiang University). In Europe, several initiatives started about twenty years ago. Clearly, deep-sea observatories are a new technological approach that has become accepted and is being implemented worldwide. Why is this?

This is due to scientific and societal needs. Indeed, only long-term observatories allow continuous observations over long-timescales (at least 10 years) of very numerous parameters collected through power intensive sensors. This capability is crucial in observations of natural processes that are either very episodic or hidden by higher-frequency noise, and thus only detected through acquisition of long data time-series. These long-term observatories are also an opportunity for society to monitor the sea in "real-time", and thus to prevent geo-hazards such as earthquakes and tsunamis. It also allows monitoring of long-term phenomena such as those caused by the global warming. The open-access policy for collected data makes them more accessible to the general public, stakeholders and socio-economic users. Furthermore, through support to those research areas where access to high-quality data is relatively rare, it widens the community that gets access to deep-sea research and attracts worldwide research cooperation. Long-term observatories are crucial for the scientific community so as to maintain the high-level reached in the research developed under the scope of past and present framework programmes.

In the following pages, we will show how observatories (i) can serve various marine research fields, in a broad sense, such as oceanography, biogeochemistry, volcanology, seismology, and geosciences and (ii) help to enlarge the vision and the understanding of marine sciences at the crossroads of these disciplines throughout Europe. The history of observatory development in Europe will be reminded in Section 2 to explain the current status, the available infrastructures and equipment. Their deployment and associated feedback as well as the expected further steps will be illustrated from on-site deployment case-studies in Section 3.

2. History of open-sea observatories in Europe

The first fixed platforms dedicated to quite long-term measurement were traditional moorings initially deployed at sea for time-span of a few months to one year according to the system autonomy in terms of energy and data storage capacity. Mooring provided unprecedented long multi-parameter time-series for oceanographic research. Nevertheless, deployment and sustainability for periods over, at least, one year imply that one has to come back to the field site, at least twice a year, for data recovery, power supply check up and eventually sensor cleaning. Under these conditions, a mooring can be maintained for several years, but a major difficulty is to get ship time at both the right periods of time and good weather conditions, which turns out to be costly. In addition, the lack of communication system during the acquisition months makes impossible to know whether the system is working correctly or not prior to data retrieval. This raised the need for development of more energy-autonomous systems able to communicate and transfer data to a ground station. These new systems should include an onboard electronic unit including, at least, a centralised 2-way communication system with a data logger. These considerations are at the origin of the observatory concept: an infrastructure equipped with several sensors and an onboard communicating electronic unit. Several types of observatory systems have, thus, been developed and deployed: let us cite upgraded landers and single acoustically-linked bottom stations further upgraded to multi-node stations, cable infrastructures and finally the observatories in use today. Two complementary review articles were published in 2006 (Person et al. 2006; Favali and Beranzoli, 2006); it appeared to us worth writing the update presented in this paper.

2.1 Equipment and infrastructure evolution until now

Several initiatives were precursors to observatory systems; let us cite the so-called "landers", which are dropped from the sea surface to free-fall to the sea bottom. The reader will find an interesting list of these systems in the final activity report of the ESONET CA project (http://www.oceanlab.abdn.ac.uk/esonet/ESONET_fullrep.pdf). Amongst the systems presented in this review paper, we will consider only those able to communicate with a ground station further to a dedicated upgrade permitting the addition of a 2-way communication unit.

2.1.1 Upgraded landers

The BOBO lander developed by the NIOZ (Van Weering et al., 2000) consists of a frame with 3 legs (2-m-high). At their foot, the lander is 4-m-wide (figure 1). As its frame was specially designed to endure stays of one year at least on the seabed, it was, thus, built with dedicated material. To prevent attacks by corrosion or electrolysis, the construction parts and connections were very carefully isolated. The instrumentation is attached to both the hexagonal frame and the lander legs. Electrical power is supplied to the instrument by a built-in battery pack housed in a glass sphere.

In Sweden, the Göteborg University (Hall et al.) has developed and operated autonomous landers since the early 1990s. Then, research institutes from France, Denmark and the USA have developed and used 5 different lander systems. Considerable efforts have also been made in the development of new sensor technology (dedicated, for example, to oxygen sensing). Later, within the framework of the EU-ALIPOR project, the consortium interconnected all sensors and systems into a modern network (a CAN network, *Controller Area Network*), which allows high-speed and high-security communication between instruments and sensors. The advantage of such a network is that new instruments and sensors can be hooked up in "plug and play" mode to the existing network and all data can be collected by a control unit and transmitted to the users network which is in wireless 2-way communication (acoustic) with the surface.

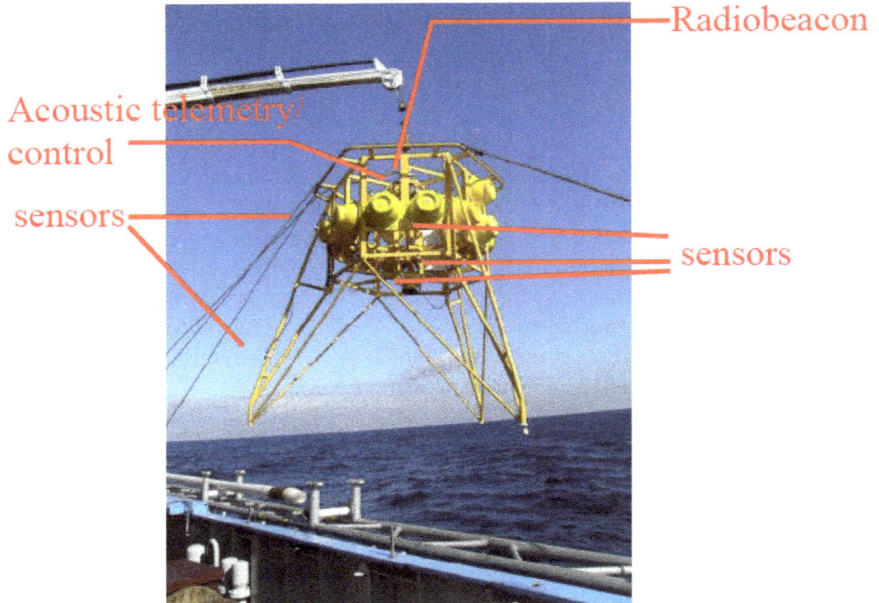

Fig. 1. BOBO lander in previous version (source: http://mars-srv.oceanlab.iu-bremen.de/infra_obser_bobo.html)

IFM-GEOMAR has operated a series of 8 landers of modular design for investigations of the deep-sea benthic boundary layer. A special launching device is used for their deployment, either in conventional free-fall mode or in targeted mode, on hybrid optical fibre or coaxial cables. The lander is accurately positioned by the launcher, and then softly deployed and quickly disconnected through activation of an electric release. The bi-directional video and data telemetry provide online video transmission as well as the power supply (< 1 kV) and surface control of various relay functions. Autonomous lander clusters connected by optical cable and ensuring data transmission to the surface, and in the future by satellite link to the shore, are envisioned as an important contribution to the sea observatories of the future. A lander cluster (figure 2) consists of various types of scientific observation-dedicated landers, power supplies and garages for small autonomous vehicles (AUV, crawler) and tethered ones (ROVs).

Fig. 2. Lander cluster developed by IFM-GEOMAR (source IFM- GEOMAR: http://www.ifm-geomar.de)

2.1.2 Acoustically linked bottom stations

The first steps towards long-term and multi-sensor bottom stations deployed in Europe by remotely controlled vehicles such as ROVs took place in the 1980s. Let us cite, for example, NADIA, SAMO and GEOSTAR systems that worked as single node platforms. Later, engineers developed multi-node platforms such as those used in the ORION and ASSEM projects.

2.1.2.1 NADIA and SAMO systems

In France, Ifremer undertook several early initiatives such as NADIA developed in 1985 and SAMO (Person et al., 2006). The former (figure 3) was an autonomous structure, moved and controlled by the Nautile submersible, and designed for deployment into Ocean Drilling Program (ODP) holes (Montagner et al., 1994). Then, the SAMO 1 & 2 stations (Abyssal Station for Oceanographic Measurements) were designed from this technology so as to continuously monitor hydrothermal vents. SAMO was a communicating infrastructure equipped with a camera for acoustic and real-time transmission of images as well as temperature and turbidity data to the surface. The first deployment took place in 1991 during the HERO campaign. Later, in 1993, the HYDROGEO structure was deployed within the framework of an Ocean Drilling Programme (ODP) to characterise temperature and pressure fields and their variations associated to the fluid dynamics in drilling holes. It consisted of 20 thermistance chains, and 3 pressure sensors. It was deployed in 1994 (June-July) in Hole 948 (15° 31'N; 56° 43'W) at a depth of 4 938 m, for more than one year, and recovered on Nadire R/V with the Nautile submarine in 1995 (November and December).

Fig. 3. NADIA observatory developed by Ifremer in 1985, © Ifremer

2.1.2.2 GEOSTAR class observatories

The Geophysical and Oceanographic Station for Abyssal Research, GEOSTAR, is a prototype of bottom stations developed under the leadership of INGV (National Italian Institute of Geophysics and Volcanology). This observatory prototype comprises a set of geophysical and oceanographic sensors for continuous, time-referenced and synchronised measurements and it is deployed from the sea surface through a dedicated and remotely controlled tethered vehicle, MODUS. GEOSTAR has been submitted to a set of pilot experiments in the Mediterranean Sea and the Atlantic Ocean prior to the development of additional seafloor observatories of different sizes and pieces of equipment. The observatory is a 4-leg marine aluminium frame of slightly less than 10 m³, which holds all of the devices constituting the scientific and operative payload (figure 4). The first phase of the project was completed in late 1998 with the 2-week deployment of a prototype in shallow waters (40 m) of the Adriatic Sea.

The success of GEOSTAR led to the subsequent development of a class of observatories designed for different applications, but which shared common solutions and infrastructures (EC ORION project, Italian projects SN-1 and MABEL). Two GEOSTAR-class observatories and associated instruments with telemetry were linked together into a submarine network known as ORION (Ocean Research by Integrated Observatory Networks, 2002-2005). Between 2003 and 2005, this network operated in the Tyrrhenian Sea at the base of the Marsili Volcano seamount (3 320 mwd), with acoustic communications of data and commands between the observatories. One of these observatories, GEOSTAR, could also communicate with a surface buoy. The system provided important results and represented significant advances in communication systems. However, it is worth noting the existence of limitations on the data rate imposed by the current performances of the acoustic telemetry. In order to increase data flow to a ground station for application addressing geo-hazards (e.g., earthquakes), INGV had the opportunity to upgrade one of the observatories to a cable configuration (Favali et al., 2006a). More information can be found in Favali et al. (2006b) and on the website: http://roma2.rm.ingv.it/en/facilities/seafloor_multidisciplinary_observatories/1/geostar.

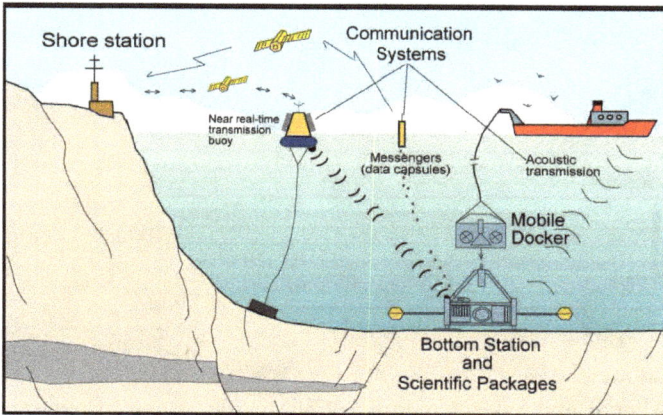

Fig. 4. GEOSTAR bottom station (from Favali and Beranzoli, 2009)

2.1.2.3 ASSEM multi-node technology: Array of Sensors for long-term Seabed Monitoring of Geohazards

ASSEM was the first application of a new observatory concept dedicated to the long-term monitoring of a limited area (some km²) and based on a network of interconnected measurement nodes. This FP5 European project was coordinated by Ifremer in 2001-2004. Its design was based on a cost-effective and light platform with a standardised layout of sensors sharing a common data and communication infrastructure. The ASSEM project enhanced some marine technologies dedicated to the real-time monitoring of the seafloor. The system main component was the COmmunication and STOrage Front end (COSTOF) permitting a local storage of data and data transmission as well as exchanges between the whole set of sensors and the external world through an underwater network. ASSEM consisted in an array of several nodes to be deployed in water and sediment by a submersible or a ROV (figure 5). Their design and sensor package (pore pressure, methane, geodesy, tiltmeter, CTD, turbidity, currents) were alike. Data were exchanged between the nodes through networked acoustic modems. One of the nodes was connected to either a surface buoy and then to the shore by satellite communication or to a ground station with a cable. ASSEM was designed as a light (less than 250 kg) and deep (4 000 m water depth-rated) platform. The first prototype was produced at Ifremer, Brest, in January 2004. Its dual objectives were: i) monitoring of a set of geotechnical, geodesic and chemical parameters distributed on a specific seafloor area in order to gain more insight into the slope instability phenomena and ii) assessment with anticipation, whenever possible, of the associated risks (Blandin et al., 2002, 2003, 2005). Two test sites were selected for validation of the basic concepts: they were located in Norway and in the Gulf of Corinth, respectively. Its use in a set of new targets, in particular within the framework of ESONET Demonstration Missions, has been fostered by the success of these field experiments. For further information refer to the ESONET NoE newsletter, ESONEWS (spring 2008).

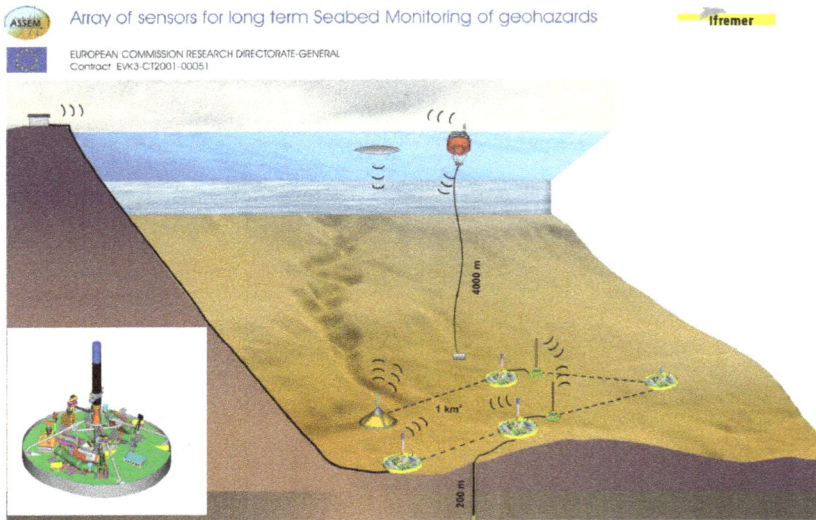

Fig. 5. An ASSEM node and a sketch of multi ASSEM node system ©Ifremer

2.1.3 Communicating moorings

The first important experiment led to the implementation, around Europe, of a network of communicating moorings managed within the framework of the ANIMATE project. ANIMATE ("Atlantic Network of Interdisciplinary Moorings And Time-series for Europe") was funded by the EU in 2002-2004 under the scope of the 5th Framework Programme and then was continued, in part, by the EU-funded projects Mersea IP and Carbo-Ocean. The project was launched in 2002 so as to develop a European carbon cycle time-series infrastructure at 4 key sites in the north-east Atlantic Ocean (figure 6a). It was based on the implementation of real-time telemetry of subsets of the prime data to be collected for immediate dissemination to the scientific community through satellite link, and via the Internet for the general public.

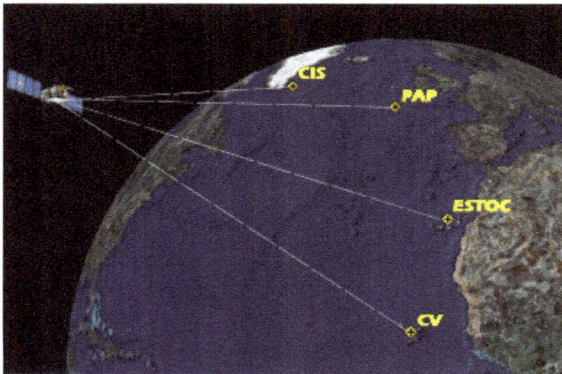

Fig. 6a. Four deep ocean mooring sites managed within the framework of ANIMATE and then EuroSITES (source: http://www.noc.soton.ac.uk/animate)

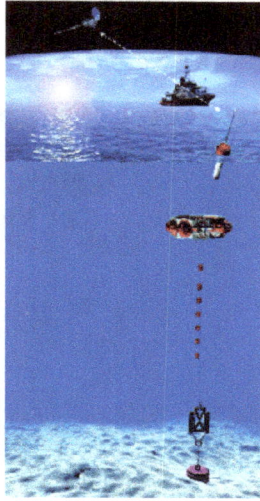

Fig. 6b. Concept of an Animate mooring (source: http://www.noc.soton.ac.uk/animate)

The 3 deep ocean mooring sites denoted as CIS, ESTOC and PAP have collected data for three years under the ANIMATE banner. Further to the validation of the original ANIMATE mooring concept (figure 6b), additional funding enabled one to enhance and refine the systems for provision of a set of long-term time-series to be included in the general effort dedicated to climate change monitoring. In 2006 a fourth mooring was developed at the Cape Verde Islands. In April 2008 the ANIMATE network became part of the EuroSITES European Ocean Observatory Network aimed at becoming a permanent infrastructure.

2.1.4 Cabled observatories

General Concept

Several layouts have been proposed for the development of cabled observatories connecting undersea sites of scientific experiments:
- Single line of Sensors or Nodes
- Sensor Rings
- Sensor Meshes

An illustration of the general concept is presented below (figure 7).

The scientific systems deployed to date have predominantly consisted in a single cable routed from the landing point to the sea, where a single sensor or several ones are connected to the cable. Power is supplied to the cable from the shore station; the return path is either a second conductor in the cable or via a seawater return that employs a seawater ground anode at the end of the cable and a ground bed on the beach. The fundamental drawback of a single cable is its susceptibility to damage (electrical shortage or cable cut).

The design of one of the very first sensor systems by Neptune Canada (University of Victoria) is based on a ring configuration (figure 8) with 2 cables routed to the same landing point via physically different routes. In the event of accidental damage to one leg, the other one should be still able to supply power to the ring and to transmit data to and from the field site.

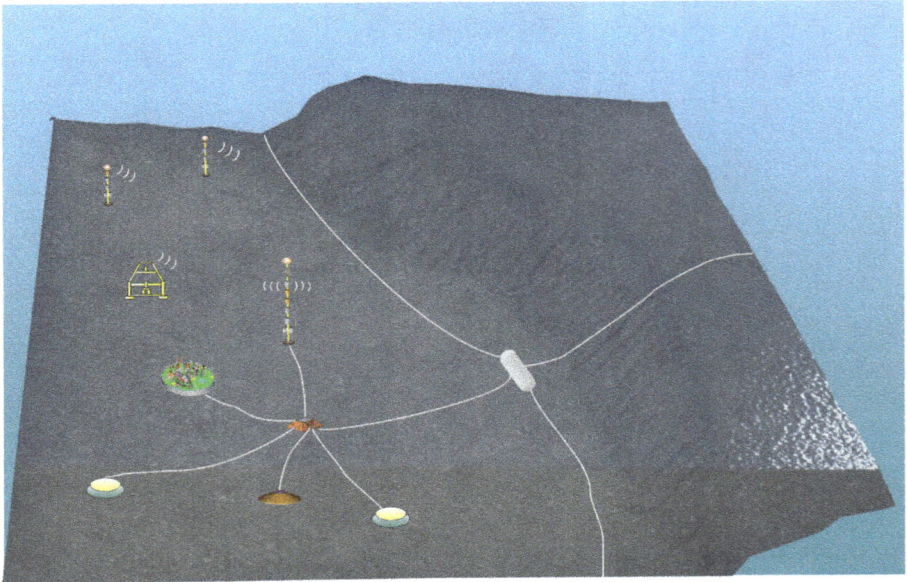

Fig. 7. Deep-Ocean observatory concept showing cabled and local wireless access from fixed and sensor systems. ©Ifremer/M. Chapon-Ashaine

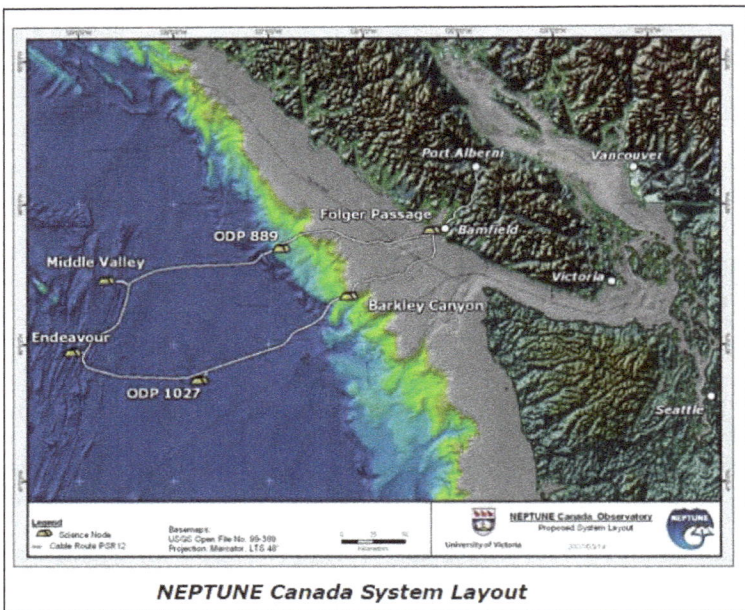

Fig. 8. Route of Neptune Canada cable infrastructure (source: http://www.neptunecanada.ca/)

Whatever the concept, the main components of a cabled observatory are concerning cables and nodes, junction boxes, permanent cabled instrumentation, short-term scientific packages, eventually associated with moored buoys with or without an electro-opto-mechanical cable. More generally, construction and maintenance of the observatory shall cover:
- Field site surveys
- Module lifting and lowering to seabed
- Cable laying and underwater connections
- Inspection and maintenance works

A common feeling is that that the main backbone of cable and cable fittings (branching unit, spur cable, main node) will be likely deployed by a fleet specialised cable installation and maintenance. However, after the initial deployment by a specialised vessel, the next interventions over the observatory operational life may be achieved by multi-purpose oceanographic vessels. This operation life includes generic operations such as those described below in the example of the ANTARES telescope about the secondary junction box and the connected permanent instrumentation.

Example of ANTARES cabled observatory

The aim of the international ANTARES (Astronomy with a Neutrino Telescope and Abyss environmental RESearch) collaboration is to detect and to study the production of high-energy neutrinos in the Universe. The ANTARES infrastructure is also a permanent marine observatory providing high-bandwidth real-time data transmission from the deep-sea for geosciences and marine environmental sciences.

The main cable lands near Toulon, a south-eastern coastal site in France, and runs along ~40 km offshore and down to2 500 m in depth, where the primary junction box is connected to the mooring network for power supply and data transfer (see figure 9). The first mooring line equipped with particle detectors was moored in early March 2006. The network was completed on the night of May 30, 2008, when the final 2 ANTARES lines were powered on, which thus brought the total number of detection lines to 12 and marked the completion of the largest undersea neutrino telescope available to date. A 13th line has been instrumented specifically for marine science research. The reader can get an overview of the system building and deployment at http://antares.in2p3.fr/News/index.html. Because of the need to extend the network, it was decided to plug a Secondary Junction Box (BJS) with an additional infrastructure.

This additional infrastructure installed on the ANTARES site was developed by Ifremer Toulon. The BJS was deployed 2 500 m deep in the Mediterranean Sea in November 2010 over a scientific cruise on the Research Vessel, *"PourquoiPas"*, and the connections were made via the Ifremer ROV, Victor6000. This upgraded network is now fully operational, with scientific modules already connected and data provided in real-time.

The BJS connected to the main *ANTARES* junction box provides Internet communication interfaces between the scientific equipment and the shore. Its 7 wet mateable ports for users allow data transmission at a rate of 100 Mb/s and supply power to the connected pieces of scientific equipment. In this way, every scientist can get access, in real-time, to the whole set of sea sensor data from the ground laboratory. Users can get access to the system from anywhere in the world and can interact with their own instruments as if they were on their workbench. The target requirements in the design of the whole BJS were high-reliability and 20-year life expectancy. Though the concept is simple, one should be aware that the physical system is much more complex.

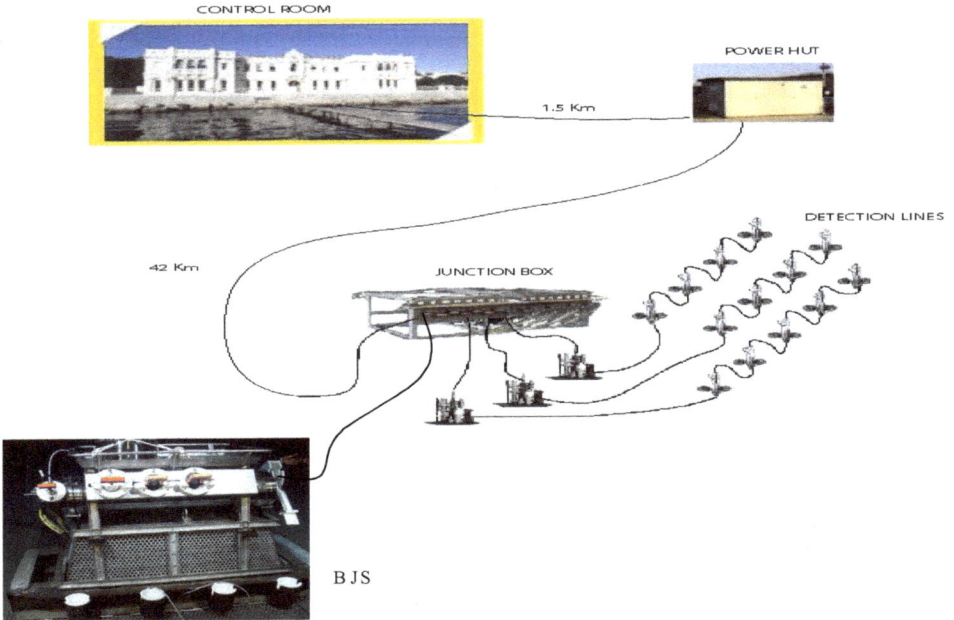

Fig. 9. Scheme of the ANTARES cable network (source: Ifremer and CNRS/CPPM)

The main design features of BJS power systems are to provide 400 V (DC) with 1 kW full power and to ensure a total protection of the power supply input, even in case of electrical fault at the output. This requirement calls for a highly accurate monitoring of the delivered current as well as detection of leakage and ground fault at the outputs. The entire system is monitored by a fully autonomous controller: the system is thus always powered and can restart the switch and the power generation in case of alarm signal. Prior to the deployment of the BJS electronic titanium housing, a heavy sled was built and laid down on the seafloor. This sled is equipped with two platforms mounted on either side so as to keep the ROV (Victor 6000) stable during the connexion operations achieved with the 2 manipulating arms.

Different scientific modules have been connected to the BJS on the 2 500 m-deep seabed, and are now delivering real-time data. Generic oceanographic sensors developed by CNRS/DT-INSU have been deployed for the real-time monitoring of key oceanographic parameters, e.g. temperature, salinity, current-velocity and -direction, pressure and oxygen concentration. Moreover, biolumiscence data have been recorded on the site with a camera. A multi-parameter seismometer module for seafloor motion observation proposed by the *Geosciences Azur* laboratory has also been connected to the BJS to meet both operational and scientific goals. A few days after the connection, the instrument recorded a rare seismic event of magnitude 4 that took place between Toulon and Corsica. The available data contributed to a better location of this event. This new infrastructure presents all the capabilities for a European demonstration mission with real-time data acquisition, in order to test recently-developed instruments and observatory technologies, e.g. a novel concept of deep-sea sensor network, or a new design of economic wet mateable connector, or a highly accurate synchronisation system. By nature open-sea observatory research is mainly interdisciplinary and has the potential to open the way for very significant advances in the field of science of concern.**Other cable observatories in Europe**

The ANTARES initiative is not isolated in Europe; a neutrino telescope has also been implemented off Sicily under the leadership of the Italian National Institute of Nuclear Physics (INFN). Similarly, in October 2002, INGV team deployed an updated version of the GEOSTAR system for long-term measurements: the SN-1 observatory, which is currently in operation offshore Eastern Sicily in about 2 100-m-deep waters and in a cabled configuration thanks to an agreement with INFN, which has made accessible a 25-km-long electro-optical cable. The GEOSTAR-class observatories have so far provided a huge amount of long-term time-series of seismological-, gravity-, magnetic-, geochemical- and oceanographic-measurements that constitute an unprecedented resource to gain more insight into the complex and maybe inter-related processes that take place in the deep sea. A cabled observatory concept is currently under development in coastal zones as shown by the MeDON initiative (Marine e-Data Observatory Network:http://www.medon.info/). Today the sensor packages are tested in Plymouth (UK) littoral zone waters, and a complete demonstrator will be deployed off Brest (France).

2.2 Toward a shared observatory infrastructure

Given that most of the most recent infrastructures are still in operation, capabilities of long-term sea observation are currently available, and it is more than likely that it will be alike in the near future. This is the case of both cabled observatories with and their modules such as GEOSTAR and ASSEM nodes, communicating moorings and communicating landers. Nevertheless, strong synergy and common governance are essential prerequisites for sustainability over decades, money- and time-saving and securing the technical know-how. This implies firstly the preparation of a state-of-the-art on existing capabilities with the assessment of their possibilities for the future, then a convergence of common procedures and pieces of equipment. This would allow the development of a European infrastructure involving joint participation at every step of the way.

An initial review of open-ocean observatory capabilities in Europe was completed in the ESONET-CA "Concerted Action" project co-funded by the EC (2002-2006) and led by Aberdeen University. Then ESONIM, a European Specific Support Action (SSA), led by the Irish Marine Institute (IMI), took the ESONET-CA plan a step further by producing a practical and flexible business plan to establish a seafloor observatory based on the ESONET Celtnet Porcupine site. In June 2007, the European Marine and Maritime Science and Technology Community defined a strategy for the community through the Aberdeen Declaration: this was the actual beginnings of the ESONET Network of Excellence (ESONET-NoE) and of the upcoming projects, EMSO and EuroSITES.

The European Seas Observatory NETwork of Excellence (ESONET-NoE, http://www.esonet-emso.org) (2007-2011) was coordinated by Ifremer (France) and co-funded by the European Commission. ESONET-NoE aimed to promote the implementation and management of a dedicated network of long-term multidisciplinary ocean observatories in the deep waters around Europe. One of its objectives was to overcome research fragmentation in Europe through unification of the various European initiatives of ocean observatory implementation in Europe. It involved 14 European countries, more than 50 institutions and SMEs, and about 300 scientists, engineers and technicians. ESONET-NoE consolidated the deep-sea-observatory community at the regional and European levels at around 11 sites across Europe. The two-step process of the ESONET-NoE submission was the occasion to strengthen the core partnership and to enlarge it to major countries such as Spain, Turkey and Norway as well as to key partners such as NOCS (UK). ESONET-NoE has been working closely on an infrastructure project entitled EMSO (European Multidisciplinary Seafloor Observatory). Its Preparatory Phase (EMSO-PP) has been co-funded by the European Commission since April 2008 (www.esonet-emso.org). The success of the EMSO proposal in getting funds was an objective of ESONET-NoE over the first year of ESONET-NoE (2007-2008). EMSO-PP project officially started one year after ESONET so as to prepare the infrastructure implementation according to ESONET initiative. While ESONET federated the community of partners to write the common technical specifications of ESONET-EMSO observatories, EMSO was working at implementing them, preparing the legal context of the infrastructure and related legal bodies and looking for funding at the national and international levels.

Throughout this collaborative work, a consensus was reached about the definition of an open-sea observatory which is an open-sea fixed station (or a network of stations), docked or moored on the seafloor *plus eventually* a set of marine sensors in communication with the shore by a 2-way system via acoustic-, satellite-, or cable- connection in real-time or near-real time. These observatories enable coherent acquisition of data relating to oceanographic, geosciences and climatic phenomena, at relative high frequency over long timescales of, at least, ten years. The two kinds of observatories that have been acknowledged are: i) "Stand-alone observatories" and ii) "Cabled observatories" (figure 10). It is worth noting that another mobile observatory has, also, been defined; it is dedicated to the monitoring of the marine environment under very particular conditions whenever it is needed but, in this chapter, focus will be only on the two previous types.

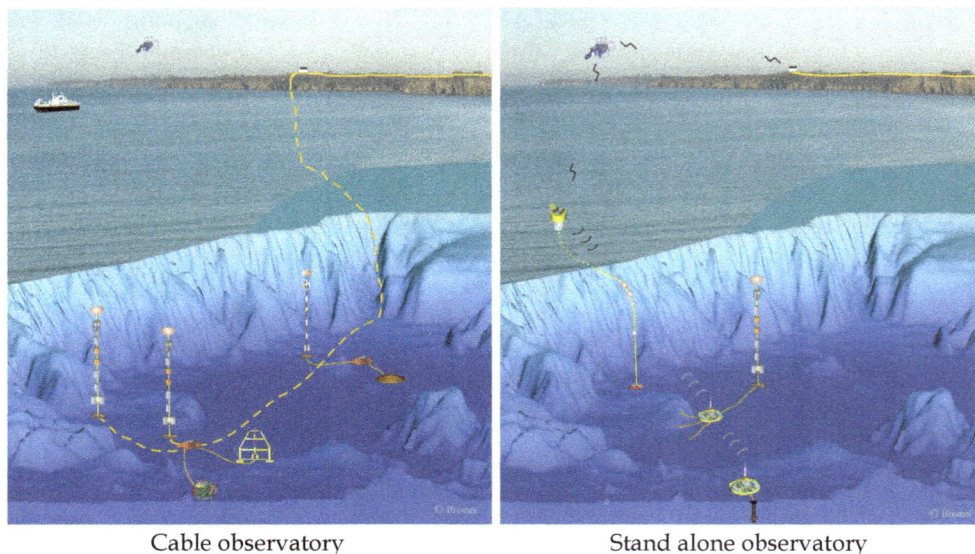

Cable observatory   Stand alone observatory

Fig. 10. The two observatory types according to the definition by ESONET EMSO and EuroSITES consortiums.© Ifremer/M. Chapon-Ashaine

Stand-alone and cabled observatories both communicate to a shore station, but through different methods. The stand-alone observatories are autonomous as regards energy, and transfer their data from a surface buoy to a data centre by satellite link. Communication below the water level is supported by acoustic link or by a cable set in-between the sensors and the surface buoy (figure 10). Cabled observatories both transfer their data to a shore station and are supplied with power via a telecommunication cable. Both kinds of observatories communicate in real-time or near real-time, but with the latter, the amount of data transmitted by satellite is restricted.

In parallel with the ESONET-NoE and EMSO-PP projects, the EC also co-funded the EuroSITES Collaborative Project (2008-2011). EuroSITES formed an integrated European network of 9 deep-ocean observatories and 3 associated sites positioned across European seas in waters off the continental shelf and at more than 1 000 m-deep in order to permit measurements of various parameters from the sea surface to the seafloor (figure 11). EuroSITES was coordinated by the National Oceanography Centre of Southampton, (NOCS, UK) and involved 13 partners across Europe and the Cape Verde Islands. This project jointly with ESONET and EMSO has contributed in a better integration across European countries of European open ocean observatories. EuroSITES project has resulted in enhancement of observatory infrastructure, from the moorings to the sensors and telecommunication facilities. Nowadays EuroSITES observatories are monitoring more atmospheric and ocean parameters (essential climate and ocean physical, biogeochemical and biological variables) than ever before. The higher rate of near real-time transmission of datasets has led to a greater use of *in situ* data in the validation of climate models.

Fig. 11. A conceptual design of a EuroSITES open ocean observatory. (Source: EuroSITES)

EuroSITES observatory infrastructure includes:

- i) a **surface buoy** used to locate the observatory and to transmit, in near real-time, data to scientists ashore via satellite. It is equipped with a platform for the attachment of the meteorological sensors in use for simultaneous monitoring of atmospheric conditions and ocean variables;
- ii) a **mooring line** from the sea surface down to the seafloor to which instruments and sensors are fixed for monitoring the ocean environment, along with buoyancy aids and weights to preserve the mooring integrity;
- iii) a suite of **devices** including:
  - the **sensor frame** equipped with sensors for measurements of various ocean parameters including temperature, salinity, chlorophyll-a, nutrients, oxygen and carbon dioxide;
  - the **sediment trap** to catch all the sinking particles from the surface (shells, animal molts, detritus and aeolian dust);
  - a time-lapse camera system such as the **bathysnap camera**, which is deployed on a lander system on the seafloor and takes photographs every 8 hours to record the benthic activity for over a year at a time.
- iv) an **anchor** to maintain the observatory in the same position.

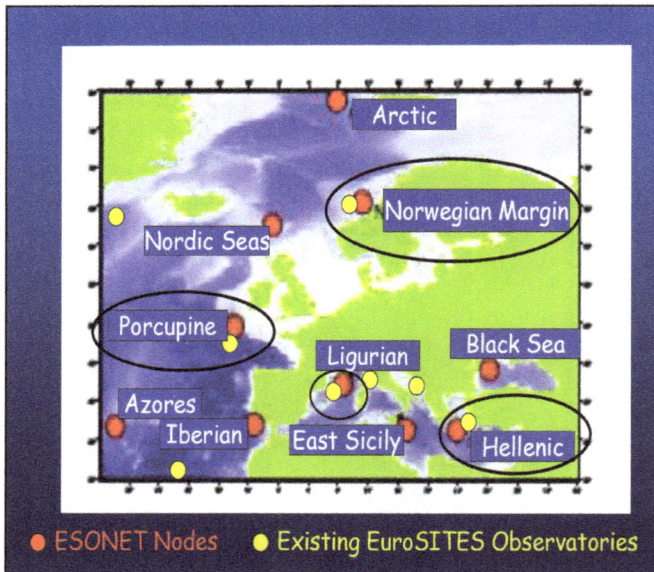

Fig. 12. Map of some European open-sea observatories, including some key sites of the European projects EuroSITES, ESONET and EMSO. This map does not include the EuroSITES Cape Verde observatory (subtropical Atlantic region), nor the Plocan observatory (Canaries islands). EMSO sites are included in ESONET ones.

Figure 12 presents an updated map of the European open sea observatory sites falling under the scope of the EuroSITES, ESONET-NoE and EMSO-PP projects developed in strong collaboration. It shows that the existing infrastructures are located in many common geographical regions of interest. The consortia of these 3 projects are moving towards a common future vision for open-/deep-ocean observatories. In the next section, 3 examples of open-sea observatories deployed around Europe in the north Atlantic Ocean (Azores sites, Porcupine Abyssal Plain) and the Mediterranean Sea (Marmara Sea) will be presented to exemplify applications in several marine science fields: e.g. ecosystems monitoring, seismology and geosciences, oceanography and biogeo-chemistry.

3. Current observatories: Examples of infrastructures in Europe and their possible future

Observatories networked at the seafloor- and water column-levels offers to earth- and ocean-scientists new opportunities for studies of multiple and interrelated processes over timescales from a few seconds to decades. These include i) episodic processes, ii) processes with time-spans from weeks to several years, and iii) global and long-term processes. Episodic processes include, for instance, eruptions at mid-ocean ridges and volcanic seamounts, deep-ocean convection at high latitudes, earthquakes, and biological, chemical and physical impacts of storm events. The second category includes processes such as hydrothermal activity and biomass variability in vent communities. The establishment of an observatory network is essential to investigate global processes such as the dynamics of the oceanic lithosphere and the thermohaline circulation in the Ocean. Any increase in sampling capability results in major advances across a range of scientific disciplines:

- Global change, physical oceanography, marine biochemistry,
- Earth sciences, geohazards and seafloor interface,
- Ecology
- And non-Living resources.

A key work was carried out by H. Ruhl et al. (2011) within the framework of the ESONET NoE project. It was aimed at reviewing the science areas in relation to ocean observatory research and introducing some of the scientific issues that observatories will help to address. The reader can refer to this article to get an overview per scientific discipline; here, focus will be only on topics relevant to the following applications.

3.1 An observatory dedicated to marine ecosystem monitoring on MOMAR site

Marine organisms are sensitive to their physical and biochemical environment. Consequently, due to the increase of anthropogenic modifications in the environment, one among the challenges of the 21st century is to evaluate the sensitivity of the ecosystem to anthropogenic changes. Indeed, given the short- and long-term impacts of climate and global changes upon the primary production, the entire trophic chain is likely to be affected in a complex way (see the review in H. Ruhl et al. (2011), section 2.4). This implies to accurately monitor the biochemical cycles and to gain more insight into the ecological functions of organisms at several levels of the trophic network according to the environment conditions and its evolutions. This should be done in several water layers, including the bottom one and the benthic one. A special focus should be made on the deepest ocean, which is scarily explored and so "well unknown". Indeed, the conduct of long-term explorations of the deep areas is difficult because this environment is hostile. The benthic zone is influenced by the processes at play within the upper water layers and by those from the seabed such as hydrothermal vents, gas hydrates, and more generally by geophysical processes. Fluid flow within and through the seabed transfers substantial amounts of mass, heat and chemical energy in relation to geophysical, biogeochemical and ecological processes. Estimation of fluxes is, thus, essential to the calculation of ocean global budgets and to enhancement in the modelling of global changes. Many of these processes are developing on relatively short timescales and can be unpredictable. Deep-sea observatories are suitable tools to monitor the changes induced by both such short-term events and long-term global changes. In addition, the interactive communication between observatories and scientists allows the latter to modify the sampling strategy whenever an episodic, and maybe unpredictable, event is detected.

3.1.1 A case study site: MoMAR Site
The MoMAR (Monitoring the Mid Atlantic Ridge) initiative is aimed at providing multidisciplinary time-series datasets about hydrothermal systems in the Azores region of the Mid-Atlantic Ridge, where 4 hydrothermal fields are currently studied (figure 13). The goal is to get some feedback of the links between volcanism, deformation, seismicity and hydrothermalism in order to gain more insight into the coupling between hydrothermal ecosystems and these sub-surface processes and into its effect upon exchanges with the ocean. The Lucky Strike hydrothermal system (figure 13) is located on the summit of a volcano and consists of a hundred of hydrothermal sources that surround a lava lake. In the close vicinity of the vent sources the environment is hostile with temperature varying from 4 to 330°C. The chemical properties of the fluid have evidenced a complex hydrothermal system. On smokers the biological communities include bivalves, often colonised by micro-organisms, and host many faunal species. In the warmest zones shrimps are the main habitants. The food network is mainly based on chemosynthesis. As hydrothermal vents are very dynamic systems, their variations induce, in the ecosystem, complex changes that need to be better understood. In the past, field-observations were organised through ship surveys conducted at nearly the same seasons and under similar or very similar conditions, and thus the temporal variations were not taken into account. Consequently it was also difficult to capture a seismologic event and to observe the associated changes in the ecosystem.

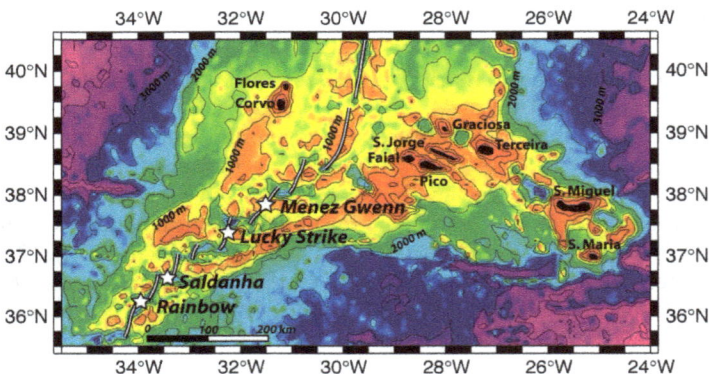

Fig. 13. Bathymetric map around the Azores islands. The Lucky Strike site is one among the 4 hydrothermal sites (white star). ©Ifremer

To fill this gap, the implementation of a permanent observatory on Lucky Strike started, in 2008, with the deployment of autonomous temperature probes at vents selected so as to build on previous experiments dedicated to ecology, geodesy and seismology. Under the scope of the ESONET Network, the MoMAR-D demonstration mission has maintained and reinforced these experiments through a stronger participation of colleagues from other European countries. The system operates 200 miles off the closest island, at a depth of nearly 1 700 m on two field sites: a geophysical node is moored in the Lucky Strike lava lake, and a geochemical/ecological node at the Eiffel Tower vent site.

3.1.2 Strategy and technology on the MoMAR site

A prerequisite to the monitoring of the effects induced by the above mentioned geophysical phenomenon is the detection and monitoring of this rather unpredictable geophysical activity, which usually takes place in short lapses. An observatory will help to capture the event and to update the sampling strategy according to the phenomenon features. This observatory must include facilities dedicated to the monitoring of the life evolution around the hydrothermal site and measurement of the impacting environmental parameters such as key physico-chemical properties associated with the geophysical activity. The system allows one to directly observe the ambient life with a camera and to measure parameters with dedicated sensors. The first part of the project was dedicated to the development and adaptation of the interfaces between the various sensors and the monitoring nodes. It was followed by an on-shore trial period that took place a few weeks before the cruise. The whole system was finally deployed during the MoMARSAT cruise with the RV *"Pourquoi Pas?"*. The website, http://www.ifremer.fr/momarsat2010/, was created to daily track the operations at sea. In the summer of 2011 a second cruise was organised for infrastructure maintenance and equipment upgrade.

The required infrastructure and pieces of equipment

The observatory infrastructure (figure 14) is composed of 2 SEA MOnitoring Nodes (SEAMON) acoustically linked to a surface relay buoy (BOREL) to ensure satellite communication to the land base station in Brest (France). These SEAMON nodes are based on a technology developed during the ASSEM project. For instance, they were upgraded for deployment from a ROV instead of being free-falling devices. One of the challenges was to prepare a system reliable for long-term deployment (several years) in the vicinity of an active hydrothermal vent (*i.e.* at less than 2 m from the hydrothermal spring) where life conditions mean important biofouling and hostile environment for sensor. In addition the system had to be ROV-operable, deployed and maintained far away from the coast. This implied that any subsystem component had to be tested and qualified prior to the final deployment.

The first SEAMON node (SEAMON Est) dedicated to large-scale geophysical studies was moored in the centre of the large lava lake present in the Lucky Strike vent field. This node hosts an Ocean Bottom Sismometer (OBS) and a Permanent Pressure Gauge (PPG).

The second SEAMON node (SEAMON West) was deployed at the base of the Eiffel Tower active edifice to study the links between faunal dynamics and variations of physico-chemical factors. This node is composed of two sensor modules (figure 15). The former is a chemical *in situ* analyser developed by NOCS for measurements of iron concentrations in diffuse hydrothermal fluids. The latter is an ecological module (TEMPO) developed by Ifremer and described hereafter. Both sensor modules were connected to SEAMON onboard and moored once attached to the SEAMON West structure.

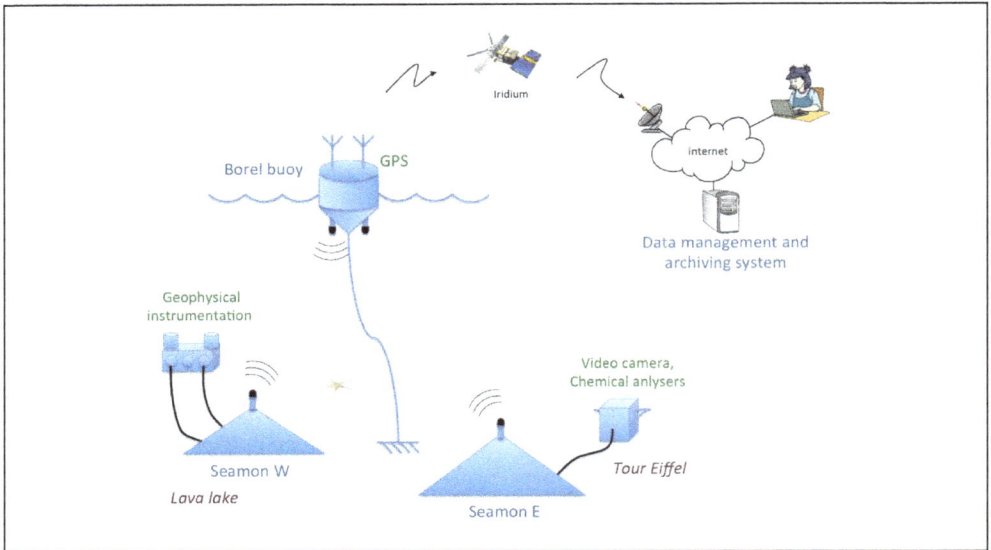

Fig. 14. MOMAR observatory concept

The TEMPO ecological module

The IronMan chemical analyser deployed on small cracks.

Fig. 15. Modules deployed on SEAMON west site (Eiffel Tower) © Ifremer/Victor 6000

TEMPO is an upgraded version of TEMPO Mini, which is an instrument package created by Ifremer for real-time monitoring of ecosystems (Auffret et al., 2009). Firstly Tempo mini integrated a 2 Megapixel streaming video camera with embedded event detection, 4 LED lights, an oxygen sensor and a temperature sensor. An efficient and innovative biofouling protection procedure was applied to the camera, lights and optical oxygen sensor (Delauney et al., 2010). TEMPO was tested and deployed on the coastal cable demonstrator, VENUS, of Neptune Canada from September 2008 to February 2009. Further to the proven evidence of its robustness, it was updated to include a chemical analyser, CHEMINI-Fe developed within Ifremer (Vuillemin et al., 2009) for measurements of total dissolved iron concentrations in the deep-sea waters. In this new version the High Definition (HD) video camera was coupled with 6 LED lights, an oxygen and temperature probe. This new version, named TEMPO, was deployed on the MoMAR site, and one photo was sent everyday to the ground station over one-year period (other video images were stored in the system).

Deployment

SEAMON nodes and the sensor modules were moored by the ROV, Victor 6000, from the RV "*Pourquoi pas?*". On Lava Lake, the sensor module was deployed near the SEAMON East and was connected underwater with wet mateable connectors by Victor6000 (figure 16).

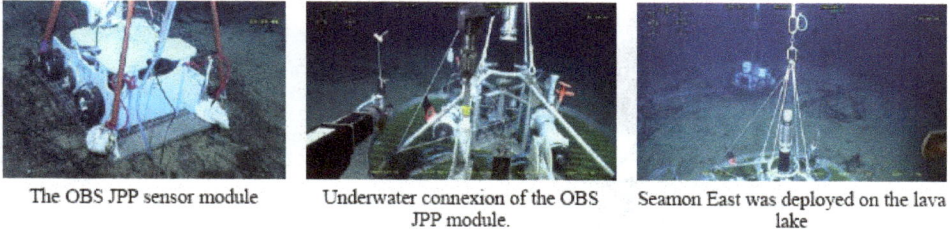

| The OBS JPP sensor module | Underwater connexion of the OBS JPP module. | Seamon East was deployed on the lava lake |

Fig. 16. Deployment and connection of the SEAMON East node, (lave Lake) © Ifremer/Victor 6000

On Eiffel tower active edifice, the SEAMON West-linked sensor modules were precisely deployed on the sea-bottom by the ROV: TEMPO was located in front of a mussel bed, and the chemical analyser was positioned on a small crack (figure 17). The deployment of TEMPO was controlled *in situ* by using a prototype of underwater WiFi link. Using this link, the first video sequences acquired by the camera during on-board deployment were helpful in the validation of the exact position of the system and in the tuning of the recorded image quality.

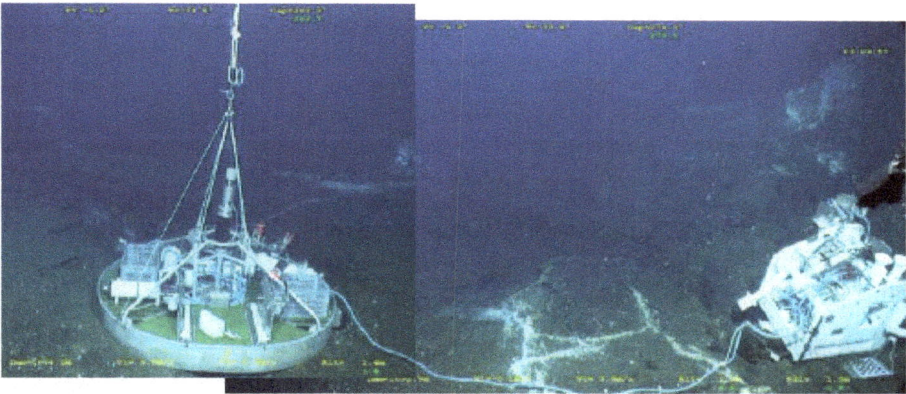

Fig. 17. SEAMON West and TEMPO after deployment and in operation on Eiffel Tower site © Ifremer/Victor 6000

The TEMPO ecological module and the IronMan chemical analyser were deployed on small cracks. The technique in use for exchanges between the two nodes and a BOREL buoy moored on the ocean surface within the acoustic range of the SEAMON stations is underwater acoustic communication. This buoy is equipped with two identical and redundant data transmission channels to ensure uninterrupted data flow. The acoustic modems were chosen further to ESONET trials and are provided by EVOLOGICS GmbH (Berlin). Scientific and technical data (including a low-resolution photo) are transmitted daily to the data centre in Brest via IRIDIUM satellite transmission. Energy is supplied to the buoy by lead batteries recharged by solar panels. SEAMON West was deployed at the base of the Eiffel Tower active edifice.

The Borel buoy was moored North of the lava lake.

Fig. 18. Autonomous surface buoy communicating with the moored system and satellite (a). © Ifremer/ J. Blandin

The 2 SEAMON nodes and the geophysics module were moored by using the vessel cable. On the bottom, Victor 6000 ROV performed the precise deployment and connection operations. The connection of the sensors to the nodes was validated by two methods tested over the cruise. A wet mateable connector was used for *in situ* connection of the geophysics-dedicated module to SEAMON West, whereas the sensors were on-board connected to SEAMON East. The Borel buoy was deployed within the acoustic range of the 2 nodes. In addition to real-time communication sensors, autonomous instruments (OBS, ocean bottom tiltmeter, current meters, particle trap, colonisation modules and temperature probes, see Table 1 below) were also deployed in the Lucky Strike vent field. They were equipped with 1-year data storage facilities.

3.1.3 Results and next steps
This observatory infrastructure has acquired a synchronised multidisciplinary data set and has enabled the development of solutions for sensor interoperability, shore-sensor interactive communication, data-management and –dissemination as well as public outreach. The first result was the validation of the procedure in use in the on-site deployment of the different components of the observatory. The two nodes were deployed on October 4 and 5, 2010. The Borel buoy was moored on October 12, 2010 and data transmission started immediately. The transmitted data are stored within Ifremer facilities. Some among the data can be viewed online at http://www.ifremer.fr/WC2en/ allEulerianNetworks. The near real-time data have been used both as support for scientific

analysis and interpretation as well as indicator of the occurrence of an event. Volcanic-events or rapid degassing of the magma chamber, tectonic (displacement along axial faults) or hydrothermal ones have been recorded. In June and July 2011, a year after the deployment, the MoMAR team went back to the field site for system recovery and upload of non real-time data for further upgrade and redeployment. Nowadays, the system is still deployed. The analysis of the provided data is in progress; however, some results are already available. For instance the TEMPO module showed a smoker that was growing of 4 to 5 cm in height in two weeks (figure 19). It evidenced an increase of the local community as well as an extension of the area colonised by bivalves and several seismic events were recorded.

Fig. 19. The TEMPO observation module was deployed in the vicinity of an active hydrothermal site at the base of the 11-m-high Eiffel Tower edifice located on the Lucky Strike vent field on the Mid-Atlantic Ridge. It recorded photos for 5 months; they were to the sea surface via acoustic communication and to the in-land Ifremer data centre via satellite. The SMOOVE camera was looking at a Bathymodiolus azoricus mussel assemblage. On the above picture, one should note the presence of many predators, e.g. the crab, Segonzacia mesatlantica. Physico-chemical measurements were made in parallel to monitor the changes in environmental conditions and to evaluate their impacts on the dynamics of the vent fauna. Copyright J. Sarrazin/PM Sarradin, Ifremer.

Future steps include a cruise planned for 2012 and targeted to temporary recovery for maintenance of the system and redeployment. Despite the autonomy of the observatory, on-site operations are needed every year. Ideally, the cable system should allow a greater data flow in real-time, especially for video streaming, but the installation of such a long cable is not economically sustainable with the technology currently available.

3.2 A multiparameter, permanent observatory dedicated to Earth sciences, geo-hazards and seafloor interface in the Marmara Sea

The solid earth interacts with the ocean through many processes such as earthquakes, slope instability and sediment failures, fluid flow and seepage of gas through sediments and gas hydrates. Gas hydrates host large quantities of solid methane in marine sediments. Seismic activity and warming can make gas hydrate unstable and result in large releases of greenhouse gases that add positive feedback to global warming. Around Europe, this can occur in the Black Sea, the Mediterranean Sea, the Gulf of Cadiz, the Nordic Sea and Arctic sites (Bohrmann et al. 2003, Woodside et al. 2006, Lykousis et al. 2009, Westbrook et al. 2009, Hustoft et al., 2009). The study of earthquake-related fluid-fault processes helps to gain more insight into gas emissions, seismic activity and oceanic warming (Etiope and Favali 2004). Earthquakes and landslides can both generate important tsunamis, phenomena that have recently become dramatically famous. These sad experiences highlight the need for, at least, enhancement and/or generalisation of early warning systems in sensitive zones. As a result, more geo-seismic deep-ocean observatories have been studied and implemented so as to improve prevention methods through a better understanding of earthquake- and tsunami-generating processes.

Cold seeps are often observed in association with active faults (e.g. Moore et al., 1990; Henry et al., 2002). This has led the scientific community to hypothesise that, at least, some of these fault channelled fluids come from deep levels within the sediments and, possibly, from the seismogenic zone in the crust. Furthermore, gas has been reported to expel from pockmarks and mud volcanoes in relation to the occurrence of earthquakes. Coupling between deformation, pore pressure transients and fluid flow may lead to post-seismic fluid release, earthquakes precursor signs and/or systematic variations in flow rates, fluid chemistry and pore pressure during inter-seismic phases. Numerous fluid vents and related features have been discovered along the North Anatolian Fault (NAF) system in the Marmara Sea.

3.2.1 A case study site: MARMARA Site

Tectonic setting

The Sea of Marmara hosts the submerged section of the highly active, 1 600-km-long, right lateral strike-slip North Anatolian Fault (NAF). According to historical evidence (Ambraseys and Jackson, 2000) a major earthquake happened in 1509 in the central part of the Marmara Sea. A series of earthquakes with estimated moment magnitudes close to, or greater than, 7 occurred in 1719, 1754, and in May and August of 1766 in the Marmara Sea region, but the distribution of damage was unable to give clues about the exact geometry of the associated segment fault ruptures (e.g. Ambraseys, 2002; Parson, 2004; Pondard et al., 2007). The next series of Mw ~ 7 events comprises 3 earthquakes in 1894, 1912 and 1999. The earthquake of the year 1894 affected the Cinarcik Basin and Izmit Gulf, but it is unclear which fault ruptured in the Cinarcik Basin. The earthquake of 1912 ruptured the Ganos fault on land and extended it further offshore (Armijo, 2005), but this distance may have been quite short (Ambraseys, 2002). Whatever the interpretations, a consensus has been reached within the scientific community about the fact that the Istanbul-Siliviri segment in the central part of the Marmara Sea is the most likely to rupture in the future. No break of this segment has been recorded since 1766.

Relations with fluids and gas emissions

In the Gulf of Izmit, repeated surveys have shown increase in the intensity of methane emissions after the earthquake of August 17, 1999 (Alpar, 1999; Kuscu et al., 2005). In the deepest parts, cold seeps with their associated manifestations, such as carbonate crusts, black patches and bacterial mats, were observed along the fault (Armijo et al., 2005). A systematic correlation was also found between active faulting and the acoustically detected gas escapes. Remarkably, the fault segment with the less acoustic anomalies found within the main fault trace corresponds to the Central High and Kumburgaz Basin area (see Fig. 1

in [Géli et al., 2008]). This segment is the most dangerous since it is the only one to have not undergone rupture since, at least, 1766. The identification of thermogenic hydrocarbons with the same geochemical signature as those found in the Thrace Basin at the top of anticline structures indicates that the North Anatolian Fault cross-cuts gas reservoirs from the southern continuation of the Thrace Basin gas field (Bourry et al., 2009). This finding opens new perspectives that were not even imaginable a few years ago. Moreover, it supports the need for monitoring gas emission activity along with seismicity: a major challenge is to determine whether gas can generate detectable signals related to the stress building process over the seismic cycle. To address this challenge, a continuous collection of geochemical and geophysical data in the immediate vicinity of the fault zone is a must. It can be done through implementation of permanent seafloor observatories in the Sea of Marmara and development of methods and tools for data processing, integration and analysis. A two-step strategy has been developed; it includes a prospecting phase aimed at getting a more accurate definition of the area prior to the implementation of a permanent system.

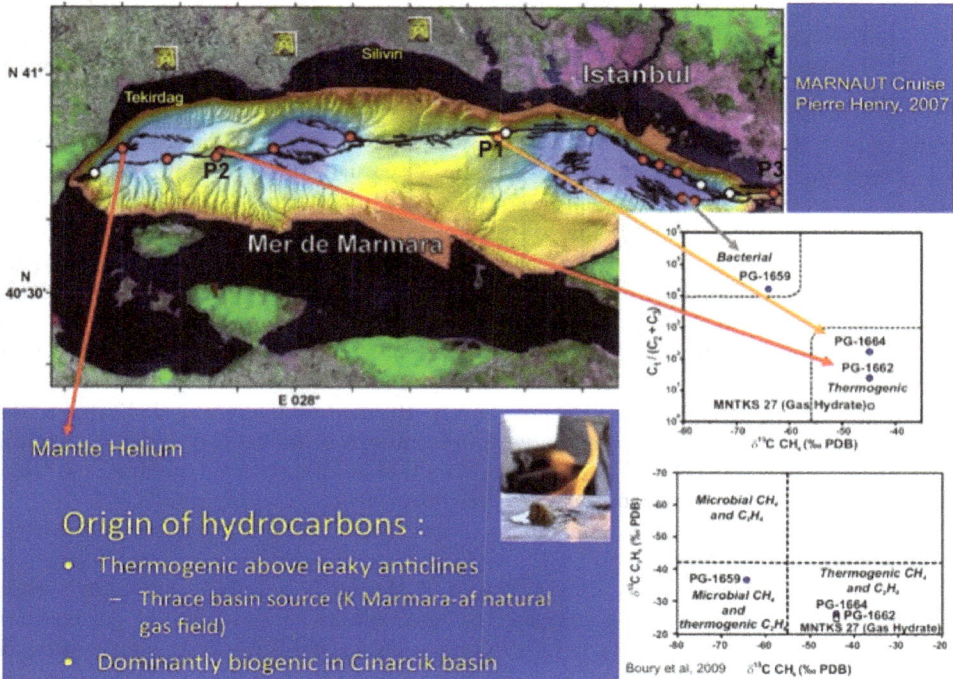

Fig. 20. Map showing the most active northern branch of the North Anatolian Fault (NAF; black line), gas emission sites (red dots). The inset diagrams show the composition plots of gases indicative of the thermogenic and deep origin (Géli et al., 2008; Bourry et al., 2009). P1, P2 and P3 indicate potential sites that were identified for multi-disciplinary seafloor observatories during the ESONET Marmara-DM Demonstration Mission.

3.2.2 Strategy on Marmara site: A prospecting step to prepare a permanent infrastructure

The Marmara Demonstration Mission (April 2008 to September 2010) was conducted within the EU-funded ESONET-NoE programme to: i) characterise the temporal and spatial relations between fluid expulsion, fluid chemistry and seismic activity in the SoM, ii) test the relevance of permanent seafloor observatories for an innovative monitoring of earthquake-related hazards, appropriate to the Marmara Sea specific environment, and iii) conduct a feasibility study for optimisation of the submarine infrastructure options (optical fibre cable, buoys with a wireless meshed network, and autonomous mobile stations with wireless messenger). The partners involved in the Marmara-DM Demonstration Mission were: Ifremer, CNRS/INSU, CNR/ISMAR, INGV, ITU and DEU (Dokuz Eylül University, Izmir). A total of 6 cruises[†] conducted within the framework of the Marmara-DM were devoted to the area exploration, which included:

- an accurate mapping of the bottom by microbathymetry measurement,
- acoustic detection of gas emissions through deployment of an acoustic gas bubble detector (named BOB and hereafter described),
- 3D and 2D high-resolution seismic survey on the Western High site,
- deployment and recovery of the multi-parameter sea-bottom SN-4 observatory of INGV (a GEOSTAR type observatory) at the entrance of the Gulf of Izmit on the fault trace and at the end of the rupture of the 1999 earthquake.

The required infrastructure and pieces of equipment

During cruises at sea, a temporary infrastructure was deployed to test the equipment and to better assess the design and specifications of the infrastructure. Significant research efforts to the development of innovative sensors for monitoring variations in the geochemical and geophysical properties of gas emissions were made during Marmara-DM. They dealt with:

- *Pore-pressure sensors.* The piezometer is a free-fall device with a 15-m-long sediment-piercing lance equipped with sensors for measurement of the differential pore pressure at 5 different depths (< 15 m) below the seafloor. This device proved to be very powerful for detection and monitoring of free gas accumulation and release in superficial sediments.

- *Methane sensor.* Based on one-year-long tests performed by INGV and dedicated to the measurement of variations in methane concentrations in the Gulf of Izmit. It uses the methane sensor developed by the German CONTROS company (HydroC™/ CH4, Hydrocarbon & Methane Sensor), which has provided encouraging results.

- Arrays of Broadband *Ocean Bottom Seismometers* (BB-OBS)[‡] working in the 0.03 – 30 Hz bandwidth

- *Gas-bubble monitoring system (BOB).* The BOB station is a bubble detection system designed from halieutic echo sonars used to detect echoes backscattered by gas bubbles. It relies on well known acoustic technology, such as high directivity single beam or multi-beam echo-sounders, used in mapping and quantification of gas bubble emissions

---

† Two Cruises with R/V Le Suroit (Ifremer, France) from November 4 to December 14, 2009, 2 other ones with R/V Urania (Italy) in September-October 2009 and September-October 2010, and the last 2 with Turkish vessels, respectively R/V Yunuz and R/V Piri Reis in March 2010.
‡ The two leading manufactures of the BB sensors in use in Ocean Bottom Seismology are: Guralp (www.guralp.com) and Kinemetrics (www.kinemetrics.com).

from the seafloor and monitoring of their temporal variability (Greinert, 2008). These echo-sounders are ideally combined with 70 to 300 kHz ADCPs systems so as to identify different seeps in the datasets and to determine the horizontal and vertical velocities of the bubbles. It can monitor an angular area of 360° in 7°steps. For each step it acquires and records acoustic data and other needed parameters such as compass of the station, temperature and pressure. The 2-m-high station frame has a surface base of 1.3 x 1.3 m² and a total weight of 635 kg (figure 21). The detection radius is 150 m and it can work at varying depth up to 1500 m deep.

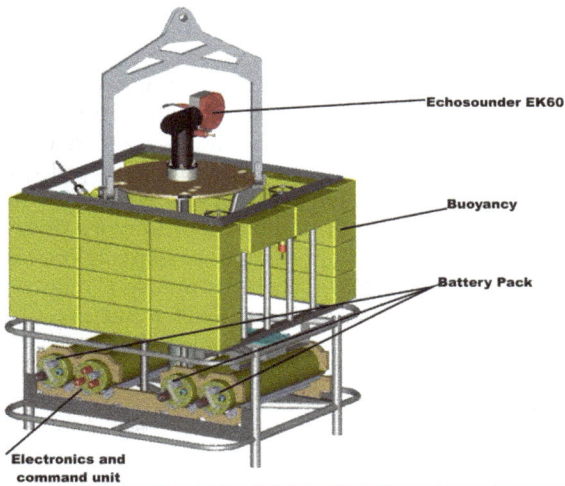

Fig. 21. Gas-bubble monitoring system (BOB module): conceptual design (left panel, ©Ifremer/RDT) and photo of the currently used BOB version (right panel, ©Ifremer/photo from M. Gouillou and O. Dugornay)

BOB was deployed by the Ifremer ROV, Victor6000, for the first time in December 2009 and off Istanbul to detect and quantify methane fluxes from the area. For its first try in deep waters, BOB recorded data over 4 days. This first *in situ* test evidenced important variations in gas flow. More recently, in summer 2011, BOB was also deployed in the Arctic area to detect and quantify methane releases under the scope of the ESONET demonstration mission AOEM (Arctic Ocean Esonet Mission) conducted between August 2 and 6, 2011. It proved to work well, and the collected data are currently analysed. BOB will be upgraded to extend the data recording duration so as to satisfy long-term observatory requirements.

3.2.3 Marmara DM results and next steps

Results and next steps for the Mamara observatory

The systematic mapping of fluid emission sites in the Marmara Sea over the Mamara-DM mission showed that, despite the association of many of them with active fault traces (Zitter et al., 2008; Géli et al., 2008), they are widespread and found in various contexts with no systematic relation to the activity of faults. The geochemistry of the fluids expelled also appears to be diverse, in particular the depth inferred for the fluid source (Bourry et al., 2009; Tryon et al., 2010). Instruments were deployed on the seafloor for durations from one month to one year. Some clusters of microseismic activity appeared to be spatially correlated with fluid migration through the crust (Tary et al., 2011). The flow-meters (Tryon et al., 2011) as well as the physical and chemical sensors, deployed with SN-4 system, measured temporal variations related to episodic fluid emissions through the seafloor.

The 6 cruises conducted in the Marmara Sea allowed the selection of the optimum sites for the future multi-parameters seafloor observatories (figure 22):

- Site P1, on the Istanbul-Silivri segment. This site is located in the seismic gap immediately south of Istanbul where intense bubbling is observed (1 km south of the main fault trace) despite the lack of evidence of fluid expulsion on the fault itself.
- Site P2, in the Western High area (also named Gas Hydrates area). The site is situated in the area where oil and gas seeps from the Thrace Basin were found and where the connections between the fluid migration conduits and the main fault system were imaged with a 3D, high-resolution seismic survey.
- Entrance of Izmit Gulf (Site P3). At this site, the principal deformation zone of the North Anatolian Fault is less than some tens of meters wide. In addition, the site is close to the western end of the surface rupture associated with the Izmit earthquake of 1999, where the next earthquake affecting the fault strand towards Istanbul may nucleate. It is a relatively accessible area, at shallow depth (200 m) and at less than 5 km from the coastline.

At sites P1 and P2, it has been proposed to install one cabled multi-disciplinary seafloor observatory on the Istanbul-Silivri fault segment that cuts an anticline characterised by the presence of numerous sites of gas emissions of thermogenic origin. The shore station will be cabled to one node, itself connected to, at least, 2 Junction Boxes (JBs) set respectively north and south of the fault (JBN and JBS). Each of them will be connected, at the most, to 10 instrument packages (Table 1). The devices deployed at each junction box will consist of: an array of 4 seismometers set at less than 500 m from the junction box, a piezometer, a bubble acoustic detector and a methane sensor. Clusters of seismometers will provide ultra-precise characterisation of earthquakes near the fault zone through use of array-based methods for hypocentre determination.

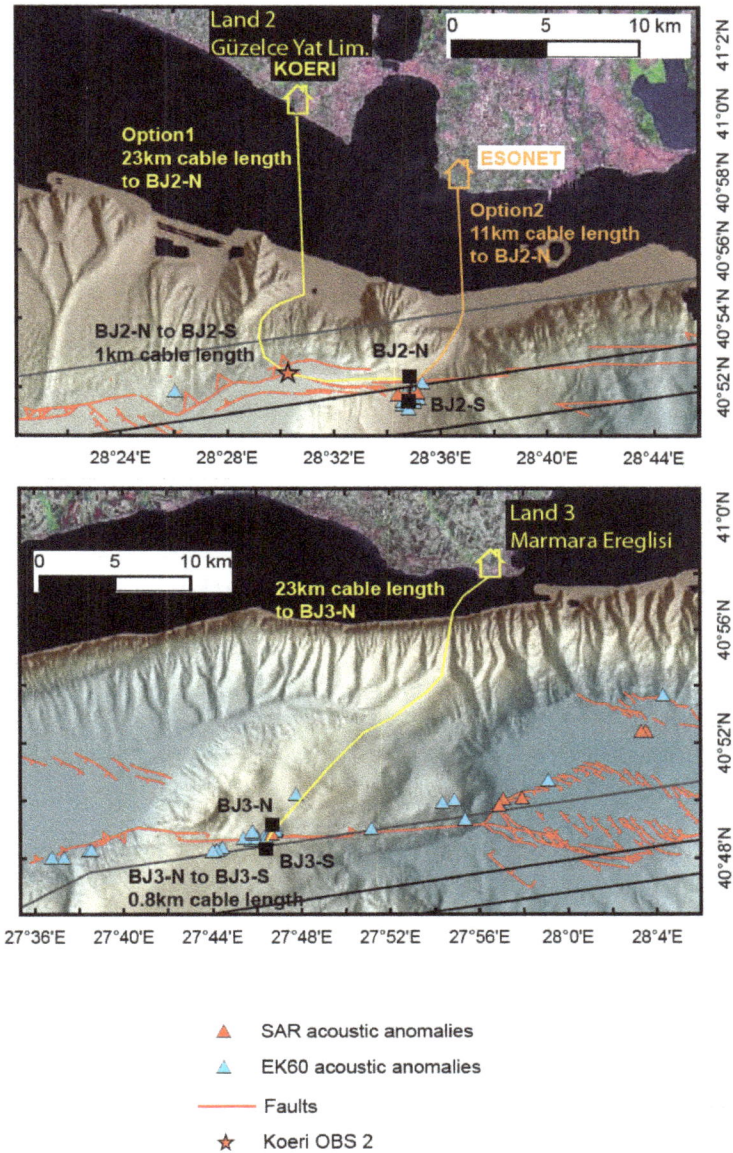

Fig. 22. View of the cable (yellow line) linking the observatory to the node at P1 (lower figure) and P2 (upper figure). The cable will be deeply buried all the way from shore to the main node. The area between the two SW-NE oriented, black lines is the separation zone between the upgoing and downgoing navigation corridor. The northern limit of the downgoing corridor is indicated as a thin grey line. The red star indicates the planned location of a cabled, permanent OBS to be deployed by KOERI in 2011.

Connectors	Supplier	Availability of instrument/ Manufacturer	Interface	Power (Approx)
1	OBS	Guralp	RS-232	< 10 W
2	Piezometer	NKE	RS-232	< 10 W
3	BOB (Bubble Observatory)	Ifremer	100BaseTX	< 10 W
4	Methane sensor	CONTROS	RS-232	< 10 W
5	Accelerometer	On-the-shelf	RS-232	< 10 W
6	Absolute Bottom Pressure Recorder	Paroscientific	RS-232	< 10 W
7	CTD/Oxygen/Turbidity	On-the-shelf	RS-232	< 10 W
8	Current meter / ADCP	On-the-shelf	RS-232	< 10 W
9	Time Lapse Camera	On-the-shelf	100BaseTX	< 10 W
10	Strong Motion Accelerometer			
11	Distance Meter	Sonardyne	RS-232	< 10 W

Table 1. List of the 11 sensor packages tested during previous EC-funded programmes, e.g.: the ESONET-NoE Programme (for slots 1 to 10) and ASSEM (for slot 11).

This infrastructure will contribute to enhance the knowledge and understanding of geophysical activity and will help scientists to make significant advances in this field. In parallel it will answer important societal needs such as seismic monitoring and prevention, which is a growing priority.

3.3 An observatory dedicated to biogeochemistry and physical oceanography on Porcupine site

Much of the transport of mass and energy from the upper ocean to the seafloor is controlled through the sinking of particles. It is, thus, essential to gain more insights into the functioning of this export. In particular the timescales of sinking and the origin and particle pathways are worth being investigated since these factors are crucial to determine the impact of a rapidly changing climate on a specific deep-sea ecosystem. The major source of particles is the photosynthetic production of phytoplankton in the euphotic zone. The matter, which finally sinks, can be either direct phytoplankton biomass or a conversion of it into certain aggregates. To investigate whether a region shows a strong vertical coupling between the surface ocean and the seafloor or whether it is dominated by lateral fluxes, the variables governing the particle export have to be analysed: this includes the characterisation of the particles themselves, the particle composition and form as well as the ability of the environment to re-suspend them, the role of ocean transport, and finally the local stratification of the water column. In regions with low stratification and deep mixing, such as the Arctic and Antarctic, a very close coupling between surface and the deep ocean has been found. It is this coupling and its variability which are the major scientific objectives of the Porcupine observatory site.

3.3.1 The Porcupine Abyssal Plain (PAP) case study

The Porcupine Abyssal Plain (PAP), 350 n.m. (nautic miles) off southwest Ireland (figure 23), is a key site for collaborative investigations within the framework of the EuroSITES and ESONET-NoE projects. The PAP site hosts the longest running multidisciplinary open-ocean time-series in Europe (http://www.noc.soton.ac.uk/pap). Over the past decade, the aim was the observation of changes in rate and state variables within the entire water column and the benthos for a wide range of biogeochemically significant features in the centre of the Porcupine Abyssal Plain. The site appears to meet many of the conditions required for analysis and interpretation of open ocean processes since it lies well away from coastal input; it is, indeed, situated in the middle of one of the biogeochemical provinces at a depth of 4 800 m where the abyssal plain is flat over large areas. Below the upper mixed layer, currents are in usually northerly and of low velocity. Winter mixing takes place at variable depth in the range 300-800 m, which facilitates investigations on the effects of the most important driving force on the upper ocean biogeochemistry: the nutrient supply.

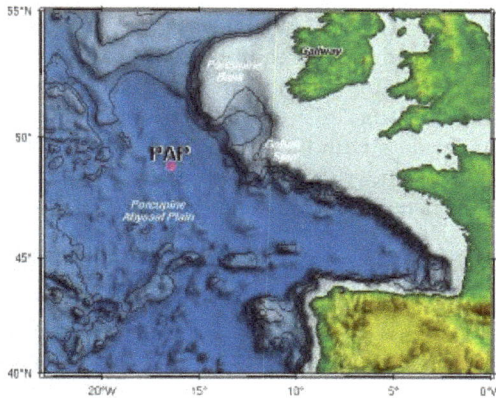

Fig. 23. Location map of the PAP Sustained Observatory (source: http://www.modoo.info/science).

The seasonal to interannual variability of downward particle flux that has been documented for the PAP site is of particular interest. Indeed, the variability has been attributed to changes in surface productivity, but longer time-series of particle fluxes have also shown a marked interannual variability, which can be related to climate modes such as the North Atlantic Oscillation (NAO). The scientific studies carried out on PAP site are targeted to an enhancement of our understanding of the complex oceanic processes from surface waters to the seafloor. They deal with the short-term variability of the oceans, including physical mixing, ecosystem dynamics and nutrient cycling. Longer-term trends in the Earth's climate are also addressed.

3.3.2 Strategy and technology on PAP site

The preliminary design for the PAP observatory was the CELTNET observatory. It involved a series of nodes linked together with sea-floor cables. However, for financial and technical reasons, it may be more appropriate that the nodes set in deeper waters work as stand-alone nodes with communication links and near real-time data transfer. Consequently, an observatory prototype was designed to demonstrate the functioning of a re-locatable system with underwater acoustic telemetry as well as surface buoy telemetry of scientific and engineering data. This was achieved within the framework of the MOdular and mobile Deep Ocean Observatory (MODOO, http://www.modoo.info), *i.e.* an observatory with real-time data access by telemetry transmission.

The MODOO project (2009-2010) was organised to take profit from the existing water-column mooring infrastructure at PAP, which forms part of the EuroSITES ocean observatories and to develop it beyond the current state-of-the-art and within the objectives of ESONET-NoE. MODOO intended to integrate the both existing water column observatory with a benthic lander observatory into a single, real-time accessible observatory and to demonstrate the efficiency of one of the deepest acoustic links from the deep seafloor to the surface. The underlying concept in the MODOO project is to link and to operate existing stand-alone observatories so that to merge them into a single observatory. MODOO was mobile (or re-locatable) to be moved to regions of interest. Its modular architecture allowed connections to other stand-alone systems.

MODOO consisted of two observatory components (figure 24): a steel/plastic wire mooring and a benthic lander. The mooring was equipped with a bi-directional telemetry buoy connected, via the mooring wire, to inductively linked instruments. One of these "instruments" was a so-called "Data Collection and Dissemination" (DCD) node. This node acoustically linked the lander data with the surface buoy via the mooring wire. The lander also had a DCD node to which lander instruments were connected. The central components of DCD nodes, mooring, telemetry buoy and lander are described hereafter.

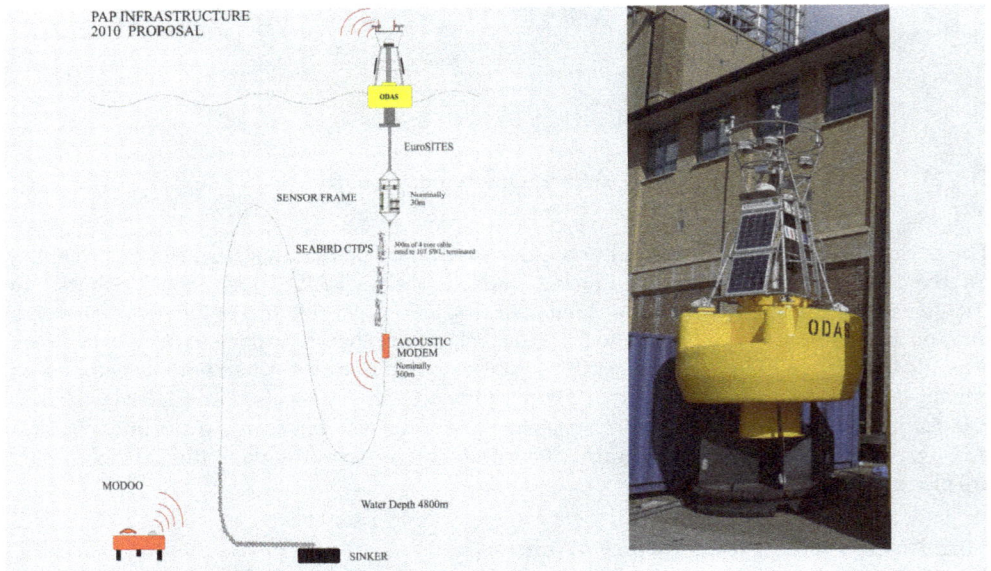

Fig. 24. Sketch of the PAP mooring as planned for the MODOO deployment (left). The PAP surface telemetry buoy at the NOCS'yard (right) (source: ESONET-NoE deliverable 45c, MODOO final report)

Data Collection and Dissemination (DCD) node

The central hardware linking the observatories components into one system was constructed around acoustic telemetry modems. These DCD nodes were expected to be used for communication between them, but also for sensor control, data storage, application of a precise time stamp to all data, data compression/uncompression for efficient transmission and checking of the transmitted data quality.

The PAP mooring

Two types of deep-sea mooring were deployed at the PAP site: they consisted of deep-water sediment trap moorings and, since 2002, a full water depth multidisciplinary mooring with surface telemetry devices. MODOO was expected to connect to the latter via an inductive coupled DCD node. The overall length of the PAP mooring is about 6 200 m; the deep sea part consists mainly in a Polyester/Polystell rope with the upper 1 300-m-long section made of Norselay® steel wire.

Surface telemetry system

The data logging and telemetry system on the PAP mooring was originally developed within the NERC-NOCS. The system is equipped with an Iridium 9522 modem, a Seabird inductive modem, a GPS receiver, a compass/pitch/roll unit and monitors for internal voltage/current/temperature. The buoy contains a solar charging controller and high-intensity LED recovery lamps that can be remotely switched on. The computer clock is regularly compared to the GPS clock and is adjusted so that it is +/- 1 second accurate. All of the data received from inductively coupled sensors are time-stamped with this clock as all the other data logged in the system. The Iridium modem is programmed to send regular position and engineering data reports by email, while scientific data are transferred through the dial-up mechanism. Commands can be sent to the unit by email according to a predefined command set for sensor control from the shore (event trigger). The system has successfully communicated with MicroCAT sensors at variable depths up to 4 960 m.

The Benthic Boundary lander (BoBo)

The BoBo lander (see §2) was developed as a stand-alone long-term monitoring system able, in its MODOO configuration, to carry a Seabird 16 CT (conductivity and temperature) probe (3 m above the bottom) with a Seapoint optical backscatter sensor (1 m above the bottom) and a new combined OBS-Fluorometer sensor (Wetlabs; 2m above the bottom) connected to it. A 1 200-kHz downward looking ADCP currentmeter is fixed to the frame 2 m above the sediment surface for measurements of bottom currents at high resolution (5 cm vertical bin size). At the same time the backscatter values of the 4 acoustic beams give additional 3D information about the amount of re-suspended material in case of correlation with the OBS sensors. In addition, a Technicap PPS 4/3 sediment trap with a rotating carousel of 12 bottles (250 ml) is mounted in the frame with the aperture (0.05 m^2) set at a height of 4 m above the bottom. A pan/tilt camera and 3 lights can be mounted in the frame at 2 m above the seabed. The camera can be programmed to record video images in 6 different settings and/or positions (these images are only accessible after retrieval). A power-unit supplies CT and ADCP probes with power and combines the RS232 connections of both sensors in a single underwater plug to establish the link to the DCD node. The Lander was expected to be placed in 5 500 m deep water and at a maximum distance, along the horizontal, of about 3 000 m from the mooring anchor to secure ship operations.

3.3.3 MODOO results and next steps

The deployment mission started at the end of May 2010 on the James Clark Ross (JCR221) R/V. Soon after the deployment of the lander and of the DCD node mooring, trials evidenced communication problems between the 2 modules. The onboard team recovered the BOBO lander to fix the problem prior to a further deployment. The second deployment ended well, but a few hours later some devices from the system were found, by accident, drifting on the sea. The team concluded that the lander spheres encapsulating the power supply had imploded. The lander was lost despite the numerous and successful previous deployments in the past. The mooring was successfully deployed, and data are transmitted daily to the ESONET and EuroSITES data portals. In order to test the DCD communication, a second deployment of the DCD system was organised in August 2010 within the framework of the MOMAR demonstration mission. This second try was successful. The next step will be the inclusion of the MODOO system in a larger infrastructure consisting of 2 moorings and 7 landers, all able to communicate with the other ones and with the outside world via a surface telemetry system (German BMBF project Molab).

4. Conclusions

The experiments and research programmes achieved from the 1980s to the present time reflect the progressive enhancement of monitoring systems in the ocean basins. During this time span, marine researchers and engineers have been witness to the achievement and strengthening of the "open sea observatory" concept and to the technical evolution of earlier, quite simple, stand-alone mono-disciplinary instrumented modules into more complex multi-parameter platforms with extended lifetime and performances. This chapter was aimed at providing evidence of these evolutions at the European scale. Nevertheless, one should be aware that ocean observatories will never replace other observing systems; they should be considered as complementary tools that help scientists to make great advances in marine science. Indeed, marine science will always need traditional equipment for prospective actions, or to monitor either inaccessible sites where an observatory cannot be installed, or sites where the deployment of a less complex and cheaper system is sufficient.

Beyond these scientific focus restricted to topics discuss in this chapter, observatories are also pushing ahead the limits of other marine disciplines such as monitoring and study of Non-Living resources: energy (renewable resources and hydrocarbons, including CO_2 sequestration), mining/deposition, but also monitoring of fisheries, biochemistry, marine acoustics, marine forecasting for instance. Observatories can answer numerous scientific questions and can also satisfy societal and economical requirements like early warning systems for seismic prevention. The scientific and society pressure is increasing the need of observatory development but it is now clear that a coordinated research effort of long-term investigations and investment is required.

5. Acknowledgements

This article gives only a snapshot of the work done by the 54 ESONET-NoE partners. Ifremer, as coordinator of this project, greatly thanks all of the staff involved during this 4-year project (see partner list on www.esonet-emso.org). The authors also acknowledge the European Commission for co-funding of this project within the framework of the 6th Framework Programme under the contract GOCE-036851. They are also grateful to Kate Larkin (NOCS) and Laura Beranzoli (INGV) for their contributions and the BTU translation office of the Brest university (Bureau de Traduction de l'Université) for support.

6. References

Alpar, B., (1999). Underwater signatures of the Kocaeli Earthquake (August 17th 1999), Turkish Journal of Marine Sciences, 5.

Ambraseys, N.N., (2002). The Seismic Activity of the Marmara Sea Region over the last 2000 years. Bull. Seismo. Soc. Am., 92, 1-18.

Ambraseys, N.N., Jackson, J.A., (2000). Seismicity of the Sea of Marmara (Turkey) since 1500, Geophys. J. Int., 141, F1 – F6.

Armijo, R., Pondard, N., Meyer, B., Mercier de Lapinay, B., Ucarkus, G. and the MARMARASCARPS Cruise Party, (2005). Submarine fault scarps in the Sea of Marmara pull-apart (North Anatolian Fault): implications for seismic hazard in Istanbul, Geochem. Geophys. Geosyst., 6, 1-29.

Auffret, Y., Sarrazin, J., Sarradin, P-M., Coail, J-Y., Delauney, L., Legrand, J., Dupont, J., Dussud, L., Guyader, G., Ferrant, A., Barbot, S., Laes, A., Bucas, K., (2009). TEMPO-Mini: a custom-designed instrument for real-time monitoring of hydrothermal vent ecosystems. OCEANS '09 IEEE Bremen, Balancing technology with future needs, May 11th – 14th 2009 in Bremen, Germany

Blandin, J., Person, R.,. Strout, J.M., Briole, P., Etiope, G., Masson, M. , Golightly, C.R., Lykousis, V., Ferentinos, G., (2002). ASSEM: Array of Sensors for long term Seabed Monitoring of geohazards. Proc. Underwater Technology, Tokyo, 16-19 April 2002, p.111

Blandin, J., Person, R.,. Strout, J.M., Briole, P., Etiope, G., Masson, M. , Golightly, C.R., Lykousis, V., Ferentinos, G., (2003): ASSEM: a new concept of observatory applied to long term SEabed Monitoring of geohazards, in Proceedings of OCEAN'2003, San Diego, 115.

Blandin J., Rolin, J.F., (2005). An Array of Sensors for the Seabed Monitoring of Geohazards, a Versatile Solution for the Long -Term Real-Time Monitoring of Distributed Seabed Parameters. Sea Technology, 46 (12).

Bohrmann, G., Ivanov, M., Foucher, J.P., Spiess, V., Bialas, J., Weinrebe ,W., Abegg, F., Aloisi, G., Artemov, Y., Blinova, V., Drews, M., Greinert, J., Heidersdorf, F., Krastel, S., Krabbenhöft, A., Polikarpov, I., Saburova, M., Schmale, O.,Seifert, R., Volkonskaya, A., Zillmer, M., (2003). Mud volcanoes and gas hydrates in the Black Sea – new data from Dvurechenskii and Odessa mud volcanoes. Geo-Marine Letters 23, 239-249.

Bourry, C., Chazallon, B., Charlou, J-L, Donval J.P, Ruffine, L., Henry, P., Geli, L., Çagatay, M.N., Sedat, İ, Moreau, M., (2009). Free gas and gas hydrates from the Sea of Marmara, Turkey: Chemical and structural characterization. Chem. Geol., 264 (1-2), 197-206. doi:10.1016/j.chemgeo.2009.03.007.

Delauney, L., Compère, C., Lehaitre, M., (2010). Biofouling protection for marine environmental sensors, Ocean Science, 6, 503–511. doi:10.5194/os-6-503-2010.

ESONEWS (Spring 2008) edited by ESONET-NoE consortium, Vol.2 (1) on:
    http://www.esonet-noe.org/content/download/20733/300246/file/ESONEWS4.pdf

Etiope, G., Favali, P. (Eds.), (2004). Geologic emissions of methane from lands and seafloor: mud volcanoes and observing systems. Environmental Geology 46, 987-1135.

Favali, P., Beranzoli, L., D'Anna, G., Gasparoni, F., Gerber, H. W, (2006a). NEMO-SN-1 The First "Real-Time" Seafloor Observatory Of Esonet. Nuclear Instruments & Methods In Physics Research Section A-Accelerators Spectrometers Detectors And Associated Equipment, 567, 462-467, Doi: 10.1016/J.Nima.2006.05.255

Favali, P., Beranzoli, L., D'Anna, G., Gasparoni, F., Marvaldi, J., Clauss, G., Gerber, H.W., Nicot, M., Marani, M.P., Gamberi, F., Millot, C. and Flueh, E.R. (2006b). A fleet of multiparameter observatories for geophysical and environmental monitoring at seafloor. Ann. Geophys., 49/2-3, 659-680.

Favali, P. and Beranzoli, L. (2006). Seafloor Observatory Science: A Review. Ann. Geophys., 49, 515-567

Favali, P., and L., Beranzoli, (2009). EMSO: European Seafloor Multidisciplinary Observatory, Nucl. Instr. and Meth. in Phy. Res. A, 602, 21-27. doi:101016/j.nima.2008.12.214.

Géli, L., Henry, P., Zitter, T., Dupré, S., Tryon, M., Çağatay, M. N., de Lépinay, B. Mercier, Le Pichon, X., Sengör, A. M. C., Görür, N., Natalin, B., Uçarkus, G., Özeren, S., Volker, D., Gasperini, L., Burnard, P., Bourlange, S., the Marnaut Scientific, Party, (2008). Gas emissions and active tectonics within the submerged section of the North Anatolian Fault zone in the Sea of Marmara. Earth Planet. Sci. Lett., 274 (1-2), 34-39.

Greinert, J., (2008): Monitoring temporal variability of bubble release at seeps: The hydroacoustic swath system GasQuant. J. Geophys. Res., 113, C07048, doi:10.1029/2007JC004704.

Henry, P., Lallemant, S.J., Nakamura, K., Tsunogai, U., Mazzotti S. and Kobayashi K., (2002). Surface expression of fluid venting at the toe of the Nankai wedge and implications for flow paths, Marine Geology, 187, 119-143.

Hustoft, S., Bünz, S., Mienert, J., Chand, S., (2009). Gas hydrate reservoir and active methane-venting province in sediments on <20 Ma young oceanic crust in the Fram Strait, offshore NW-Svalbard. Earth and Planetary Science Letters, 284 (1-2), 12-24 doi:10.1016/j.epsl.2009.03.038.

Kuşçu, I., Okamura, M., Matsuoka, H., Gokasan, E., Awata, Y., Tur, H. and Şimşek, M. (2005). Seafloor gas seeps and sediment failures triggered by the August 17, 1999 earthquake in the Eastern part of the Gulf of Izmit, Sea of Marmara, NW Turkey. Marine Geology, 215: 193-214.

Montagner, J.P., Karczewski, JF., Romanowicz, B., Bouarich, S., Lognonne, P., Roult, G., Stutzmann, E., Thirot, JL. , Brion, J., Dole, B., Fouassier, D., Koenig, JC., Savary, J., Floury, L., Dupond, J., Echardour, A. and Floc'h, H., (1994). The French Pilot Experiment OFM-SISMOBS: first scientific results on noise level and event detection. Physics of The Earth and Planetary Interiors, 84, 321-336.

Moore, J.C., Orange, D. and Kulm, L.D., (1990). Interrelationship of Fluid Venting and Structural Evolution – Alvin Observations from the Frontal Accretionary Prism, Oregon. J. Geophys. Res., 95 (B6), 8795-8808.

Lykousis, V., Alexandri, S., Woodside, J., de Lange, G., Dählmann, A., Perissoratis, C., Heeschen, K., Ioakim, Chr., Sakellariou, D., Nomikou, P., Rousakis, G., Casas, D., Ballas, D., Ercilla, G., (2009). Mud volcanoes and gas hydrates in the Anaximander mountains (Eastern Mediterranean Sea). Marine and Petroleum Geology 26, 854-872.

Person, R., Aoustin, Y., Blandin, J., Marvaldi, J., Rolin, J. F. ,(2006). From bottom landers to observatory networks, annals of Geophysics, 49, 2-3, 581-593.

Parson T., (2004). Recalculated probability of M>7 earthquakes beneath the Sea of Marmara. J. Geophys. Res., 109. doi:10.1029/2003JB002667.

Pondard, N., Armijo, R., King, G. C. P., Meyer, B., Flerit, F. (2007). Fault interactions in the Sea of Marmara pull-apart (North Anatolian Fault): earthquake clustering and propagating earthquake sequences, Geophys. J. Int., 171, 1185-1197. doi :10.1111/j.1365-246X.2007.03580.x.

Ruhl, H., Karstensen, J., Géli, L., André, M., Beranzoli, L., Namik Çağatay, M., Colaço, A., Cannat, J., Dañobeitia, J., Favali, P., Gillooly, M., Greinert, J., Hall, P., Huber, R., Lampitt, R., Lykousis, V., Miranda, M., Person, R., Priede, I., Puillat, I., Thomsen, L., and Waldmann, C., (2011). Societal need for improved understanding of climate change, anthropogenic impacts, and geo-hazard warning drive development of ocean observatories in European Seas, Progress In Oceanography, 91, (1), 1-33. doi:10.1016/j.pocean.2011.05.001.

Tary, JB, Geli, L, Henry, P, Natalin, B, Gasperini, L, Comoglu, M, Cagatay, N, Bardainne, T., (2011). Sea-Bottom Observations from the Western Escarpment of the Sea of Marmara. Bulletin of the Seismological Society of America, 101 (2), 775-791. doi: 10.1785/0120100014

Tryon, M.D., Henry, P., Çağatay, M.N., Zitter, T.A.C. Géli, L., Gasperini, L,. Burnard P., Bourlange, S., Grall, C., (2010). Pore fluid chemistry of the North Anatolian Fault Zone in theSea of Marmara: A diversity of sources and processes. Geochemistry, Geophysics, Geosystems, 11. doi:10.1029/ 2010GC003177.

Van Weering, T.C.E., Koster, B., Heerwaarden, J., Thomsen, L., Viergutz, T., (2000).New technique for long term deep seabed studies. Sea Technology, 2, 17–25.

Vuillemin, R., Le Roux, D., Dorval, P., Bucas, K., Sudreau, J.P., Hamon, M., Le Gall, C. and Sarradin, P.M., (2009). CHEMINI: A new in situ CHEmical MINIaturized analyzer. Deep Sea Research Part I: Oceanographic Research Papers, 56 (8), 1391-1399. doi:10.1016/j.dsr.2009.02.002.

Westbrook, G.K., Thatcher, K.E., Rohling, E.J., Piotrowski, A.M., Pälike, H., Osborne, A., Nisbet, E.G., Minshull, T., Lanoisellé, M., James, R.H., Hühnerbach, V., Green, D., Fisher, R.E., Chabert, A., Bolton, C., Beszczynska-Möller, A., Berndt, C., Aquilina, A., (2009). Escape of methane gas from the seabed along the West Spitsbergen continental margin. Geophysical Research Letters 36, L15608. doi:10.1029/2009GL039191.

Woodside, J.M., David, L., Frantzisc, A., Hooker, S.K., (2006). Gouge marks on deep-sea mud volcanoes in the eastern Mediterranean: Caused by Cuvier's beaked whales? Deep-Sea Research I 53, 1762-1771.

Zitter, T.A.C., Henry, P., Aloisi, G., Delaygue, G., Çağatay, M.N., Mercier de Lepinay, B., Al-Samir, M.F., Fornacciari, M., Pekdeger A., Wallmann, K., Lericolais, G., (2008). Cold seeps along the main Marmara fault in the Sea of Marmara (Turkey), Deep Sea Research, Part I, 55, 552-570.

An Introduction to FY-3/MERSI, Ocean Colour Algorithm, Product and Application

Sun Ling[1], Hu Xiuqing[1], Guo Maohua[2],
Zhu Jianhua[3], Li Sanmei[1] and Ding Lei[4]
[1]*National Satellite Meteorological Centre,*
[2]*National Satellite Ocean Application Service,*
[3]*National Ocean Technology Centre,*
[4]*Shanghai Institute of Technical Physics,*
China

1. Introduction

Ocean colour is the water-leaving radiance in the visible and near-infrared just above the ocean surface owing to selective absorption and scattering by phytoplankton and its pigments such as chlorophyll, as well as dissolved organic matter and suspended particulate matter in the subsurface ocean waters. Ocean colour carries useful information concerning biogeochemical properties of the water body. Ocean colour remote sensing can be applied in investigating the optical properties of upper ocean layers, biological productivity, global carbon and biogeochemical cycles in the oceans.

When an ocean-colour sensor measures the radiance backscattered by the ocean-atmosphere system, the signal it received is largely dominated by the atmosphere. Ocean colour retrieval from satellite measured top-of-the-atmosphere (TOA) radiances over the oceans requires removal of the atmospheric and ocean surface interfering effects, a process termed atmospheric correction which is the key step in ocean colour data processing. Water-leaving radiance is the basic product in ocean colour remote sensing, and can be used to derive the water inherent optical property (IOP) and constituent's concentration, photosynthetically active radiation (PAR), red-tide index. Other oceanic applications such as primary productivity and global carbon can be investigated further. Besides water-leaving radiance, concentration of phytoplankton pigments such as chlorophyll a is another standard ocean colour product. Diffuse attenuation coefficient, fluorescence line height, photosynthetically available radiation, concentrations of particulate organic and inorganic carbon, chromophoric dissolved organic matter and suspended particulate matter can also be provided.

With the advantage of spatial coverage and frequent overpass, satellite instruments are widely used in ocean research at local and regional scales. Since the launch of the Coastal Zone Colour Scanner (CZCS) in 1978, ocean colour remote sensing has been focused on studying the spatial and temporal evolution of phytoplankton in open oceanic areas. The main purpose is to improve our understanding of the carbon cycle and the role of the ocean in climate change. After CZCS, instruments with improved spatial and spectral characteristics have been deployed on space borne satellites. The Sea-Viewing Wide Field-of-View Sensor (SeaWiFS) provides the oceanographic community an unprecedented opportunity to retrieve ocean colour on a global scale. Then, the Moderate Resolution Imaging Spectrometer (MODIS) with a total of 36 channels continues the ocean colour observation. It is worth mentioning some ocean colour instruments, e.g. Ocean Colour Temperature Scanner (OCTS), Modular Optoelectronic Scanner (MOS), Medium Resolution Imaging Spectrometer (MERIS), Ocean Colour Monitor (OCM) and Global Line Imager (GLI). Geostationary Ocean Colour Imager (GOCI) launched in 2010 allows more frequent data provision.

FengYun-3 (FY-3) is the second generation of Chinese polar-orbit meteorological satellite operating in a near polar, sun-synchronous orbit at an altitude of 836 km. The morning satellite FY-3A (launched on May 27, 2008) with local equator-crossing time of 10:30 A.M. (descending southward) and the afternoon satellite FY-3B (launched on Nov 5, 2010) with local equator-crossing time of 1:30 P.M. (ascending northward) have been in operation. Medium Resolution Spectral Imager (MERSI), a major sensor among its 11 payloads, is a MODIS-like sensor covering visible to infrared spectral region. FY-3/MERSI is capable of making continuous global observations for a broad range of scientific studies of the Earth's system, and ocean colour application is one of its main targets.

In this chapter, an overview of MERSI instrument is presented in section 2. Quantitative remote sensing such as ocean colour is sensitive to sensor's radiometric performance. MERSI can not realize the onboard absolute radiometric calibration in the reflective solar spectral region. Various calibration techniques are adopted to monitor the radiometric degradation and assure calibration accuracy, such as absolute calibration using China Radiometric Calibration Site (CRCS) with in-situ measurements, multi-sites calibration tracking and relative calibration with onboard calibrator. The sensor performance related with ocean colour and primary calibration results in reflective solar spectral bands are briefly introduced in section 3. FY-3A/MERSI ocean colour products consist of water-leaving reflectance retrieved from atmospheric correction algorithm, chlorophyll a concentration, pigment concentration, total suspended mater concentration and absorption coefficient of CDOM and NAP from global and Chinese regional empirical models. The FY-3A/MERSI ocean colour product specification, atmospheric correction algorithm based on lookup tables (LUT) and ocean colour components concentration estimation models are described in section 4. The ocean colour product has been primarily validated against in situ data and the comparison result is given in section 5. Ocean colour product is helpful in understanding the ocean variability and changes with their effects on climatic processes. In section 5, some ocean colour application cases using MERSI data such as algae bloom monitoring and coastal suspended sediment variation are also demonstrated.

2. Overview of MERSI instrument

MERSI is manufactured by Shanghai Institute of Technology and Physics (SITP), Chinese Academy of Sciences (CAS). First serial MERSI (called MERSI-1) onboard the firstly three FY-3 satellites has 20 spectral bands, of which 19 are the reflective solar bands (RSBs) covering the wavelength range of 0.4-2.1 μm and one is the thermal emissive band (TEB) covering 10-12.5 μm. FY-3/MERSI is a cross-track scanning radiometer. There are two calibrator systems inside: a Visible Onboard Calibrator (VOC) for RSBs and a Blackbody (BB) for TEB. Figure 1 shows the instrument photo. MERSI makes earth view observations via a single-sided 45° scan mirror in concordance with a K mirror (de-rotation) over a scan angle range of ±55° about nadir. It provides a swath of 2900 km cross track by 10 km (at nadir) along track for each scan with multi-detectors (10 or 40), enabling a complete global coverage in one day. MERSI has a nominal ground instantaneous field of view (GIFOV) of 250 m, or 1000 m at nadir. The general information and spectral band specifications are listed in Table 1 and 2. The global moderate-resolution narrow-band observations in 20 spectral bands have provided useful data for scientific studies and applications in land, ocean, and atmosphere.

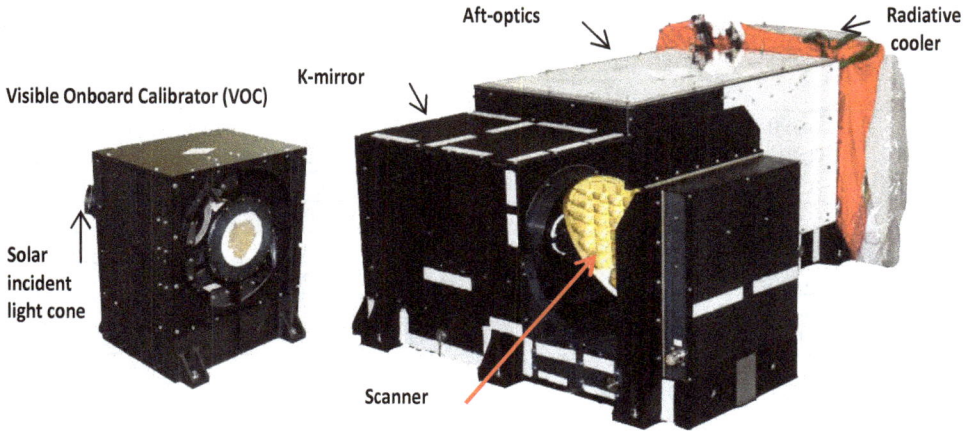

Fig. 1. Instrument photo of MERSI (the left is the VOC and the right is the instrument main component).

Parameters	Specification
Earth scanning	±55.1°±0.1°
Quantization	12 bits
Scanner speed	40 rotations/minute
Scanning stability	<0.5 IFOV (1000m)
Sampling pixels of each scan	2048(1000-m bands); 8192(250-m bands)
Response degradation rate	<20%/3 years
Spectral characterization accuracy	Bias of central wavelength < 10%×band width; out of band response < 3%
Inter-band co-registration	<0.3 pixel
Bright target recovery	≤6 pixels(1000-m bands) ≤24pixels(250-m bands)
MTF	≥0.27(1000-m bands); ≥0.25(250-m bands)
Radiometric calibration accuracy	Visible bands <7%; thermal band<1K(270K)
Detector consistency within one band	Unconsistency≤5-7%

Table 1. General information of MERSI.

Band	Central wavelength(μm)	Band width (μm)	Resolution (m)	NEΔρ(%)/ NEΔT (300K)	Dynamic range (maximum ρ or T)
1	0.470	0.05	250	0.45	100%
2	0.550	0.05	250	0.4	100%
3	0.650	0.05	250	0.4	100%
4	0.865	0.05	250	0.45	100%
5	11.25	2.5	250	0.54K	330K
6	1.640	0.05	1000	0.08	90%
7	2.130	0.05	1000	0.07	90%
8	0.412	0.02	1000	0.1	80%
9	0.443	0.02	1000	0.1	80%
10	0.490	0.02	1000	0.05	80%
11	0.520	0.02	1000	0.05	80%
12	0.565	0.02	1000	0.05	80%
13	0.650	0.02	1000	0.05	80%
14	0.685	0.02	1000	0.05	80%
15	0.765	0.02	1000	0.05	80%
16	0.865	0.02	1000	0.05	80%
17	0.905	0.02	1000	0.10	90%
18	0.940	0.02	1000	0.10	90%
19	0.980	0.02	1000	0.10	90%
20	1.030	0.02	1000	0.10	90%

Table 2. Spectral band specification of MERSI.

Scene radiant flux reflects from the continuously rotating scan mirror at 40 rpm. Energy from the scan mirror strikes the primary mirror (the entrance pupil), goes through a field stop and then onto the secondary mirror. Radiance reflected from the secondary mirror is transmitted to K mirror which is used to remove the rotation of the image due to the rotation by 45° scanner and multi-detector parallel. The K mirror rotates at exactly half the rate of the scanner and uses alternating mirror sides on successive telescope scans. Immediately after K mirror there is a dichroic beam splitter assembly (consisting of three beam splitters) that directs the energy through four refractive objective assemblies and then onto the four focal plane assemblies (FPAs) with their individual band-pass filters. The beam splitters are used to achieve spectral separation, dividing the MERSI spectral domain into four spectral regions: visible (VIS) (412 to 565 nm), near infrared (NIR) (650 to 1030nm), short wavelength infrared (SWIR) (1640 to 2130nm), and long wavelength infrared (LWIR) (12250nm). The SWIR and LWIR FPAs are cooled to approximately 90 K by a passive radiative cooler. All instrument bands, each with multiple detectors (40 and 10 along-track detectors for 250-m and 1000-m band separately) are aligned in the scan direction on focal planes. The VIS and NIR FPAs utilize p-i-n photovoltaic silicon diodes. The SWIR bands have photovoltaic HgCdTe detectors and LWIR band uses photoconductive HgCdTe. The outputs from multiple detectors in each band are added using a time delay and integration (TDI) technique to improve the signal-to-noise ratios (SNRs). The VOC is mounted on the side of the instrument, allowing the scanner to view the VOC's exit when passing over the South Pole. Space view (SV) port is used to provide a zero reference.

3. In-flight performance of solar bands

3.1 Radiometric response change using VOC

MERSI VOC is the first onboard visible calibration experimental device for FengYun series sensors. It is composed of a 6-cm diameter integrating sphere with interior lamp and sunlight import cone, an export beam expanding system with a flat mirror and a parabola to create a collimated beam, and absolute radiance trap detectors (4 detectors with the same filter designs as MERSI bands of 470nm, 550nm, 650nm and 865nm, and one panchromatic detector with no filter). The export parallel light from the expanding system fills the entrance aperture of MERSI and is viewed each scan. Although the VOC can not realize the absolute radiometric onboard calibration, it can be used as a radiometric source to monitor the radiometric response degradation of MERSI.

Figure 2(a) shows the digital number (DN) variation of 5 VOC trap detectors at different time. The degradation of interior lamp illumination appears. Response degradation rates of 19 MERSI solar bands are then derived using the scanning DN and lamp illumination degradation. Figure 2(b) reveals the great degradation of band 1, 8, 9, 10 of MERSI since launch. The greatest degradation is in band 8 (412nm), more than 15% during one year. It has been found that there exist signal anomaly jumps at band 6 and 7. The instrument vendor explained that it was induced by the electronic gain anomaly jump of MERSI SWIR bands. So the degradation status of band 6 and 7 have to be re-evaluated based on the electronic gain levels.

3.2 Vicarious radiometric calibration tracking

China Radiometric Calibration Site (CRCS) for satellite calibration in VIS/NIR bands is located at Dunhuang Gobi desert, centered at 40.65°N, 94.35°E. Annual field campaigns have been routinely carried out at CRCS, and the vicarious calibration (VC) based on synchronous in-situ measurements is the baseline operational calibration approach for Chinese FengYun (FY) series satellites (Hu et al., 2010). Annual CRCS field campaign for MERSI has been conducted since Sep. 2008.

Although the CRCS based vicarious calibration has the accuracy of ~5%, the limited data amount is not enough for frequent and stable in-flight calibration coefficient updates. Multi-sites with stable surface properties have been chosen for radiometric calibration tracking. Gobi and desert targets such as Dunhuang, Libya1(24.42°N, 13.35°E), Libya4(28.55°N, 23.39°E) and Arabia2(20.13°N, 50.96°E) recommended by CEOS/WGCV are used, as well as an ocean site at Lanai(MOBY, 20.49°N,°-157.11E). Surface directional reflectance is calculated using MODIS BRDF products for land sites or taken from MOBY measurements for the ocean site. Aerosol optical depth is taken from MODIS monthly aerosol product. With the V6S model, TOA reflectance (Ref) can be calculated. Data within certain days from 5 sites are used to get the calibration coefficient (slope):

$$ARef_i = Ref_i \, (d0/d)^2 \cos(SolZ) = Slope_i \, (DN_i\text{-}SV_i), \tag{1}$$

where $ARef_i$ is the TOA apparent reflectance for band i, $SolZ$ is solar zenith angle, $(d0/d)^2$ is earth-sun distance correction factor, $Slope$ is calibration slope, DN and SV are digital numbers of earth observation and space view respectively. Based on the calibration coefficient series, a linear model is used to describe the varying trend of calibration slopes:

$$Slope_i = a_i DSL + b_i, \tag{2}$$

where DSL is the day number since launch (May 27, 2008), a reflects the degradation rate of response gain ($1/Slope$).

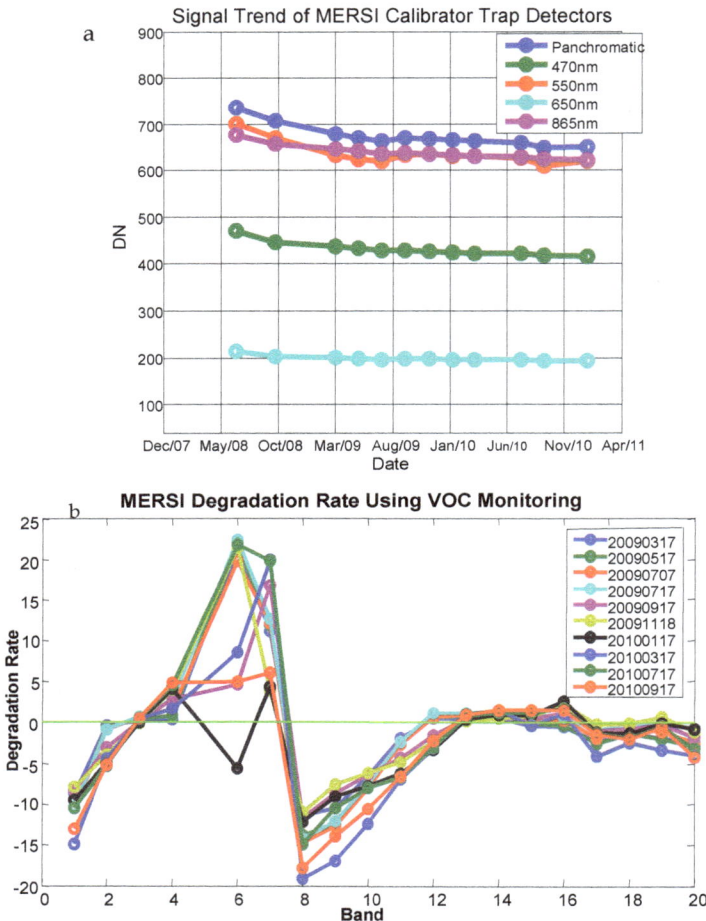

Fig. 2. MERSI VOC's output monitoring by trap detectors (a) and derived response degradation of 19 reflective solar bands (b) from VOC's observation at different dates.

Typical examples of band 1, 2, 3 and 4 are shown in Fig. 3, in which the linear fitting model is shown with black solid line. Band 6 and 7 are omitted because of anomaly electronic gain jumps possibly induced by electrostatic discharge. It can be seen that the calibration coefficients present a linear trend with DSL, and seasonal periodicity exists especially in the short-wave and the water vapor channels. The calibration tracking results are listed in Table 3 using data from Aug. 2008 to Dec. 2010. The short-wave channels have large degradation, especially band 8 with the annual decay rate up to 14%. In the red and near-infrared bands (600 ~ 900nm), e.g. band 3, 4, 13, 14, 15 and 16, the calibration coefficients almost have no change with the annual decay rate below 1%. The uncertainty (2σ/mean) for the trend analysis is below 5% except for water vapor bands (17, 18 and 19).

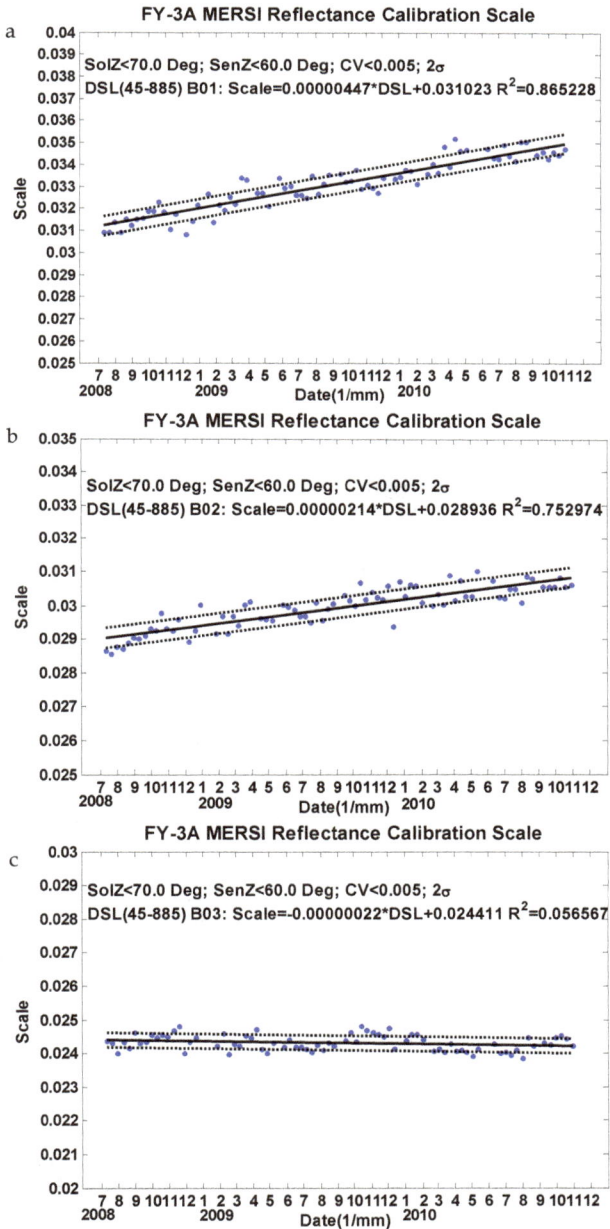

a

FY-3A MERSI Reflectance Calibration Scale

SolZ<70.0 Deg; SenZ<60.0 Deg; CV<0.005; 2σ
DSL(45-885) B01: Scale=0.00000447*DSL+0.031023 R^2=0.865228

b

FY-3A MERSI Reflectance Calibration Scale

SolZ<70.0 Deg; SenZ<60.0 Deg; CV<0.005; 2σ
DSL(45-885) B02: Scale=0.00000214*DSL+0.028936 R^2=0.752974

c

FY-3A MERSI Reflectance Calibration Scale

SolZ<70.0 Deg; SenZ<60.0 Deg; CV<0.005; 2σ
DSL(45-885) B03: Scale=-0.00000022*DSL+0.024411 R^2=0.056567

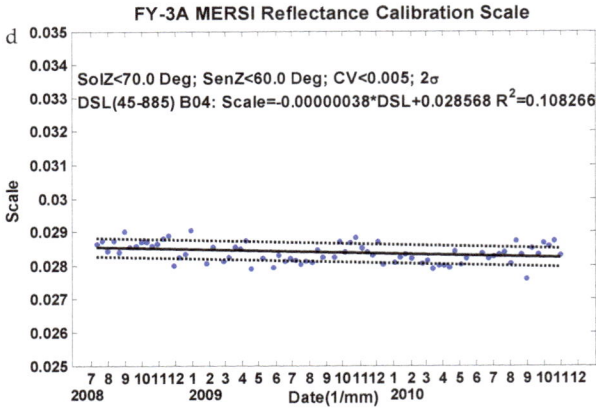

FY-3A MERSI Reflectance Calibration Scale

d

SolZ<70.0 Deg; SenZ<60.0 Deg; CV<0.005; 2σ
DSL(45-885) B04: Scale=-0.00000038*DSL+0.028568 R²=0.108266

Fig. 3. Calibration coefficient trend of band 1(a), 2(b), 3(c) and 4(d) from multi-sites calibration.

Band	a	b	2σ/mean(%)	Annual Decay Rate(%)
1	4.47E-06	0.0310	2.6694	5.2561
2	2.14E-06	0.0289	2.0388	2.6947
3	-2.19E-07	0.0244	1.8415	-0.3282
4	-3.83E-07	0.0286	1.9638	-0.4895
8	8.61E-06	0.0217	4.3801	14.4623
9	4.78E-06	0.0237	3.4290	7.3481
10	2.87E-06	0.0245	2.4686	4.2702
11	1.91E-06	0.0199	2.2112	3.5070
12	1.16E-06	0.0225	1.7599	1.8896
13	-5.91E-08	0.0223	1.9964	-0.0970
14	-2.39E-08	0.0217	1.8287	-0.0402
15	6.73E-07	0.0276	1.9260	0.8887
16	5.12E-08	0.0211	1.3659	0.0884
17	2.36E-06	0.0243	6.1333	3.5431
18	7.25E-06	0.0262	19.7859	10.0894
19	2.48E-06	0.0233	6.3302	3.8836
20	2.99E-06	0.0253	2.1729	4.3028

Table 3. FY-3A/MERSI solar bands calibration tracking results with multi-sites method.(σ: The standard deviation of difference between calibration coefficients and the linear regression line).

4. Ocean colour product

4.1 Product specification

Several quantitative parameters are provided in FY-3A/MERSI ocean colour product:

- Water-leaving reflectance ($\rho_w = L_u(0^+)/L_d(0^+)$) for band 8 to 16 retrieved from atmospheric correction algorithm based on LUTs,
- Chlorophyll a (Chla) concentration (CHL1) and pigment concentration (PIG1) from global empirical models,
- Chlorophyll a concentration (CHL2), total suspended mater concentration (TSM), absorption coefficient of CDOM and NAP at 443nm band (YS443) from Chinese regional empirical models.

FY-3A ocean colour product specification is shown in Table 4. The product file format is HDF5. Examples of ocean colour product are shown in Fig. 4.

Type	Projection	Coverage	Spatial Resolution
Day	Geographic Longitude/Latitude	Global, 10°×10° per breadth	0.01°×0.01°
Ten days	Ditto	Global	0.05°×0.05°
Month	Ditto	Global	0.05°×0.05°

Table 4. FY-3A ocean colour product specification.

4.2 Atmospheric correction algorithm

For the ocean-atmosphere system, TOA reflectance measured by the satellite sensor in a spectral band centered at a wavelength λ, $\rho_t(\lambda)$, can be written as a linear sum of various contributions (angular dependencies are omitted):

$$\rho_t(\lambda) = [\rho_w(\lambda)+\rho_{wc}(\lambda)+\rho_g(\lambda)]T_g(\lambda)T_{r+a}(\lambda)+\rho_{atm}(\lambda), \tag{3}$$

$$\rho_{atm}(\lambda) = [\rho_{mix}(\lambda)-\rho_r(\lambda)]T_g(\lambda)+\rho_r(\lambda)T_g{'}(\lambda), \tag{4}$$

where ρ_w is water-leaving reflectance, ρ_{wc} is whitecap reflectance, ρ_g is sun glint reflectance, T_g is gaseous absorption transmittance, $T_g{'}$ is gaseous absorption transmittance excluding water vapor, T_{r+a} is scattering transmittance of the Rayleigh and aerosol mixing atmosphere, ρ_{atm} is intrinsic atmospheric reflectance reaching the sensor, ρ_{mix} is reflectance for the Rayleigh and aerosol mixing atmosphere not accounting for absorptive gases, and ρ_r is reflectance for a pure Rayleigh atmosphere not accounting for absorptive gases.

None water-leaving effects are calculated and removed from ρ_t to extract ρ_w. ρ_r is calculated using look-up tables with angles and surface pressure (Sun et al., 2006). T_g together with $T_g{'}$ is calculated using formulas based on simulations with angles and gases amounts (Sun & Zhang, 2008). ρ_{wc} is estimated using an empirical model with wind speed (Gordon, 1997). ρ_g is estimated using the Cox & Munk model with wind and angles. T_{r+a} is calculated using look-up tables with certain aerosol model, aerosol optical thickness τ_a and angles (Sun, 2005). To determine ρ_{mix}, mainly the aerosol contribution because ρ_r can be calculated in advance, $\tau_a \leftarrow\rightarrow \gamma$ ($\gamma=\rho_{mix}/\rho_r$) lookup tables with various aerosol models and angles are used (Sun & Guo, 2006). Seven candidate aerosol models from OPAC database are adopted. The aerosol scattering phase function $P_a(\Theta)$ and normalized aerosol optical thickness $\Delta\tau_a$ are separately shown in Fig. 5.

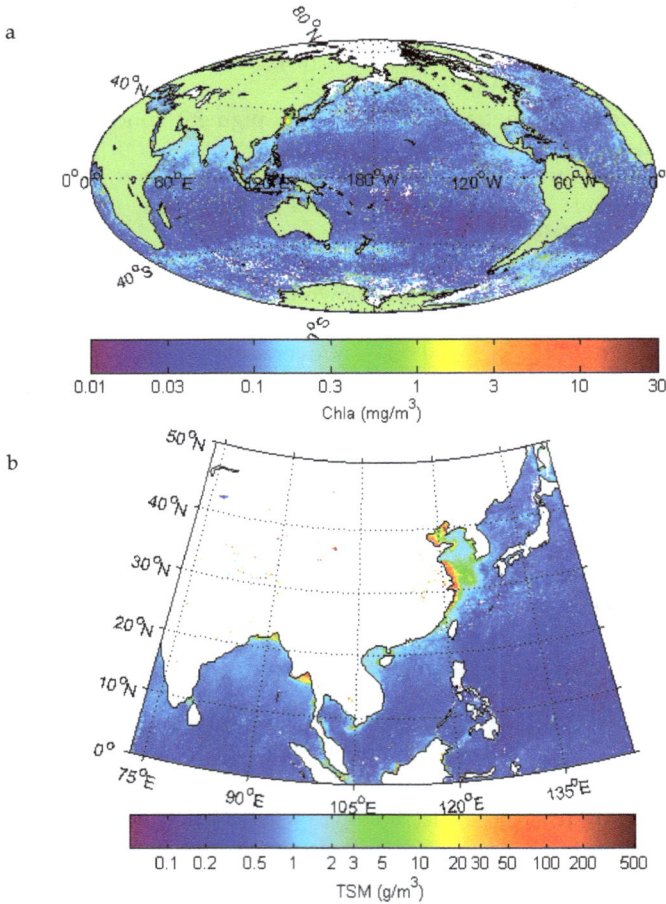

Fig. 4. Monthly mean global Chl-a (a) and TSM (b) around China coast of Feb. 2010.

The water-leaving reflectance at two NIR bands (centered at 765 and 865nm) are assumed zeros. Thus, $\gamma(765)$ and $\gamma(865)$ are obtained from ρ_t. $\tau_a(865)_i$ for 7 aerosol models are calculated using $\tau_a \leftrightarrow \gamma$ lookup tables, and $\tau_a(\lambda)_i$ are extrapolated with the normalized optical thickness $\Delta\tau_a(\lambda)_i$, and $\gamma(765)_i$ are reversely calculated using $\tau_a \leftrightarrow \gamma$ lookup tables. Then, two aerosol models (represented with $i1$ and $i2$) most similar to the actual one are selected according to $\gamma(765)_{i1} < \gamma(765) < \gamma(765)_{i2}$, and the mixing ratio X for interpolation between two aerosols is calculated as $[\gamma(765)-\gamma(765)_{i1}]/[\gamma(765)_{i2}-\gamma(765)_{i1}]$. And then, for the visible and NIR bands, $\gamma(\lambda)_{i1}$ and $\gamma(\lambda)_{i2}$ are calculated from $\tau_a(\lambda)_{i1}$ and $\tau_a(\lambda)_{i2}$, and $\gamma(\lambda)$ is estimated as $(1-X)\gamma(\lambda)_{i1}+X\gamma(\lambda)_{i2}$, and $\rho_{mix}(\lambda)$ can then be calculated. Till now, ρ_w can be calculated from equation (3).

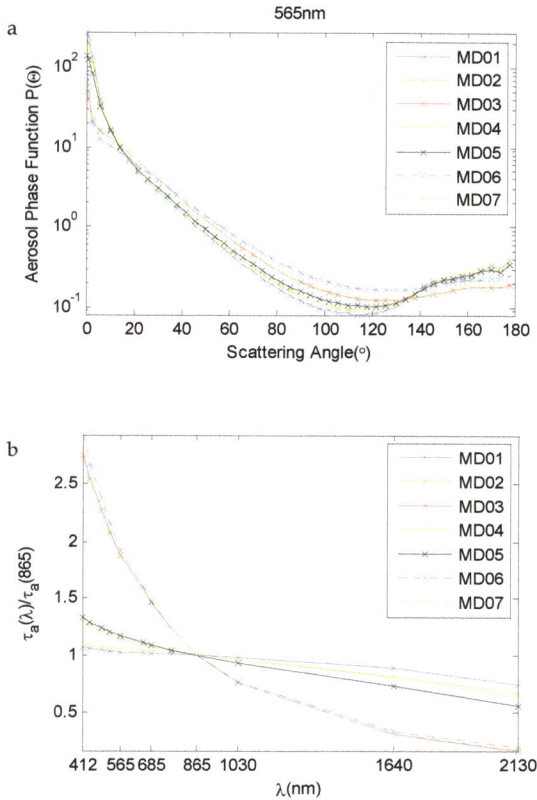

Fig. 5. Aerosol scattering phase function for 565nm band (a) and normalized aerosol optical thickness (b).

4.3 Water constituents concentration estimation models

The MODIS OC3 model is used in global chlorophyll a concentration (CHL1) estimation.

$$\text{Log(Chl}a) = A_0 + A_1 X_c + A_2 X_c^2 + A_3 X_c^3 + A_4 X_c^4, \tag{5}$$

in which,

$$X_c = \text{Log(MAX}(R_{rs}(443), R_{rs}(488))/R_{rs}(551)),$$

$$A = [0.283, -2.753, 1.457, 0.659, -1.403].$$

The gobal pigment concentration (PIG1) estimation model is derived from the SeaBAM dataset:

$$\text{Log(PIG)} = A_0 + A_1 X_p \quad (N=156, R^2=0.92), \tag{6}$$

in which,

$$X_p = \text{Log}(R_{rs}443/R_{rs}565),$$

$$A = [0.3560, -1.6198].$$

With independent data validation (N=56), the data percentage within relative error of 30% (calculated in log scale) is 71%.

The local models in China coastal region have been constructed with in situ measurements during 2003 spring and fall cruise in the Yellow Sea and the East China Sea organized by the National Satellite Application Service (Tang et al., 2004).

$$\text{Log(Chl}a) = A_0 + A_1\text{Log}(X_c) + A_2\text{Log}^2(X_c) \quad (R^2=0.87, N=40), \tag{7}$$

in which,

$$X_c = (R_{rs}(443)/R_{rs}(565))(R_{rs}(412)/R_{rs}(490))^b,$$

$$b=-0.75, A=[0.054, -1.46, 0.879].$$

The maximum, minimum and mean Chl$a$ are 6.83, 0.54 and 1.55 mg/m^3, respectively.

$$\text{Log(TSM)} = A_0 + A_1\text{Log}(X_s) + A_2\text{Log}^2(X_s) + A_3\text{Log}^3(X_s) \quad (R^2=0.94, N=40), \tag{8}$$

in which,

$$X_s = (R_{rs}(565) + R_{rs}(685))(R_{rs}(490)/R_{rs}(565))^b,$$

$$b=-2, A=[4.309, 4.324, 1.592, 0.212].$$

The maximum, minimum and mean TSM are 340.1, 0.6 and 16.8 g/m^3, respectively.

In China coastal region, the concentration of suspended particle is very high caused by terrestrial inputs, and in-situ data show that the absorption coefficient of none pigment particles is obviously larger than yellow substance.

$$\text{Log(YS443)} = A_0 + A_1\text{Log}(X_{dg}) + A_2\text{Log}^2(X_{dg}) + A_3\text{Log}^3(X_{dg}) \quad (R^2=0.93, N=40), \tag{9}$$

in which,

$$X_{dg} = (R_{rs}(565) + R_{rs}(685))(R_{rs}(490)/R_{rs}(565))^b,$$

$$b=-2, A=[2.709, 4.286, 1.711, 0.233].$$

The maximum, minimum and mean YS443 are 8.072, 0.052 and 0.583 m^{-1}, respectively.

Independent data are used for validation (N=40). The maximum, minimum and mean for Chl$a$, TSM and YS443 is 3.76, 0.51 and 1.40 mg/m^3; 182.3, 0.5 and 16.6 g/m^3; 6.712, 0.080, and 0.604 m^{-1}, respectively.

The mean relative error

$$1/N\sum_{i=1}^{N}|(X_i^{\text{Mod}} - X_i^{\text{Mea}})/X_i^{\text{Mea}}|$$

is 19.3%, 19.3%, and 26.1%, and the data percentage within relative error of 30% is 67%, 80% and 57% for Chl$a$, TSM and YS443 respectively.

5. Product validation and application

5.1 Primary validation

By comparing with in situ data collected in Feb. 2009, the FY-3A/MERSI ocean colour product was preliminarily validated. But only two match-ups were available, whose time differences were less than 5 minutes. The RMS difference and percentage difference values were given in Table 5. The ρ_w at 443, 520 and 565 nm was systematically overestimated, while ρ_w at 490 nm was underestimated.

| | ρ_w | | | | | Chl$a$ | TSM |
	412	443	490	520	565		
RMSD	0.005	0.004	0.004	0.003	0.003	0.023	0.095
RMSPD(%)	21.3	18.9	20.9	30.2	48.2	9.4	10.1

Table 5. FY-3A ocean colour product primary validation result.

5.2 Algae bloom monitoring

FY-3A/MERSI is a nice sensor for algae bloom monitoring. In the late spring and early summer 2008, a severe *enteromorpha prolifera* bloom erupted in the Yellow Sea and the sea area of Qingdao Olympic regatta was severely interfered. A massive algae clearing action was organized in which FY-3A/MERSI played an important role in determining algae distribution, moving route, and directing in situ salvage. Figure 6 showed one false colour MERSI image (RGB with band 3/4/1) on June 28, 2008 (Sun et al., 2010).

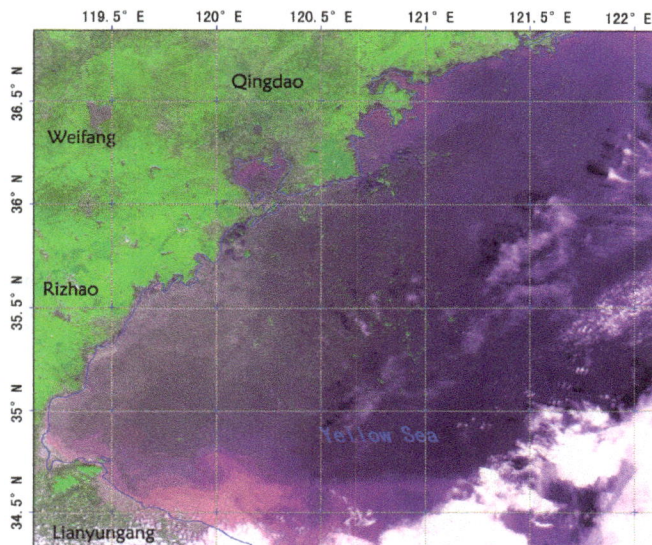

Fig. 6. Algae bloom captured on June 28, 2008 around Qingdao.

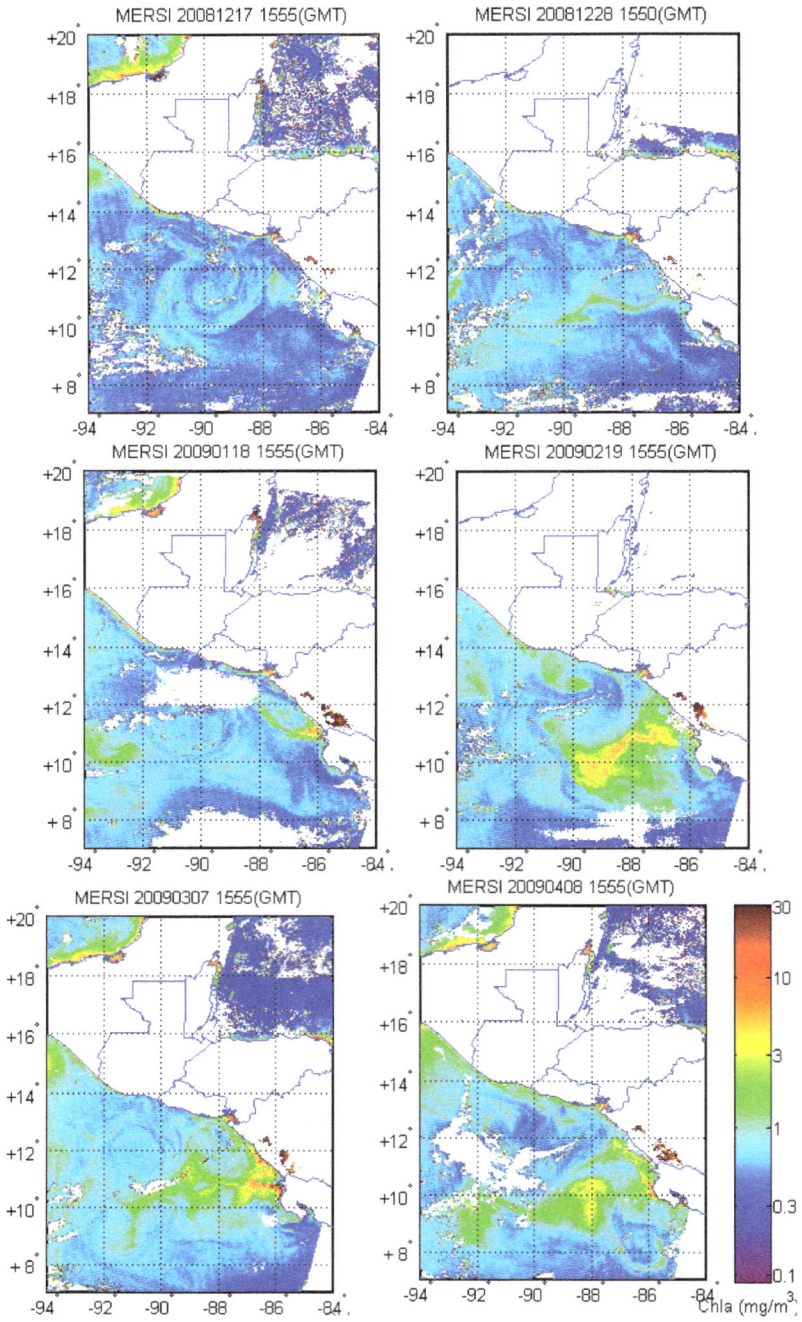

Fig. 7. Red tide process in Central America.

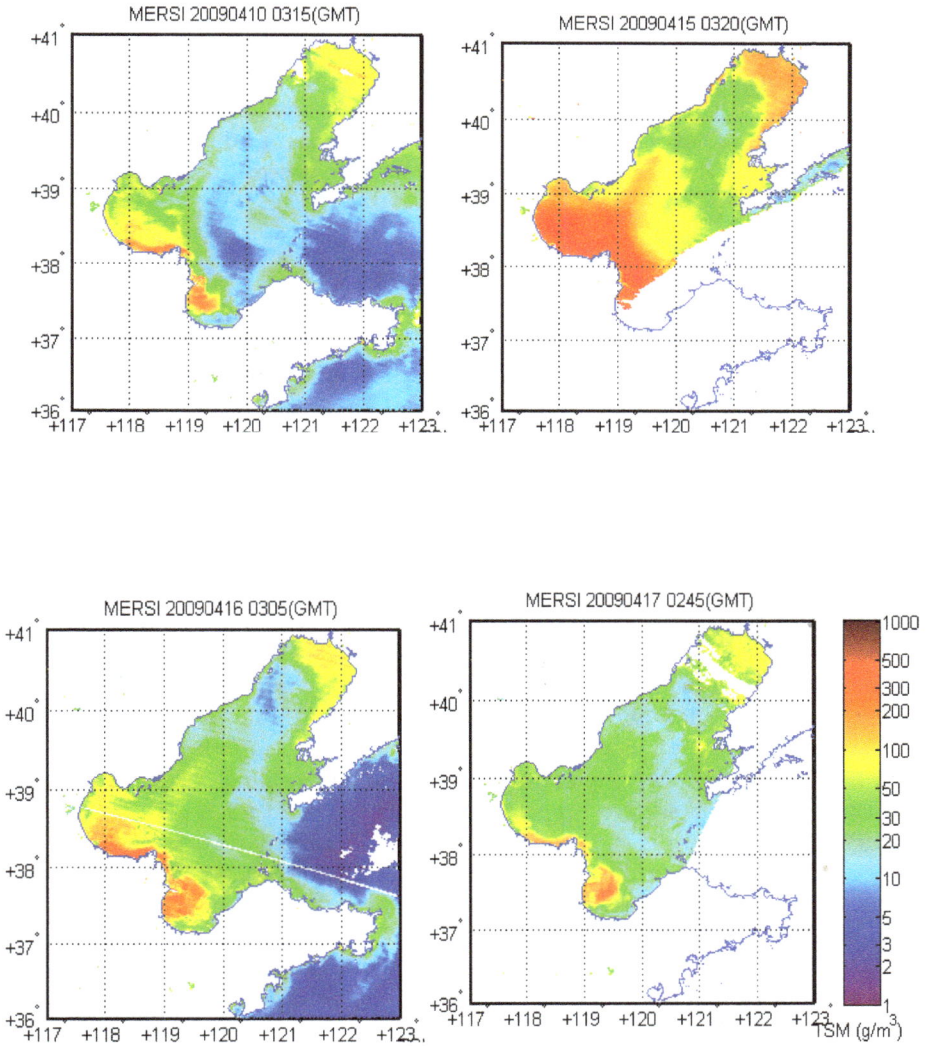

Fig. 8. Suspended particle concentration variation caused by a storm surge at Bohai Sea.

From the end of 2008, a serious HAB attack the Pacific coastal region of the Central America, especially around Costa Rica, Guatemala, El Salvador and Nicaragua. Figure 7 showed this red tide developing process using FY-3A Chl$a$ product.

5.3 Suspended sediment variation

In the early morning of Apr.15, 2009, a large storm surge suddenly happened in Bohai Sea, the eleven to twelve level wind rapidly resuspended the marine sediment and changed the suspended particle concentration distribution in the Bohai Sea, especially Bohai Bay and Laizhou Bay. Figure 8 showed this process using the FY-3A TSM product.

6. Conclusion

FY-3A/MERSI is the milestone sensor for Chinese meteorological satellite beyond it's previous instruments. It adopted a special 45° mirror scanning and de-rotation techniques with K mirror which can do multiple detectors scanning. One of the instrument's most outstanding features is that it can provide no gap global observation with its large coverage swath as well as five 250-m bands. The MERSI's global mission will make a major contribution to scientific projects that seek to understand the role of the land, atmosphere and ocean in the climate system and will further increase our ability to forecast change through modeling.

FY-3A/MERSI is the first Chinese sensor having the ability to make ocean colour global coverage. Ocean colour remote sensing is sensitive to sensor's radiometric performance. The MERSI VOC is the first experimental onboard calibrator for visible bands and provides a good way to monitor the relative radiometric response change at visible bands. The based-line calibration approach is VC based on synchronous in-situ measurements, and has been conducted annually using Dunhuang CRCS site in summer since 2008. Although the CRCS based vicarious calibration has the accuracy of ~5%, the limited data amount is not enough for frequent and stable in-flight calibration coefficient updates. Multi-sites with stable surface properties have been chosen for radiometric calibration tracking. It reveals that the short-wave bands have large degradation, especially band 8 with the annual decay rate up to 14%. In the red and near-infrared bands (600 ~ 900nm), e.g. band 3, 4, 13, 14, 15 and 16, the calibration coefficients almost have no change with the annual decay rate below 1%.

FY-3A/MERSI ocean colour products consist of water-leaving reflectance, chlorophyll a concentration (CHL), total suspended mater concentration (TSM) and absorption coefficient of CDOM and NAP (YS443) derived from atmospheric correction algorithm based on lookup tables (LUT) and global and Chinese regional empirical models for ocean colour components concentration estimation. By comparing with in situ data, the FY-3A/MERSI ocean colour product has been preliminarily validated. The ρ_w at 443, 520 and 565 nm was systematically overestimated, while ρ_w at 490 nm was underestimated. The RMS percentage difference for Chl$a$ and TSM are 9.4% and 10.1%, respectively.

Ocean colour product is helpful in understanding the ocean variability and changes with their effects on climatic processes. MERSI data have shown good application ability at ocean colour aspect, such as monitoring of algae bloom and coastal suspended sediment variation.

Although various calibration methods have been used to assure data accuracy, there is still much work to be done to acquire high quality radiometric data. And more validation work is needed to be carried out in the near future. The FY-3A/MERSI data (from L1B to L3) is public, and can be searched and ordered from the website (http://fy3.satellite.cma.gov.cn).

7. Acknowledgment

The authors would like to thank Song Qingjun and all participants in the 2009 Nanhai Cal/Val Campaign conducted by National Satellite Ocean Service for the in-situ dataset. This research is supported by National Key Basic Research Science Foundation ("973" project) of China under contracts No. 2010CB950803 and 2010CB950802, the National Natural Science Foundation of China under contract No. 40606043 and Meteorological Special Project under contract No. GYHY200906036.

8. References

Gordon, H. R. (1997). Atmospheric correction of ocean colour imagery in the earth observing system era. *J. Geophys. Res.*, Vol.102, No.D14, pp. 17081-17106, ISSN 0148–0227

Hu, X., Liu, J., Sun, L., Rong, Z., Li, Y., Zhang, Y. , et al. (2010). Characterization of CRCS Dunhuang Test Site and Vicarious Calibration Utilization for Fengyun (FY) Series Sensors, *Can. J. Remote Sensing*, Vol. 36, No. 5, (October 2010), pp. 566–582, E-ISSN 1712-7971

Sun, L. & Guo, M. (2006). Atmospheric correction for HY-1A CCD in Case 1 waters. *Proceedings of SPIE, Remote Sensing of the Environment: 15th National Symposium on Remote Sensing of China*, Vol. 6200, pp. 20-31, ISBN 0-8194-6256-X, Guiyang, China, August 19-23, 2005

Sun, L. & Zhang, J. (2008). Influence analysis of gaseous absorption on "HY-1A" CZI data processing: Simulation and correction for Rayleigh scattering. *ACTA Oceanologica Sinica*, Vol.27, No.6, (December 2008), pp. 102-114, ISSN 0253-505X

Sun, L. (2005). Atmospheric correction and water constituent retrieval for HY-1A CCD. PhD dissertation (in Chinese). Qingdao: Institute of Oceanology, Chinese Academy of Science, 67-69

Sun, L., Guo, M., Li, S., & Zhao, W. (2010). Enteromorpha Prolifera monitoring with FY-3A MERSI around the sea area of Qingdao. *Remote Sensing Information*, No.1, (February 2010), pp.64-68, ISSN 1000-3177

Sun, L., Zhang, J. & Guo, M. (2006). Rayleigh lookup tables for HY-1A CCD data processing. *J. Remote Sens.*, Vol.10, No.3, (June 2006),pp. 306-311, ISSN 1007-4619

Tang, J., Wang, X., Song, Q., Li, T., Chen, J., Huang, H. & Ren, J. (2004). The statistic inversion algorithms of water constituents for the Huanghai Sea and the East China Sea. *ACTA Oceanologica Sinica*, Vol.23, No.4, (August 2004), pp. 617-626, ISSN 0253-505X

5

Seismic Oceanography: A New Geophysical Tool to Investigate the Thermohaline Structure of the Oceans

Haibin Song[1], Luis M. Pinheiro[2], Barry Ruddick[3] and Xinghui Huang[1]
[1]Key Laboratory of Petroleum Resources Research,
Institute of Geology and Geophysics, Chinese Academy of Sciences, Beijing,
[2]Departamento de Geociências and CESAM, Universidade de Aveiro, Aveiro,
[3]Department of Oceanography, Dalhousie University, Halifax, Nova Scotia,
[1]China
[2]Portugal
[3]Canada

1. Introduction

Seismic oceanography is a new cross discipline between seismology and physical oceanography. It consists of the application of the multichannel seismic reflection method, commonly used in the oil industry to image the subsurface geological structure, to the investigation of the thermohaline fine structure of the oceans. The application of the seismic reflection method for this purpose was first reported by Gonella and Michon (1988), but that work remained largely unknown, and it was not until its rediscovery and the publication of the work of Holbrook et al. (2003) that this new method became widely established.

Seismic reflection sections provide very high resolution images of the oceans structure, both vertical and, in particular, horizontal, and complement conventional physical oceanography CTD/XBT data (e.g. Ruddick et al., 2009). These images consist of seismic reflections that occur and are recorded whenever a seismic wave travelling in a heterogeneous media encounters interfaces between different water masses with different acoustic impedances (the product of density by sound speed) and is reflected back to the surface. Nandi et al. (2004) and Nakamura et al. (2006) have shown that the reflectors imaged correspond indeed to oceanic thermal structures and, more recently, Ruddick et al. (2009) have shown that temperature variations have the dominant contribution to acoustic impedance contrasts and that salinity variations strengthen impedance contrasts by O(10%). Since the salinity variations are highly correlated with temperature variations on the scales that reflect sound, they enhance but do not change the appearance of reflectors. Therefore, these authors further demonstrated that seismic images of the water column are primarily images of vertical temperature gradient smoothed over the resolution scale of the seismic source wavelet, typically ~10m. Ocean "fine-structures" of that order of dimension are well-known in the ocean and are associated with a variety of physical phenomena: internal waves, thermohaline intrusions, double-diffusive layering, mixed water patches, vortical modes, and others (Ruddick et al., 2009).

Compared to conventional methods for investigating the ocean used in physical oceanography, seismic oceanography has the advantages of high lateral resolution and fast imaging (typical horizontal sample rate is about ten meters but time sample rate is far less, although with lower vertical resolution than conventional physical oceanographic methods). Holbrook and Fer (2005) have made quantitative inferences about internal wave energy levels near sloping ocean bottom that may eventually link to internal wave reflection properties and near-bottom ocean mixing. As noted by Ruddick et al. (2009), similar to satellite images that clearly show mixing events around the edges of structures like the Gulf Stream, Warm Core Rings, and eddies, seismic images allow us to synoptically see the relationships between finescale structures and the mesoscale features (like eddies) that produced them. Images that show the links from mesoscale features to finescale features that are associated with mixing allows hypotheses about the causes and consequences of mixing to be developed and tested in ways not previously possible. As noted also by Ruddick et al. (2009), being images of temperature gradient, seismic images are closely analogous to Schleiren images, which revolutionized laboratory fluid dynamics by showing how small-scale details relate to larger structures. This is the most exciting promise of seismic oceanography: synoptic visualization of features such as eddies and their associated fine structures allows the relationship between them to be explored in a new way. Since mixing generally passes energy from mesoscale features to fine scales, then to turbulence and molecular dissipation, this visualization tool provides a new insight into important stages in the energy cascade (Ruddick et al., 2009).

The method has thus far been successfully applied to image water mass fronts, currents, boundaries, mesoscale features such as cyclones, intrathermocline eddies, Meddies and the Mediterranean Undercurrent, staircases and internal waves (e.g. Holbrook et al., 2003; Holbrook et al., 2005; Tsuji et al, 2005; Schmitt et al., 2005; Nakamura et al., 2006; Biescas et al, 2008, Krahmann et al., 2008; Song et al., 2009, 2010; Dong et al., 2009; Hobbs et al., 2009; Klaeschen et al., 2009; Sheen et al., 2009; Pinheiro et al., 2010)).

In the development of seismic oceanography, inversion of the seismic data to obtain quantitative physical properties, such as sea water sound speed, temperature and salinity is now one of the key problems. In order to build benchmark calibration between reflection seismic and oceanographic datasets and make further progresses in seismic oceanography, EU launched the interdisciplinary GO (Geophysical Oceanography) project in 2006 for seismic and oceanographic joint investigation (http://www.dur.ac.uk/eu.go/general_public/project_info.html). This has allowed for the first time to get a dedicated dataset with near simultaneous acquisition of seismic and oceanographic data which provided essential for the first inversion tests (Papenberg et al. 2010; Huang et al., 2011). This work introduces the basic principles of the seismic method, shows results of its application to obtain high spatial resolution images of ocean eddies, lateral intrusions and internal waves, and presents preliminary results of CTD/XBT-controlled wave impedance inversion to derive the detailed thermohaline structure. In this paper, we briefly review the principles and limitations of the multichannel seismic technique as applied to the water column (section 2), show several examples of physical oceanographic phenomena that can be imaged with the technique (section 3), and discuss inversion techniques used to obtain highly detailed temperature and salinity sections from combined seismic and hydrographic data (section 4).

2. The multichannel seismic reflection method

Multichannel reflection seismology has been used for decades as an efficient exploration method, both in fundamental research and in the oil industry, because it allows a detailed acoustic imaging of the subsurface geology, both onshore and offshore (see, for example, Sheriff and Geldart, 1995). Offshore, the seismic (or acoustic) signal is normally generated by an array of airguns, towed behind a ship, a few meters below the water surface. These airguns release sudden bubble-pulses of compressed air into the water, as the ship navigates (Figure 1). This acoustic energy propagates downward in the water layer, and a fraction of it is reflected back to the surface whenever it encounters a contrast in the acoustic impedance between water masses with different temperature, salinity and/or density. The upward propagating reflected acoustic energy is detected at the surface by an array of piezoelectric sensors (hydrophones), commonly known as a *streamer of hydrophones* and is recorded digitally; these hydrophones are organized into groups (active sections) in which the signals are summed together to get a better signal-to-noise-ratio; each group forms a channel and common 2D systems can have between 96 and more than 500 channels (Figure 1). The processed signal conveys the necessary information to derive structural and physical information of the layers through which the acoustic energy propagated.

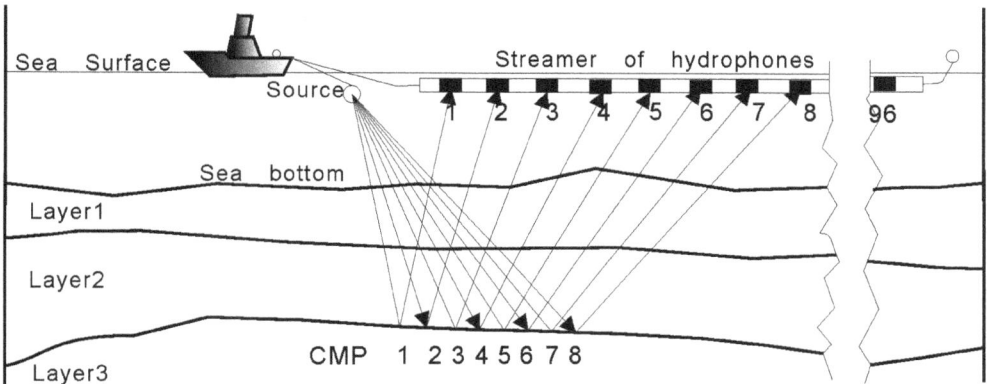

Fig. 1. Schematic illustration of the marine multichannel reflection seismic acquisition method.

In order to enhance the signal-to-noise ratio, obtain a near vertical reflection signal and derive the sound speed in the different propagation medium layers, the so-called *Common Mid Point Method* (CMP) is generally used. The returned reflected signal from each shot is recorded in all the active channels and this constitutes a *Shot Gather* composed of the seismic traces recorded at each channel. After editing bad seismic traces, correcting for wave amplitude decay with distance (spherical divergence correction) and filtering undesirable noise, all the traces from different shots that correspond to reflections from a *Common Mid Point* (CMP–points with a common half distance, or *offset*, between the source and the receiver) are ordered into *CMP Gathers*, in function of increasing *offset*. Then, all the reflections in each trace that correspond to common reflection points are corrected for varying offset distance and travel time (*Normal Moveout Correction – NMO*) and summed (stacked) together, to obtain a *Stacked Section* with a much higher signal-to-noise ratio than any single channel section. This stacked section represents the response that would be generated by waves that travel vertically downward and are reflected to a virtually coincident source and receiver location. If the seismic sections include reflection events with a significant dip, this effect should be corrected, as the simplistic approach explained above is only correct for horizontal layering. This can be done through a processing step called *Migration* of the seismic section, which can be performed before or post-stack (for a detailed discussion on these topics, the reader is referred to Sheriff and Geldart, 1995 and Yilmaz, 2001). When reflector dips are not too significant, as in seismic oceanographic data, the effect of migration is small and, as pointed out by Holbrook et al. (2003), the artifacts often produced by migration algorithms sometimes degrade the seismic image; therefore, seismic oceanographic data is sometimes presented unmigrated.

Normally, the seismic sections are plotted with the vertical axis in "two-way travel time (TWT)", because this is the measured parameter (similar to the fact that physical oceanographers normally plot their data against in pressure, rather than in depth). Conversion from TWT to depth can be done using the velocity/depth functions derived from the hyperbolic velocity analysis carried out during the application of the *Normal Moveout Correction*. As a simplistic first approach these sections can be converted from time to depth assuming a constant sound speed of 1500 ms^{-1} so that 1000 ms TWT=750m water depth. The individual seismic traces in a section are plotted side by side, usually using a two-color palette to show positive and negative reflection peaks producing the seismic image of the underlying structure. A simplified description of the methodology, assumptions and approximations of reflection seismology in the ocean, including limitations on vertical resolution set by the sound source, the effect of source wavelet side lobes, Fresnel Zone limitations on horizontal resolution, and the application of migration techniques to compensate dip, are described in Ruddick et al. (2009); for a more generalized discussion on the seismic reflection method, the reader is again referred to Sheriff and Geldart (1995) and Yilmaz (2001).

3. Some examples of thermohaline fine structure imaged with the reflection seismic method

3.1 Imaging of Mediterranean Outfow Eddies (Meddies)

Multichannel seismic sections have proved an excellent tool to investigate eddies in the oceans (Biescas et al., 2008; Pinheiro et al., 2010). Figure 2 shows the location of an approximately E-W and 326-km long multichannel seismic line acquired in the Tagus Abyssal Plain off west and south Iberia (Figure 2), in 1993, in the scope of the Iberian Atlantic Margins (IAM) Project, under the JOULE Programme, funded by the European Commission (Banda et al., 1995). This seismic line was acquired with a 4.8 km long analogue streamer, with 192 channels and a group interval of 25 meters, towed at an average depth of 15 m. The shot interval was 75 m and the sampling interval was 4 ms. The near offset was 254 m. The seismic source consisted of a 36 airgun array with a total volume of 7524 ci. Figure 3 shows an example to the type of detailed imaging of the thermohaline structure of several mesoscale features within the water column that can be obtained with the seismic reflection method (see location of the seismic line in Figure 2). This image was obtained in 2004 (Song and Pinheiro, unpublished data, 2004) and a refined processed and interpreted version was published by Pinheiro et al. (2010).

The spacing between CMP's is about 12.5m, which provides a very high horizontal resolution, allowing subtle lateral variations within the eddy and lateral intrusions to be observed; which could not be observed with conventional physical oceanographic data. Besides a large meddy observed in the eastern most portion of the section (Figures 3 and 4) and a cyclone observed in the western portion, the central portion of this line shows a complex structure within a meddy whose origin is discussed in Pinheiro et al. (2010). Images with such a high lateral resolution provide new insights into the fine structure in the oceans and will hopefully contribute to a deeper understanding of the detail of mixing processes in the ocean.

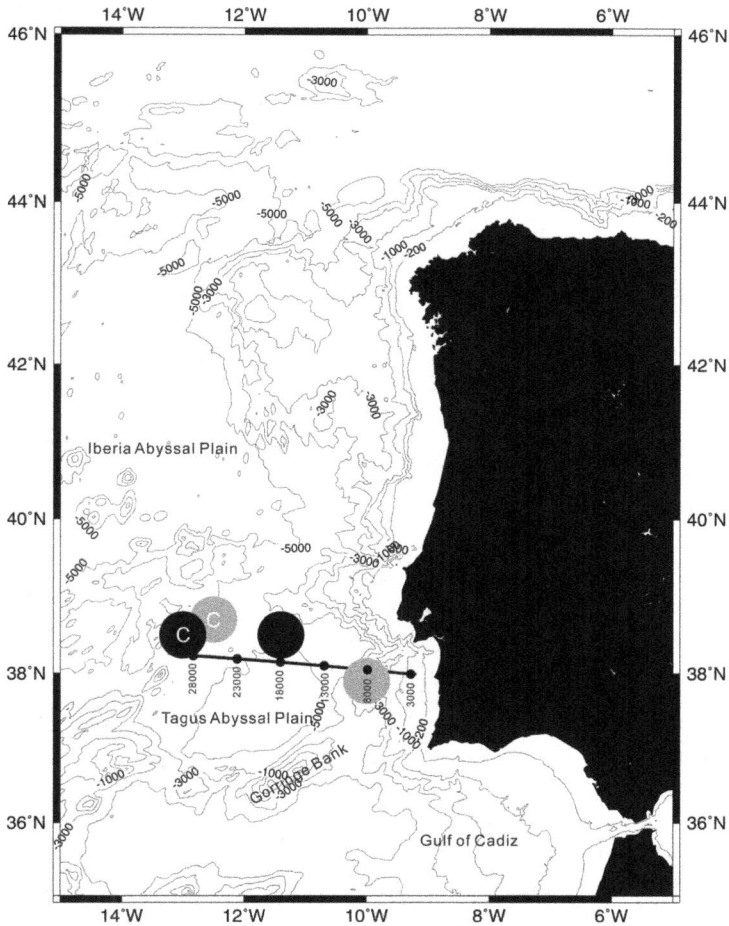

Fig. 2. Bathymetric map of the west Iberian margin showing the location of the multichannel seismic Line IAM-5 from the IAM cruise. Bathymetry from the GEBCO 1´ compilation grid. Also shown the location of Meddy-9 and the cyclone C from Richardson et al. (2000) that were used to confirm the seismic interpretation; the positions of these features for August 1993 are represented as grey circles and those corresponding to September are represented as black circles. The two westernmost circles correspond to the cyclone and the two easternmost circles to Meddy-9. Also shown the location of the CMPs along the line, for reference.

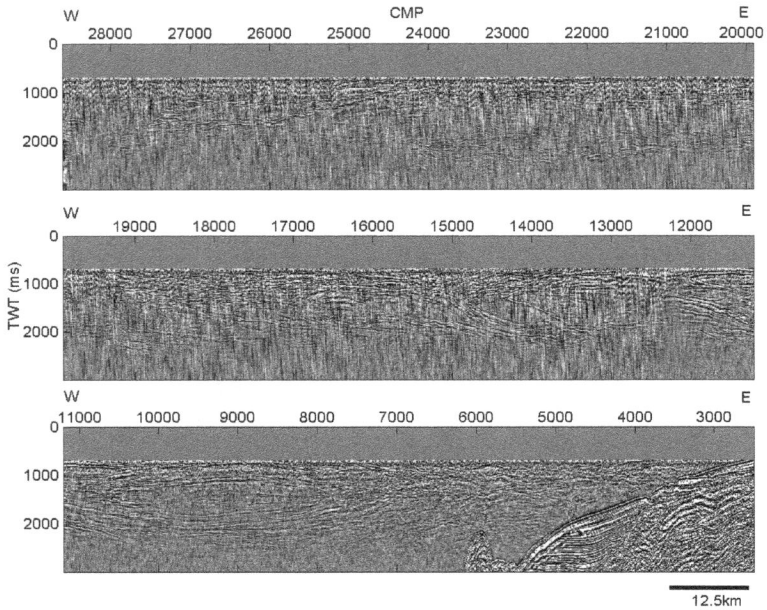

Fig. 3. Complete stacked seismic section along the processed Line IAM-5 (2004 processed version of Pinheiro et al., 2010). The vertical scale is in Two-Way Time (seconds) and the numbers in the horizontal scale correspond to CMP locations.

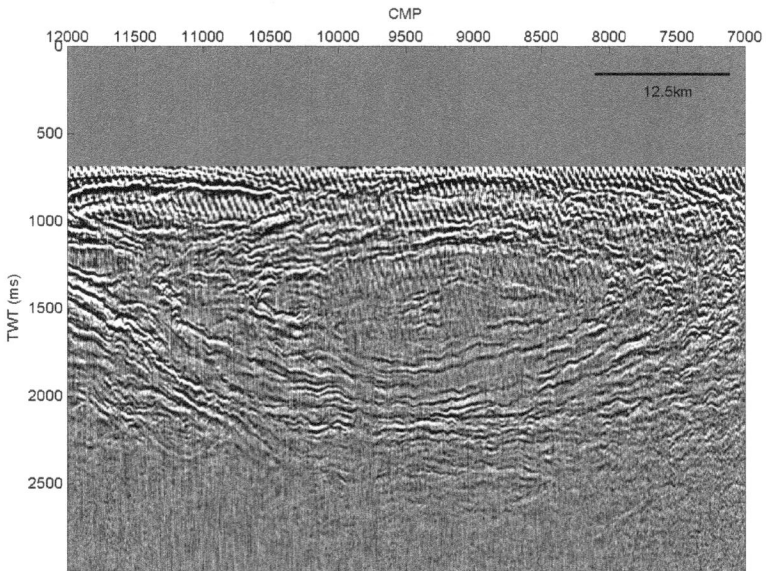

Fig. 4. Detail of Figure 3, showing a high resolution seismic image of a meddy (Song and Pinheiro, unpublished data, 2004). Several smaller lenses are also observed above the main eddy.

3.2 Imaging of internal waves in the South China Sea

Another successful application of the seismic reflection method is to image internal wave patterns with great detail, as first shown by Gonella and Michon (1988) and later by Holbrook and Fer (2005) and Blacic and Holbrook (2009). Figure 5 shows the location of a 463 km long multichannel seismic profile acquired in the Luzon Straight area, in the Northeast South China Sea (SCS); some portions of this line, highlighted in red, in Figure 5, are depicted in Figure 6. This multichannel seismic line was acquired in the framework of the National Major Fundamental Research and Development Project of China (No. G20000467), in 2001, using the R/V Tanbao of the Guangzhou Marine Geological Survey. The seismic signals were recorded by a 240-channel streamer, with a 12.5m group interval, and were sampled at 2ms. The total record length is 10s (Two-Way Travel Time – TWT) and the source to near trace offset is 250 m. The seismic source used was a 3000 in³ air-gun array. These sections show in great detail undulating seismic reflectors, coherent over vertical and horizontal distances of order 1 km, that correspond to internal waves. They provide detailed images of the lateral and vertical continuity of internal waves, allow the calculation of horizontal, as well as vertical spectra of internal waves (Song et al., in prep.) and therefore contribute to a better understanding of these phenomena.

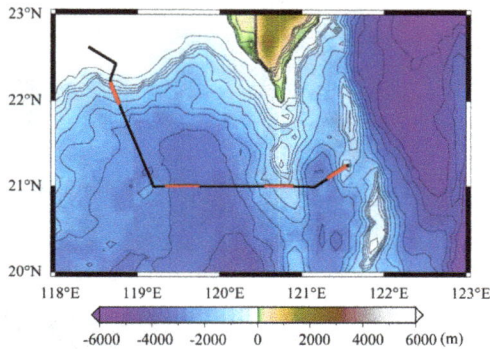

Fig. 5. Location of one seismic line acquired in the Luzon Strait, in northeastern South China Sea, whose seismic sections highlighted in red are depicted in Figure 6.

4. Thermohaline structure inversion of seismic data

Inversion for oceanographic parameters of temperature and salinity from seismic data is a very important research field in seismic oceanography. In its early stage, seismic oceanography paid more attention to imaging boundaries between water masses in the ocean (e.g. Holbrook et al., 2003; Nandi et al., 2005; Nakamura et al., 2006) and discriminating and interpreting reflection events in the seismic sections, rather than inverting acoustic impedance for physical oceanographic parameters. This was mainly due to the fact that there were not many situations in which both seismic and oceanographic data had been acquired simultaneously. As stated above, after analyzing XBT (Expendable Bathythermograph) and XCTD (Expendable Conductivity-Temperature-Depth) derived oceanographic data combined with seismic data, Nandi et al. (2005) found that reflections in the seismograms and thermohaline fine structures in the ocean were strongly correlated with each other, demonstrated that reflectors can originate from temperature changes as small as 0.03 degree, and anticipated that high-resolution spatial distribution of temperature could be derived from the seismic sections. Tsuji et al. (2005) also confirmed that seismic sections could image fine structure of the Kuroshio Current and, through analysis of the amplitude of the seismic signal, found the maximum change of temperature observed in that area was about 1 degree. These results therefore inferred that temperature could be inverted from seismograms. Subsequently, Paramo et al. (2005) retrieved temperature gradients in the Norwegian Sea using the AVO (Amplitude Versus Offset) method, and Wood et al. (2008) applied full wave inversion to synthetic seismograms and real seismic data, with good results; however, both of these studies were restricted to 1-D. More recently, Papenberg et al. (2010) presented high-resolution 2-dimensional temperature and salinity distributions derived from inversion applied to combined seismic and physical oceanographic experiment conducted in the scope of the GO (Geophysical Oceanography) European project, and showed that seismic data can indeed provide reliable estimates of high resolution temperature and salinity, provided it is constrained by physical oceanographic data.

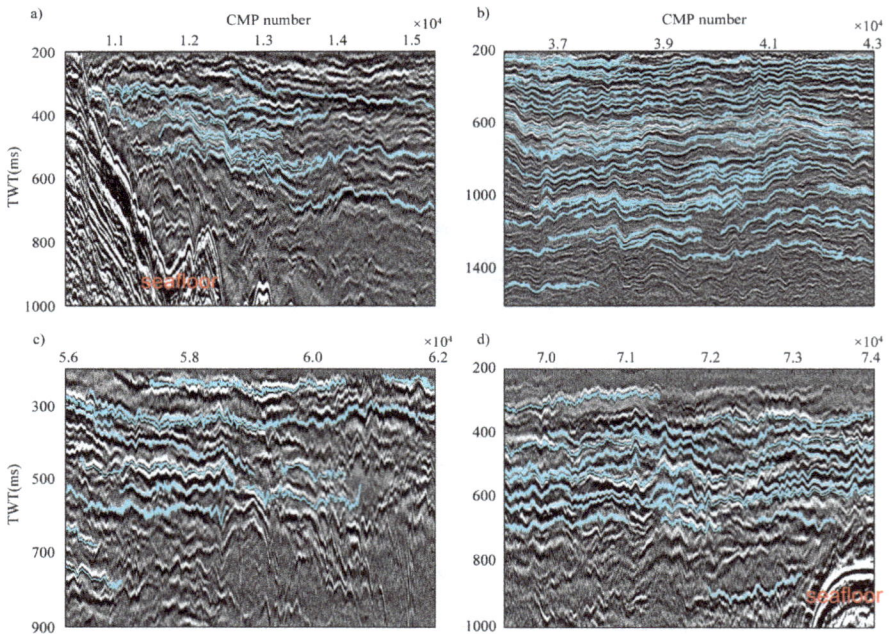

Fig. 6. Seismic sections along the seismic profile in the northeastern SCS shown in red, in Figure 5, and digitized undulate seismic reflectors (blue curves). a),b),c) and d) show sections across the continental slope, abyssal basin, Hengchun ridge and Luzon volcanic arc respectively. Vertical scale in Two Way Traveling Time (ms), CMP interval is 6.25m.

Here, we present a CTD/XBT constrained thermohaline structure inversion method (Huang et al., 2011). Using this method and synthetic seismograms Song et al. (2010) first demonstrated that it was possible to derive 2D high-resolution temperature and salinity sections from seismic data by using only a few CTD data to constrain the inversion (the

CTDs acted here as control wells in conventional seismic inversion for geological structure). This method was applied to low frequency seismic data from one multichannel seismic (MCS) line (line GOLR-12), acquired in the scope of the European GO Project, with simultaneous acquisition of XBT and CTD (Conductivity-Temperature-Depth) data. The Post-Stack Constrained Impedance Inversion method was used to derive temperature and salinity distributions of seawater and to demonstrate that this method can indeed provide reliable temperature and salinity distribution profiles every 6.25m along the seismic line, with resolutions of 0.16°C and 0.04 psu respectively.

4.1 Inversion method for temperature and salinity

The inversion for temperature and salinity distributions of seawater from combined seismic and oceanographic data is divided into two steps: the first step is the inversion for sound speed or impedance (product of sound speed and density) distribution from combined seismic and oceanographic data, using a post-stack constrained impedance inversion method; the second step is to convert the inverted velocity distribution into temperature and salinity distributions. There are some differences between the application of the wave impedance inversion method in the water layer and conventional impedance inversion in oil and gas exploration. The former needs forward calculations of the variations of the sound speed, density, and impedance from temperature and salinity changes with depth, based on CTD/XBT data and using the sea water state equations (Fofonoff and Millard, 1983).

Post-stack constrained inversion is a relatively mature technique that is widely used in the oil and gas industry in the inversion for geological parameters, and there are several software packages available for this purpose. The inversion method used is model based, and aims to build the best-fitted model to the real seismic data. The initial impedance model is built from seismic data (interpreted seismic horizons) and constraining CTD/XBT data (equivalent to constraining wells in inversion for geology). Each time a synthetic seismogram is derived from the impedance model, the model is adjusted after comparing it with the real seismic data. This procedure is then repeated until a best-fitted model is reached and the final inversion result obtained.

The inversion process consists of the following main steps: (a) importing the CTDs/XBTs and the seismic data; (b) extracting the wavelet from the seismic data using a statistical approach; (c) picking horizons or importing picked horizons; (d) doing CTD/XBT correlation with the seismic data and extracting the wavelet again using the first wavelet estimate from the seismic as the initial guess (sometimes it is also necessary to adjust the depths of the CTD/XBT to get the best correlation with the seismic data); (e) building an initial acoustic impedance model and optimizing to bet fit the observations (seismic and CTD/XBT); (f) convert the impedance model into a velocity model, using a density function for the area.

Converting velocity to temperature and salinity is also an important task. This was done by an iterative process, using the empirically derived formula for the relationship between velocity and temperature, salinity and depth (Wilson, 1960):

$$v_p = 1492.9 + 3(T - 10) - 6 \cdot 10^{-3}(T - 10)^2 - 4 \cdot 10^{-2}(T - 18)^2 +$$
$$1.2(S - 35) - 10^{-2}(T - 18)(S - 35) + Z / 61$$

Where v_p is the sound speed (m/s), T is temperature (°C), S is salinity (psu), and Z is the depth (m).

The iteration process is as follows: (a) given an initial salinity of 36 psu, the temperature is derived using the formula above; (b) then, the salinity is adjusted according to the T-S relationship for the region, based on CTD casts; (c) the new salinity is introduced in the formula above to get a new temperature; (d) this process is repeated until a convergence in T and S is reached. This approach works because the variation scales of temperature and salinity and the T-S relationship lead to a unique pair of temperature and salinity for a given acoustic velocity. More detailed information on this procedure can be found in Song et al. (2010), Dong (2010) and Huang et al. (2011).

4.2 Application of the inversion method to a case study - the GO data

In 2007, the GO (Geophysical Oceanography) project, briefly described above, carried out a two-month combined seismic and physical oceanography survey in the Gulf of Cadiz, and acquired more than 40 seismic lines (high frequency, intermediate frequency and low frequency), and more than 500 simultaneously XBT measurements and 43 CTDs (Hobbs et al., 2009; http://www.dur.ac.uk/eu.go/general_public/ project_info.html). This acquisition campaign was most successful and a wide range of oceanographic features were observed in the study area. This included several Meddies (Mediterranean Outflow eddies).

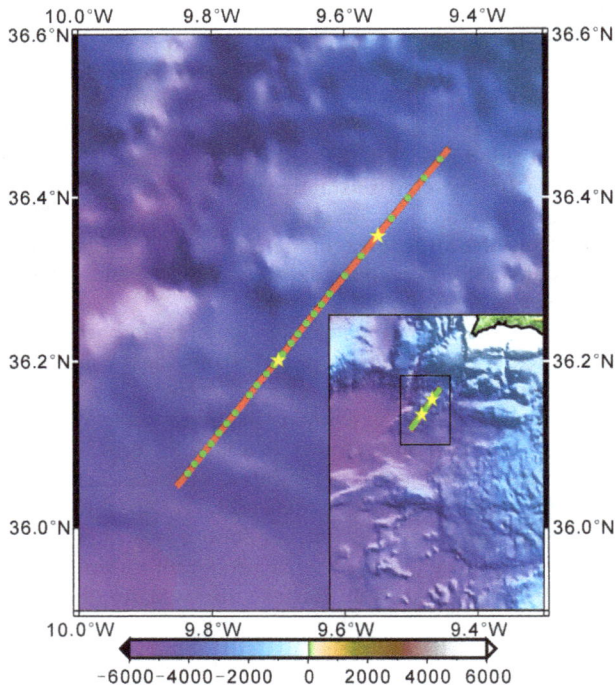

Fig. 7. Location and seafloor topography in the study area, SW of Portugal, with the location of the seismic line GOLR-12, and of the XBT and CTD positions. The red line represents the seismic line, the green solid points represent XBT measurements, and the yellow stars represent CTD locations.

These meddies detach from the main vein of the Mediterranean Outflow (the warm and saline seawater that spills out of Mediterranean from the Strait of Gibraltar and flows along the south margin of Iberian Peninsula) as it passes near the Portimão Canyon and other bathymetric features, and drift away from the continental slope at the depth of neutral buoyancy.

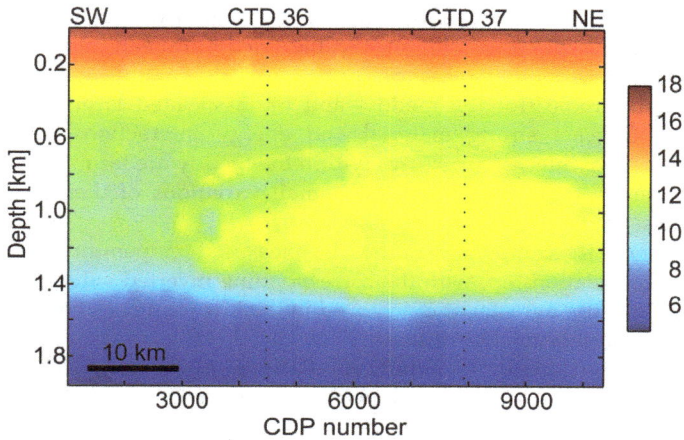

Fig. 8. Inverted temperature distribution for the seismic line GOLR-12. Dot lines represent CTD locations.

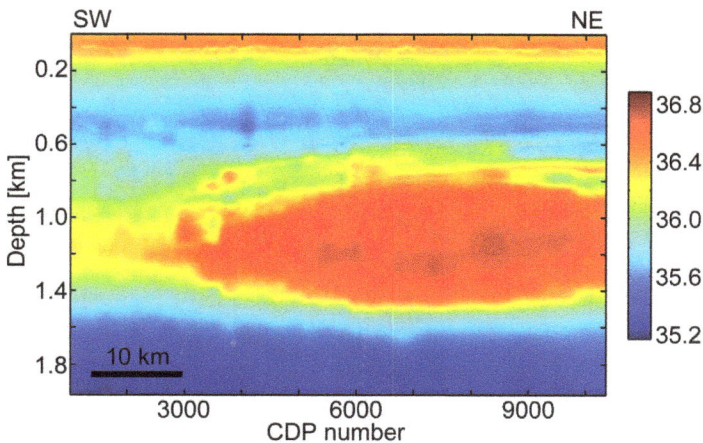

Fig. 9. Inverted salinity distribution for the seismic line GOLR-12.

Here we show results of the inversion of one of the acquired low frequency seismic lines (Line GOLR-12), which was kindly made available for this study (courtesy Dirk Klaeschen and Richard Hobbs). The acquisition parameters for this seismic line were: (a) seismic source: a 1500L BOLT air-guns system with a main frequency band of 5-60 Hz, towed at a depth of 11m; (b) shot interval: 37.5m; (c) receiver: a 2400 m long SERCEL streamer, towed 8 m below the sea surface, with 192 traces (12.5 m spacing); (d) near offset: 84 m. 24 XBT and 2 CTD profiles were acquired simultaneously and the XBT data was used to constrain in the inversion procedure. Locations of the seismic line, XBTs and CTDs are depicted in Figure 7.

After applying the inversion method described above to the seismic line, we got the inverted sections for temperature and salinity depicted in Figures 8 and 9, respectively (Huang et al., 2011). These sections show an elliptical region with the temperatures and salinities characteristic of the warm and saline Mediterranean outflow water, and therefore this elliptical structure can be interpreted as a meddy. Many fine structures can be detected at the boundary of the meddy, and the associated large temperature and salinity gradients indicate strong material and energy interactions. In contrast, the temperature and salinity internal structures within the meddy are far more homogeneous, although some fine structure associated with small variations of these parameters can nevertheless be observed.

Figures 10 and 11 show the comparison of the XBT-measured temperature and XBT-derived salinity using the T-S relationship, with the corresponding inverted values. As can be seen from these figures, a better agreement is achieved in the low frequency components than in the high frequency components. Quantitatively calculated results indicate that mean square errors of temperature and salinity are 0.16°C and 0.04psu respectively. Taking into account the normal variation ranges of these parameters, the inversion result for temperature is better than that for salinity, which is acceptable, because previous studies have shown that the relative contribution of salinity contrasts to reflectivity is approximately 20% (Sallares et al., 2009), i.e. far less than of the contribution from temperature (reflectivity is fairly insensitive to salinity change). The fact that the salinity values used for the inversion were derived from the T-S relationship and not measured by CTDs also certainly affected the inversion results.

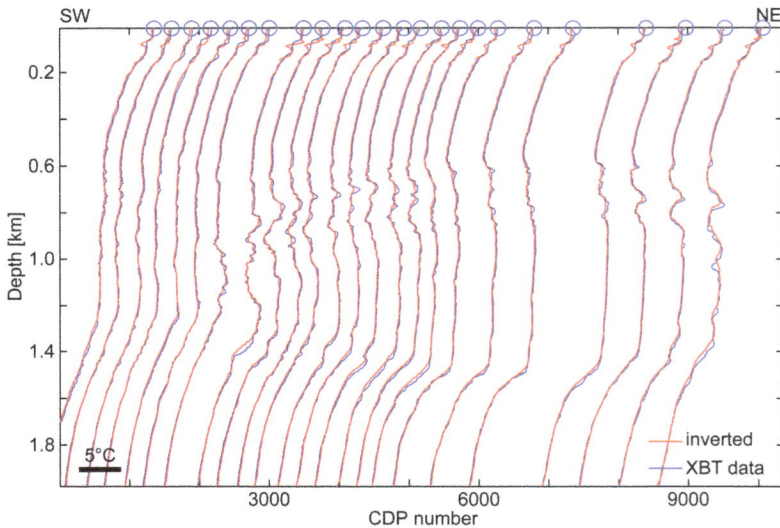

Fig. 10. Inverted and XBT temperature distribution. Blue lines represent XBT data; red lines represent inverted data.

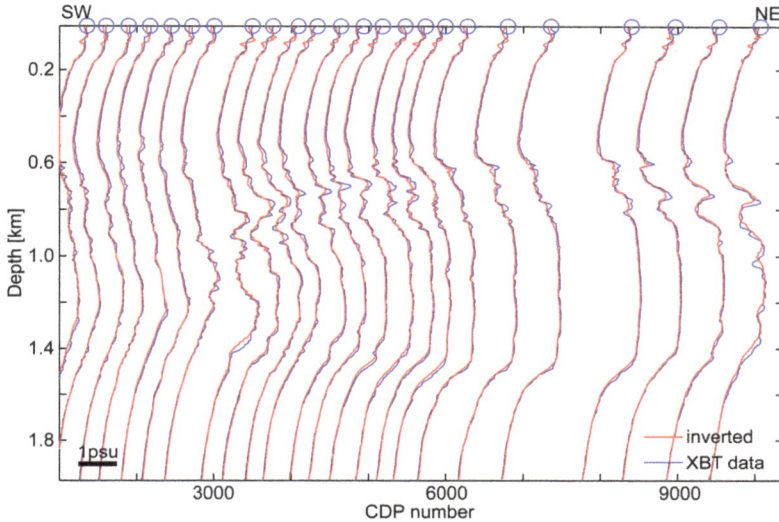

Fig. 11. Inverted and XBT-derived salinity distribution. Blue lines represent XBT-derived data from the temperature and salinity relationship; red lines represent inverted data.

In summary, from the comparisons of the measured and inverted data, we can see that although inversion results can represent detailed temperature and salinity distributions and thermohaline fine structures very well, some errors are nevertheless always inevitable. First, the seismic data processing flows may introduce some errors, since the removal of the direct wave and filtering have some effects on the amplitude and shape of signal. Also, there are always some inaccuracies in the velocity model derived from the acoustic inversion due to the process itself and to noise in the data. It should be also noted that, in the region of strong water mass interaction, the T-S relationship is more complex than the one used here, which is basically an average, and therefore highly accurate results are very difficult to achieve. Finally, the main frequency of the seismic data used is so low and therefore fine structures with the scale less than 15m cannot be detected.

Comparing currently published inversion results, the temperature and salinity inversion resolutions of Papenberg et al. (2010) are 0.1°C and 0.1psu respectively, and those presented here, from Huang et al. (2011) are 0.16°C and 0.04psu. It should be noticed that in their method to derive temperature and salinity, Papenberg et al. (2010) used the inversion result from the seismic data for the high frequency component and the XBT-derived information for low frequency component; hence, their inversion errors are very close to the high frequency component inverted from the seismic data, and the higher resolution can be explained. In summary, the results obtained from combined seismic and oceanographic data inversion using a post-stack constrained impedance inversion method are consistent with the observed XBT data and information derived from them. With the constraints of the seismic data it is possible to achieve high inversion resolution for temperature and salinity in the area between XBTs and with this method it is possible to obtain reliable temperature and salinity distribution profiles every 6.25m along the seismic line, which can be used for analysis on small scale oceanographic features.

5. Conclusions

Physical oceanographic research is based on a large amount of observational data. Mesoscale and small scale thermohaline fine structures are very difficult to be observed using the conventional observation methods because of their low lateral resolution and long observation time. Seismic oceanography makes up for these deficiencies, and makes it possible to obtain fast imaging of a large research area with high lateral resolution; if this acquisition is complemented with simultaneous conventional physical oceanographic measurements such as XBTs or CTDs, the seismic data can be inverted to temperature and salinity with high lateral and reasonably good vertical resolution (although lower than with conventional CTD/XBT measurements). In its early stage, seismic oceanography was mostly used for discriminating seawater boundaries in the ocean and they have been shown to be smoothed images of temperature gradients.

As shown above, two-dimensional temperature and salinity structure sections, with a high lateral resolution, can be obtained from seismic data by using CTD/XBT-controlled seawater wave impedance inversion. The application of this method to the low frequency seismic data of GOLR12, combined with simultaneously acquired XBT and CTD data, derived temperature and salinity distributions of seawater with resolutions of 0.16°C and 0.04psu respectively which demonstrate that seismic data can be used to extract two dimensional temperature and salinity distributions, and hence provide high lateral resolution data for physical oceanographic research. Although the inversion result is affected by the quality of seismic data, data processing and the complexity of the thermohaline mesoscale and fine structures in research area, it can be anticipated that with its development, seismic oceanography will play in the near future a more and more important role in physical oceanographic research.

Besides these post-stack inversion studies, studies have also been carried out to invert physical parameter contrasts from the north-eastern South China Sea, using the AVO (Amplitude Vesus Offset) technique (Dong, 2010), and also to invert seismic data from the same area and obtain the 1D velocity structure of seawater at three CMP (Common Mid-Points) locations, using full waveform inversion (Dong, 2010). Also, a new technique is being developed to explore the 2D thermohaline structure of seawater based on a hybrid inversion method using both pre and post-stack seismic data (Dong, 2010), which should help to solve the problem of the lack of constraints of simultaneous thermohaline data for seismic data.

As shown here and also by Papenberg et al. (2010), thermohaline structure sections with high lateral resolution can be obtained from seismic inversion of multichannel seismic reflection data, constrained by a few CTD/XBT data. This method therefore is highly complementary to conventional physical oceanographic measurements and can overcome some of the difficulties with the conventional measurement methods used in physical oceanography, in particular those concerning the low lateral resolution with high penetration, and can provide vast high resolution oceanographic data for ocean science studies. As a final remark, it should be noted that the inversion studies for the thermohaline structure carried out until present are still preliminary and continued research in this field is expected to provide more accurate determinations in the near future.

6. Acknowledgment

This research was co-financially supported by China NSF(No. 41076024), the European funded GO Project (Geophysical Oceanography - FP6-2003-NEST 15603) and the National

Major Fundamental Research and Development Project of China (No. 2011CB403503). The seismic data used here for inversion were collected as part of the GO-project supported by the EU project GO (15603) (NEST), the United Kingdom Natural Environment Research Council and the German research agency DFG (KR 3488/1-1). We are grateful to Dr. Richard Hobbs and Dr. Dirk Klaeschen for their kind permission to use seismic and oceanographic data from the GO project to apply post-stack inversion.

7. References

Banda, E., Torné, M. and the Iberian Atlantic Margins Group. (1995). Iberian Atlantic Margins Group investigates deep structure of ocean basins. *EOS, Trans. Am. Geophys. Union*, 76(3), 25, 28–29.

Biescas, B., Sallarès, V., Pelegrí, J. L., Machín, F., Carbonell, R., Buffett, G., Dañobeitia, J. J., Calahorrano, A. (2008). Imaging meddy finestructure using multichannel seismic reflection data. *Geophysical Research Letters*, 35, L11609.

Blacic, T. M., Holbrook, W. S. (2009). First images and orientation of internal waves from a 3-D seismic oceanography data set. *Ocean Sci. Discuss.*, 6, 2341–2356.

Dong, C. Z., Song, H. B., Hao, T. Y., Chen, L., Song, Y. (2009). Studying of oceanic internal wave spectra in the Northeast South China Sea from seismic reflections. *Chinese J. Geophys.* (in Chinese), 52(8), 2050-2055.

Dong, C. Z. (2010). Seismic oceanography research on inversion of sea-water thermohaline structure and internal waves in South China Sea. Ph.D thesis, Institute of Geology and Geophysics, Chinese Academy of Sciences. 128 pp.

Fofonoff, N. P., Millard, Jr. R. C. (1983). Algorithms for computation of fundamental properties of seawater, Tech. Pap. Mar. Sci. 44, UNESCO, Paris.

Gonella, J., et Michon D. (1988). Ondes internes profondes révélées par sismique réflexion au sein des masses d'eau en Atlantique-*Est. Comptes Rendus Acad. Sci. Paris, T.* 306, 781-787.

Hobbs, R. W., Klaeschen, D., Sallarès, V., Vsemirnova, E., Papenberg, C. (2009). Effect of seismic source bandwidth on reflection sections to image water structure, *Geophys. Res. Lett.*, 2009, 36.

Holbrook, W. S., Fer, I. (2005). Ocean internal wave spectra inferred from seismic reflection transects. *Geophys Res Lett*, 2005, 32, L15604, doi:10.1029/2005GL023733

Holbrook, W. S., Páramo, P., Pearse, S., Schmitt, R. W. (2003). Thermohaline fine structure in an oceanographic front from seismic reflection profiling. *Science*, 301, 821-824

Huang, X. H., Song, H. B., Pinheiro, L. M., Bai, Y. (2011). Ocean temperature and salinity distributions inverted from combined reflection seismic and hydrographic data. *Chinese J. Geophys.* (in Chinese), 2011, 54(5), 1293-1300.

Klaeschen, D., Hobbs, R. W., Krahmann, G., C. Papenberg, and E. Vsemirnova. (2009). Estimating movement of reflectors in the water column using seismic oceanography. *Geophys. Res. Lett.*, 36, L00D03, doi:10.1029/2009GL038973.

Krahmann, G., Brandt, P., Klaeschen, D., Reston, T. J.. (2008).Mid-depth internal wave energy off the Iberian Peninsula estimated from seismic reflection data. *Journal of Geophysical Research*, 113, C12016, doi:10.1029/2007JC004678.

Nakamura, Y., Noguchi, T., Tsuji, T., Itoh, S., Niino, H., Matsuoka, T. (2006). Simultaneous seismic reflection and physical oceanographic observations of oceanic fine structure in the Kuroshio extension front. *Geophys. Res. Lett.*, 33, L23605, doi:10.1029/2006GL027437

Nandi, P., Holbrook, W. S., Pearse, S., Paramo, P., Schmitt, R. W. (2004). Seismic reflection imaging of water mass boundaries in the Norwegian Sea. *Geophys. Res. Lett.*, 31, L23311, doi:10.1029/2004GL021325

Papenberg, C., Klaeschen, D., Krahmann, G., Hobbs, R. W. (2010). Ocean temperature and salinity inverted from combined hydrographic and seismic data. *Geophys. Res. Lett.*, 37, L04601, doi:10.1029/2009GL042115

Paramo, P., Holbrook, W. S. (2005). Temperature contrasts in the water column inferred from amplitude-versus-offset analysis of acoustic reflections. *Geophys. Res. Lett.*, 32, L24611, doi:10.1029/2005GL024533

Pinheiro, L. M., Song, H. B., Ruddick, B., Dubert, J., Ambar, I., Mustafa, K., Bezerra, R.(2010). Detailed 2-D imaging of the Mediterranean outflow and meddies off W Iberia from multichannel seismic data. *Journal of Marine Systems*, 79, 89-100

Richardson, P. L., Bower, A. S., Zenk, W., (2000). A census of Meddies tracked by floats. *Prog. Oceanogr.*, 45 (2), 209–250.

Ruddick, B., Song, H. B, Dong, C. Z., Pinheiro L. (2009). Water column seismic images as maps of temperature gradient. *Oceanography*, 22(1), 192-205

Sallares, V., Biescas, B., Buffett, G., Carbonell, R., Dañobeitia, J. J., Pelegrí, J. L.(2009). Relative contribution of temperature and salinity to ocean acoustic reflectivity. *Geophys. Res. Lett.*, 36, L00D06, doi:10.1029/2009GL040187

Sheen, K. L., White, N. J., Hobbs, R. W. (2009). Estimating mixing rates from seismic images of oceanic structure. Geophys. Res. Lett., 36, L00D04, doi:10.1029/2009GL040106.

Sheriff, R.E., Geldart, L. P. (1995). *Exploration Seismology*. Cambridge University Press, Trumpington Street, Cambridge CB2 1RP. Second Edition, 592 pp., ISBN 0-521-46826-4.

Schmitt, R.W., Nandi, P., Ross, T., Lavery A., Holbrook, S. (2005). Acoustic detection of thermohaline staircases in the ocean and laboratory. *Oceans 2005, Proc. MTS/IEEE*, V. 2, 1052-1055

Song, H. B., Bai, Y., Dong, C. Z., Song, Y. (2010). A preliminary study of application of Empirical Mode Decomposition method in understanding the features of internal waves in the northeastern South China Sea. *Chinese J. Geophys.* (in Chinese), 53(2), 393-400

Song, H. B., Pinheiro, L. M., Wang, D. X., Dong, C. Z., Song, Y., Bai, Y. (2009). Seismic images of ocean meso-scale eddies and internal waves. *Chinese J. Geophys.* (in Chinese), 52(11), 2775-2780

Song, Y., Song, H. B., Chen, L., Dong, C. Z., Huang, X. H. (2010). Study of sea water thermohaline structure inversion from seismic data. *Chinese J. Geophys.* (in Chinese), 53(11）, 2696-2702.

Tsuji, T., Noguchi, T., Niino, H., Matsuoka, T., Nakamura, Y., Tokuyama, H. (2005). Two-dimensional mapping of fine structures in the Kuroshio Current using seismic reflection data. *Geophys. Res. Lett.*, 32, L14609, doi:10.1029/2005GL023095

Wilson, W. D. (1960). Equation for the speed of sound in seawater. *J. Acoust. Soc. Am.*, 32, 1357.

Wood, W. T., Holbrook, W. S., Sen, M. K., Stoffa, P. L. (2008). Full waveform inversion of reflection seismic data for ocean temperature profiles. *Geophys. Res. Lett.*, 35, L04608, doi:10.1029/2007GL032359

Yilmaz, O. (2001), *Seismic Data Analysis: Processing, Inversion, and Interpretation of Seismic Data*, vol. II, Invest. Geophys., vol. 10, 2nd ed., 2027 pp., Society for Exploration Geophysics, Tulsa, Okla.

Part 2

Physical Oceanography

6

Bodies of Water Along the Coast of a Tideless Sea in Areas with Young Pleistocene Accumulation from Scandinavian Glaciers (Baltic Sea)

Roman Cieśliński and Jan Drwal
University of Gdańsk, Institute of Geography,
Department of Hydrology, Gdańsk
Poland

1. Introduction

Oceanographers usually investigate coastal areas in terms of how they affect various processes taking place in the sea including wave action, high tide, low tide as well as flora and fauna. On the other hand, coastal areas may also be investigated in terms of the sea affects the strip of land along the coast. This strip is often called the coastal zone (Rotnicki, 1995). Water circulation in the interior and the action of the sea both affect water systems in the coastal zone depending on geographic conditions, which may help produce temporary flooding, seawater intrusions, increased water salinity and the formation of marshes.

Key geographic determinants include climate type, geological structure, relief and the resulting potamic discharge regime. Hydrography itself may also be considered a determinant. Key marine determinants include high tide, extent of high tide, short-term changes and sudden changes in sea level.

Half-closed seas are a special case, which occurs in the humid climate of the northern hemisphere, where Scandinavian shelf ice used to cover the area during the Pleistocene. The Baltic Sea is a half-closed sea. The southern shore of the Baltic is made of Pleistocene and Holocene clastic sediments with varying degrees of cohesion (Tomczak, 1995). The Polish section of the Baltic coast includes sandbars (79%) with dunes between 2 and 35 m high, cliffs (18%) up to 30 m high as well as alluvial coastlines (less than 3%). These characteristics make it difficult for discharge to take place along 75% of the Polish coastline (Drwal, 1995). This results in large marshy sandbars and grassy alluvial plains with a variety of bodies of water.

The virtually inland Baltic Sea is connected to the North Sea via the Straits of Denmark. This results in very small tides (15 cm) in the western part of the Baltic and even smaller tides (2 – 5 cm) in the southern part of the Baltic (Sztobryn et al., 2005). Their hydrological effects, therefore, should be negligible. In spite of this, some aspects typical of open seas may be observed along the southern Baltic coast (Drwal, 1995; Cieślinski, Drwal, 2005, Drwal, Cieśliński, 2007). The rationale for this may be found in climate conditions. Zaidler et al. (1995) argue that wind conditions resulting from pseudo-monsoon circulation characteristic of middle latitudes of the northern hemisphere cause a permanent exchange of air masses leading to significant daily and annual variability in wind direction and speed. This results in occasional storm surges along the western and southern Baltic coastline, reaching 300 cm above the sea's average level (Dziadziuszko, 1994).

The coastal zone of the southern Baltic Sea features young accumulation from Pleistocene glaciers and possesses a large variety of hydrographic entities, which may be affected by marine effects, as earlier research has shown. Sandbars with marshy deflation basins stretch along the Baltic Sea – separating the sea from lowlands featuring large wetland systems – many of which include polders, lakes, lagoons, deltas and estuaries. In some cases, these bodies of water form hydrographic systems (Cieśliński, 2004).

Changes in water chemistry are one indicator of marine impact on coastal bodies of water. Chloride concentration is often used to assess the influx of seawater into coastal bodies of water along the southern Baltic coast. Changes in all of the above conditions yield the current state of the hydrographic network along the southern Baltic.

2. Research subject

The research covered single hydrographic entities such as lagoons, lakes, mouth sections of rivers and wetlands as well as entire hydrographic systems along the Polish section of the southern Baltic coast (Fig. 1). The research sites were selected in a way that would capture any potential differences in environmental processes.

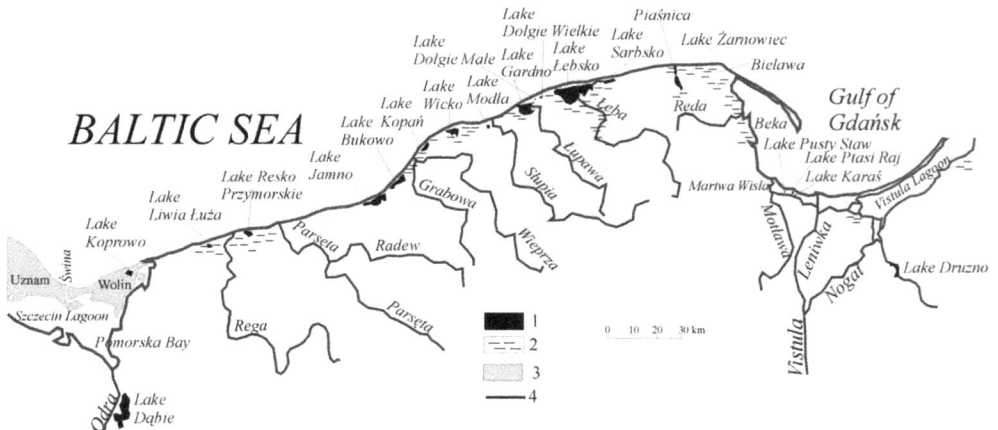

1 – lakes, 2 – wetlands, 3 – islands, 4 - rivers

Fig. 1. Location of objects investigation.

Fieldwork and library research were done in the period 2001-10. Fieldwork included hydrographic mapping and the collection of water samples for chemical analysis.

Chloride was selected as the best hydrochemical indicator, as it migrates well in the natural environment and does not react with other chemical entities (Hem, 1989). The chloride ion is also used in comparative papers for the southern Baltic coast (Cieśliński, Drwal, 2005).

The reference salinity level for brackish water varies from paper to paper. Appelo and Willems (1987) define it as 100 mg Cl⁻ dm⁻³, while Davies and DeWiest (1966) define it as 200 mg Cl⁻ dm⁻³. The reference level according to the Venetian Classification System is 500 mg Cl⁻ dm⁻³. The reference level assumed in this paper for the southern Baltic coast in Poland is 200 mg Cl⁻ dm⁻³.

3. Seawater intrusions in bodies of freshwater

Occasional storm surges occur along the southern Baltic coast. Low water levels occur periodically in the southern Baltic coastal zone. Both types of events cause seawater intrusions into bodies of freshwater along the Baltic coast. The Polish and international research literature contains a large quantity of descriptive information on seawater intrusions along the southern Baltic. Halbfass (1901, 1904) and Kunisch (1913) investigated two coastal lakes along the southern Baltic (Gardno and Łebsko) and found that the concentration of chloride decreased in canals linking the lakes to the sea with increasing distance to the sea. They explained this in terms of the influx of water from the Baltic Sea but found that it is quickly pushed out by freshwater. Kunisch (1913) also found that temporarily high concentrations of chloride in Lake Gardno may have been caused by sand blocking the lake's sole outlet – the Łupawa Canal. Kunisch also stated that the elevated concentration of chloride did not last long because it was reduced by the influx of large quantities of freshwater from the lake's drainage basin.

Szopowski (1962) wrote about water exchange taking place between Lake Łebsko and the Baltic Sea in 1956-58. He was the first to link the lake's influx of seawater with hydrometeorological factors. Szopowski based his analysis of changes in water levels in Lake Łebsko on measurements made at the Rąbka gauging site. He also used seawater data from a water gauge located in the Port of Łeba. His paper did not include a calculation of the quantity of seawater flowing into Lake Łebsko.

Mikulski, Bojanowicz and Ciszewski (1969) investigated the exchange of water between Lake Druzno and Vistula Bay and proposed a new method of calculating the influx of seawater into Lake Druzno based on differences in average water levels in the coastal lake and the neighboring Baltic Sea. The three researchers calculated channel cross sections as well as the slope and discharge for the river linking the lake with the bay. The length of river supplying water to the lake was calculated based on the ratio of discharge and influx time. If the calculations indicated that the theoretical river length was larger than the actual length of the Elbląg River, then it was inferred that brackish water from Vistula Bay was entering Lake Druzno.

Łomniewski and coworkers (1972) evaluated temporal and spatial changes in water salinity in the Vistula Delta. Majewski (1972) evaluated water exchange between the Baltic Sea and lakes Łebsko and Jamno. Majewski performed his research as part of a project on coastal lakes as transitional estuaries and assessed the number of seawater intrusions into Lake Łebsko (1972) during the period 1958-65.

Cieśliński and Drwal (2005) analyzed quasi-estuary processes along the Polish Baltic coastline and their impact on human activity. Drwal and Cieśliński (2007) analyzed seawater intrusions and their effects on selected coastal lakes with special attention being paid to the reasons for differences between the selected lakes.

Coastal areas are places where human life, safety and economic conditions often depend on phenomena and processes characterized by great intensity and dynamics. Seawater intrusions are some of the most common events taking place during extreme weather conditions. The effects of seawater intrusions have been observed in coastal lakes (Saeijs,

Stortelder, 1982; Tiruneh, Motz, 2003), lagoons (Ishitobi et al., 1999; Tanaka et al., 2005), wetlands (Glover, 1959; Flynn, McKee, 1995) and mouth sections of rivers (Foster, 1980; Giambastiani et al., 2007). Seawater intrusions cause an increase in water salinity (van der Thuin, 1990) and abruptly increased water levels (Haslett, 2008). Seawater intrusions are easily detectable in lakes (Bear et al., 1999; Pulido-Leboeuf, 2004). The subject of seawater intrusions into coastal lakes has been covered by a number of researchers worldwide. The following citations are just a sample of the literature on this subject and were deemed most relevant to the theme of this paper.

Bowden (1967) identified the mechanism of water exchange in selected estuaries. Folk (1974) analyzed changes in the concentration of calcium and magnesium resulting from the influx of seawater. Davidson et al. (1991) characterized estuaries in Great Britain. Ishitobi et al. (1999) described physical effects resulting from seawater intrusions into the coastal lake Shinja. Godo et al. (2001) described the effects of wind on the mixing of water from two coastal lakes in Japan. Murray (2002) wrote about the natural environment and its effect on an estuary-type lake. Spagnoli et al. (2002) described the hydrological and sedimentation characteristics of the Laguna di Varano along the northern Gargano coast in Italy. Whittecar et al. (2005) wrote about seawater intrusions into a hollow along the Chesapeake Bay in the United States. Mosquera et al. (2005) determined the effect of the wind on seawater intrusions into Venice Bay.

Hsing-Juh et al. (2006) described the structure and function of a tropical lagoon experiencing minor seawater intrusions. Macdonald et al. (2006) wrote about surface runoff and its effects on the chemistry of sediments in estuary-type lakes. Sanderson and Baginska (2007) determined the influx of seawater into coastal lakes in New South Wales, Australia, caused by fluctuations in ocean water levels. A number of researchers have ascribed estuary-type characteristics to various lakes (Beletsky et al., 1999; Piasecki, Sanders, 1999).

Seawater intrusions along the southern Baltic coast vary in extent and hydrochemical effects. The nature of the effect depends on the body of water and whether the given body of water is isolated or is part of a larger system of bodies of water.

4. Types of hydrographic entities experiencing seawater intrusions

The following types of bodies of water have been found to experience seawater intrusions: lagoons, standing water, mouths of rivers, canals linking lakes with the sea, wetlands, hydrographic systems. The research literature offers insight into the course of seawater intrusions and their effects.

4.1 Lagoons
Vistula Bay is connected to the Baltic Sea via the Strait of Pilawa (Fig. 2). The capacity of the bay is estimated at 2.3 km^3. The water salinity level in the nearby Bay of Gdansk is three times higher than that in Vistula Bay. The maximum depth of Vistula Bay is only 5.1 m, which favors the mixing of water from top to bottom. Water salinity in the bay can be classified in terms of salinity zones (Fig. 3). The lowest chloride concentrations are found in the western part of the bay (site no. 8). The western part of the bay is affected by the inflow of Nogat River. The highest chloride concentrations are found in the northeastern part of the bay (site nos. 1, 2, 3, 4). The salinity zones shift based on the rate of water exchange with the Baltic Sea and the magnitude of river water inflow.

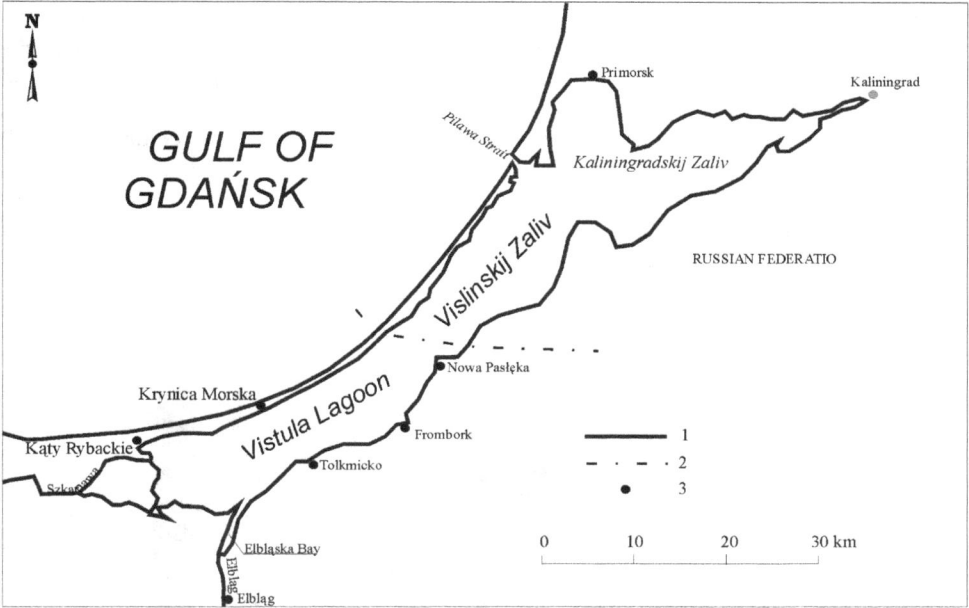

1 – rivers, 2 – border of Russian Federation and Poland, 3 – more important town

Fig. 2. Vistula Lagoon.

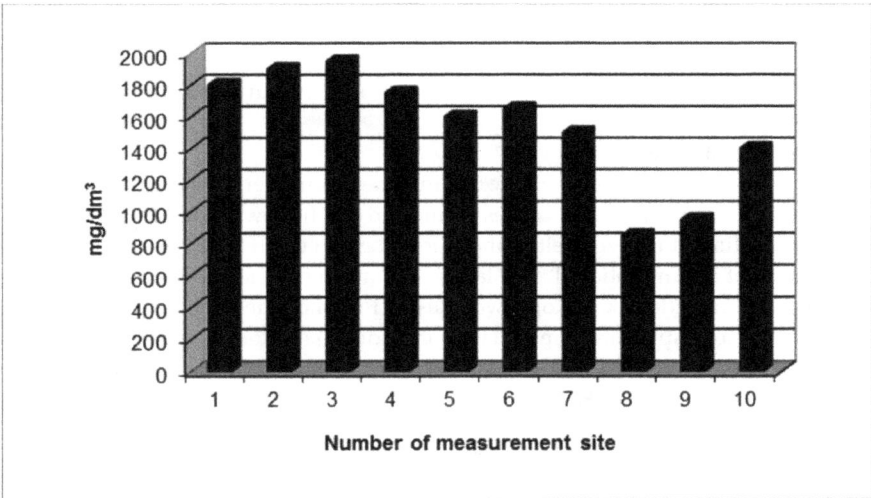

Fig. 3. Mean concentrations of chloride in the surface waters of Vistula Bay in 1996.

The mean concentration of chloride in the period 1996-98 ranged from 1,373 to 1,991 mg/dm³. Extremely low and extremely high concentrations were noted in 1998 (121 mg/dm³) and 1997 (3,025 mg/dm³) (Elbląg WIOS[1] data). Low concentrations of chloride were detected during the spring surface runoff season in the drainage basin. High concentrations of chloride were detected at low bay water levels during the summer and during the autumn (mainly October) when Baltic sea storms are more common (Fig. 4), which results in more seawater intrusions. This is shown by measurements performed in 1996-98, which indicate elevated chloride concentrations at measurement site A (1,005 – 1,495 mg/dm³). The concentration of chloride varied substantially at measurement site B ranging from 132 to1,783 mg/dm³.

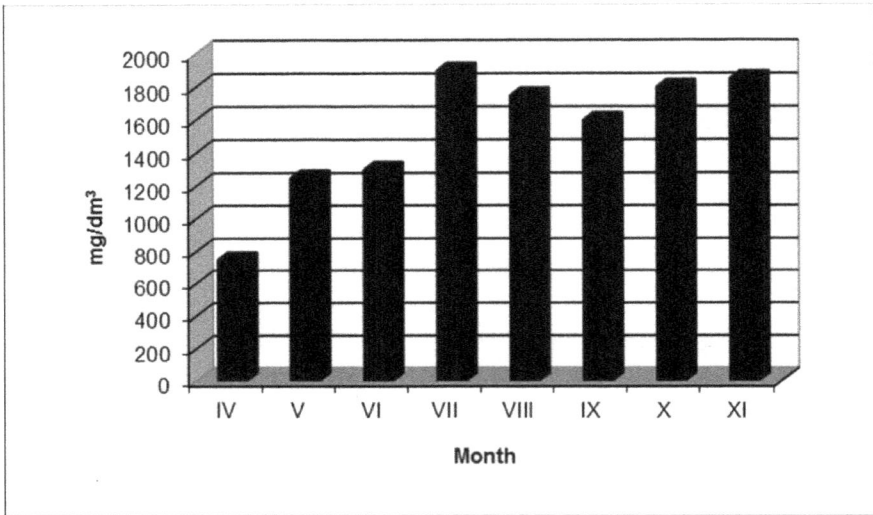

Fig. 4. Mean monthly concentrations of chloride (1996) in the surface waters of Vistula Bay.

4.2 Bodies of standing water

The coastal area along the southern Baltic Sea includes a number of bodies of standing water such as large lakes as well as small lakes formed in old hollows resulting from an uneven distribution of sediment in river deltas and due to being cut off from larger bodies of water thanks to sediment accumulation. Large lakes with mean chloride concentrations exceeding 200[2] mg Cl⁻ dm⁻³ include lakes Koprowo, Resko Przymorskie, Jamno, Bukowo, Gardno, Łebsko (Fig. 5). Corresponding small lakes include lakes Ptasi Raj and Karaś, (Fig. 5). Research in the period 2002-07 has shown that the concentration of chloride does not fall below 200 mg Cl⁻ dm⁻³ in lakes Koprowo, Resko Przymorskie, Bukowo and Łebsko (Tab. 1). It may, therefore, be inferred that these lakes constantly experience seawater intrusions. The weakest intrusions affect Lake Koprowo (320 - 780 mg Cl⁻ dm⁻³).

---

[1] Provincial Inspectorate of the Environmental Protection
[2] 200 mg Cl- dm-3 is a boundary value signifying the impact of marine water

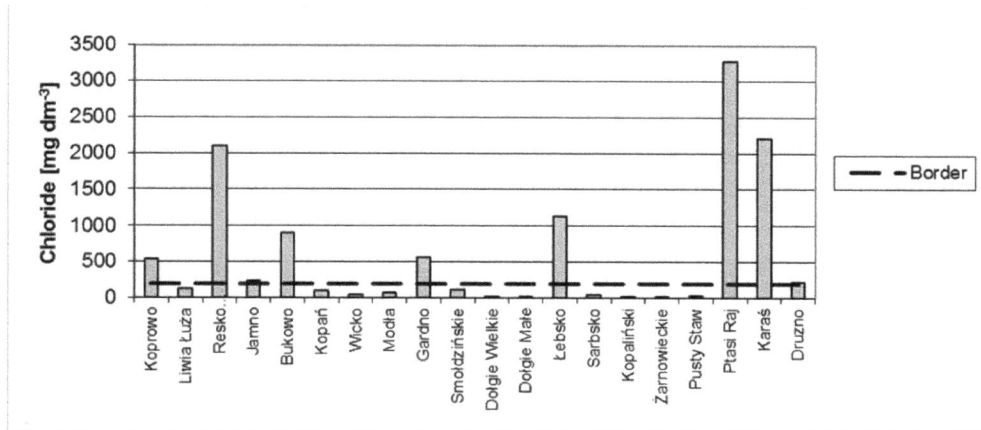

Fig. 5. Mean concentration of chloride in bodies of standing water in 2002-07

Name of lake	Minimum	Maximum
Koprowo	320,0	780,0
Liwia Łuża	90,0	170,0
Resko Przymorskie	1560,0	2700,0
Jamno	70,0	698,0
Bukowo	531,0	1188,0
Kopań	85,2	112,1
Wicko	38,2	66,6
Modła	26,4	152,0
Gardno	13,9	1512,0
Smołdzińskie	81,7	240,0
Dołgie Wielkie	13,0	18,9
Dołgie Małe	9,1	11,9
Łebsko	409,0	1970,0
Sarbsko	21,0	87,7
Kopalińskie	10,0	30,0
Żarnowieckie	8,3	28,6
Pusty Staw	33,4	40,9
Ptasi Raj	2311,0	4090,0
Karaś	1830,0	2703,0
Druzno	39,0	652,0

Table 1. Extreme chloride concentrations in bodies of standing water in 2002-07.

The concentration of chloride did not exceed 200 mg Cl⁻ dm⁻³ in the following lakes: Liwia Łuża, Kopań, Wicko, Modła, Dołgie Wielkie, Dołgie Małe, Sarbsko, Kopalińskie, Żarnowieckie and Pusty Staw. This indicates that each of the above lakes is being constantly supplied by freshwater from the drainage basin (Tab. 1).

Lake Gardno is yet another type of case of seawater intrusion. The concentration of chloride is very high in Lake Gardno for most of the year. The maximum recorded concentration is 1,512 mg Cl⁻ dm⁻³. However, the lake's salinity decreases substantially during certain periods of time and can reach as low as 13.9 mg Cl⁻ dm⁻³ (Fig. 6). While the Baltic Sea is the dominant factor in the lake's salinity level, freshwater influx from the lake's drainage basin can also play a role in some situations.

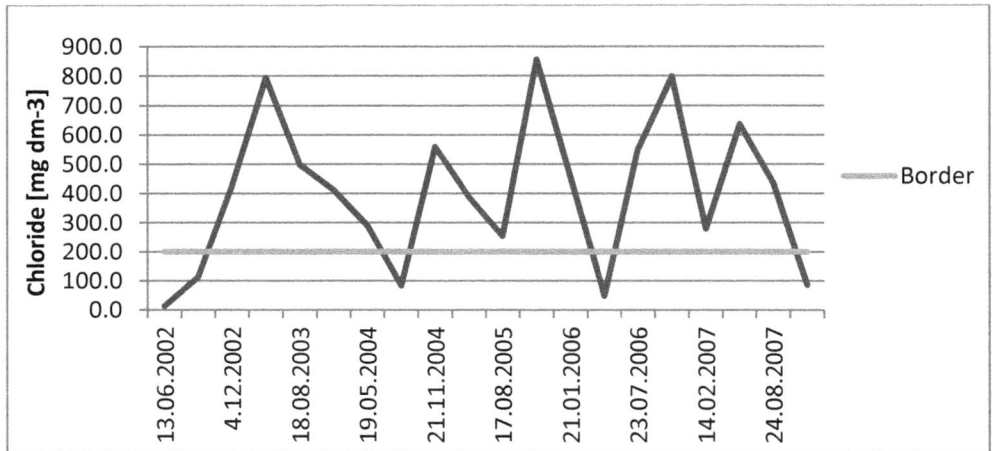

Fig. 6. Changes in chloride concentration in Lake Gardno for the period 2002-07.

The concentration of chloride in Lake Jamno generally remains below 200 mg Cl⁻ dm⁻³ and freshwater chloride levels tend to be common. The minimum concentration of chloride detected in Lake Jamno was 70 mg Cl⁻ dm⁻³. Seawater intrusions in Lake Jamno can cause abrupt increases in chloride concentration across the entire lake or in certain parts of the lake (698 mg Cl⁻ dm⁻³).

Two good examples of the Baltic Sea impacting the chemistry of a small lake originating as a puddle due to uneven sediment accumulation in a delta are Lake Ptasi Raj (mean concentration: 3,284 mg Cl⁻ dm⁻³) and Lake Karaś (mean concentration: 2,212 mg Cl⁻ dm⁻³). The lowest chloride concentration recorded during the summer in Lake Ptasi Raj was 2,311 mg Cl⁻ dm⁻³, while the highest concentration was 4,090 mg Cl⁻ dm⁻³. The lake is permanently affected by water from the Baltic Sea.

4.3 Canals linking lakes with the sea

Lake Gardno is linked with the sea by a canal 1 km long and 15-20 m wide (gradient: 0.3 ‰) (Fig. 7). The rate of water flow from the lake to the sea ranged from 6.3 to 11.8 m³ s⁻¹ during the study period. The direction of flow becomes reversed with winds from the north and a higher water level in the sea versus the lake. In 15 of 16 cases observed during the 2002-07 study period, the concentration of chloride ranged from 250 mg Cl⁻ dm⁻³ to just under 2,000 mg Cl⁻ dm⁻³, which indicates seawater intrusions (Fig. 8). The chloride concentrations in the lake were close to those in the Baltic Sea itself.

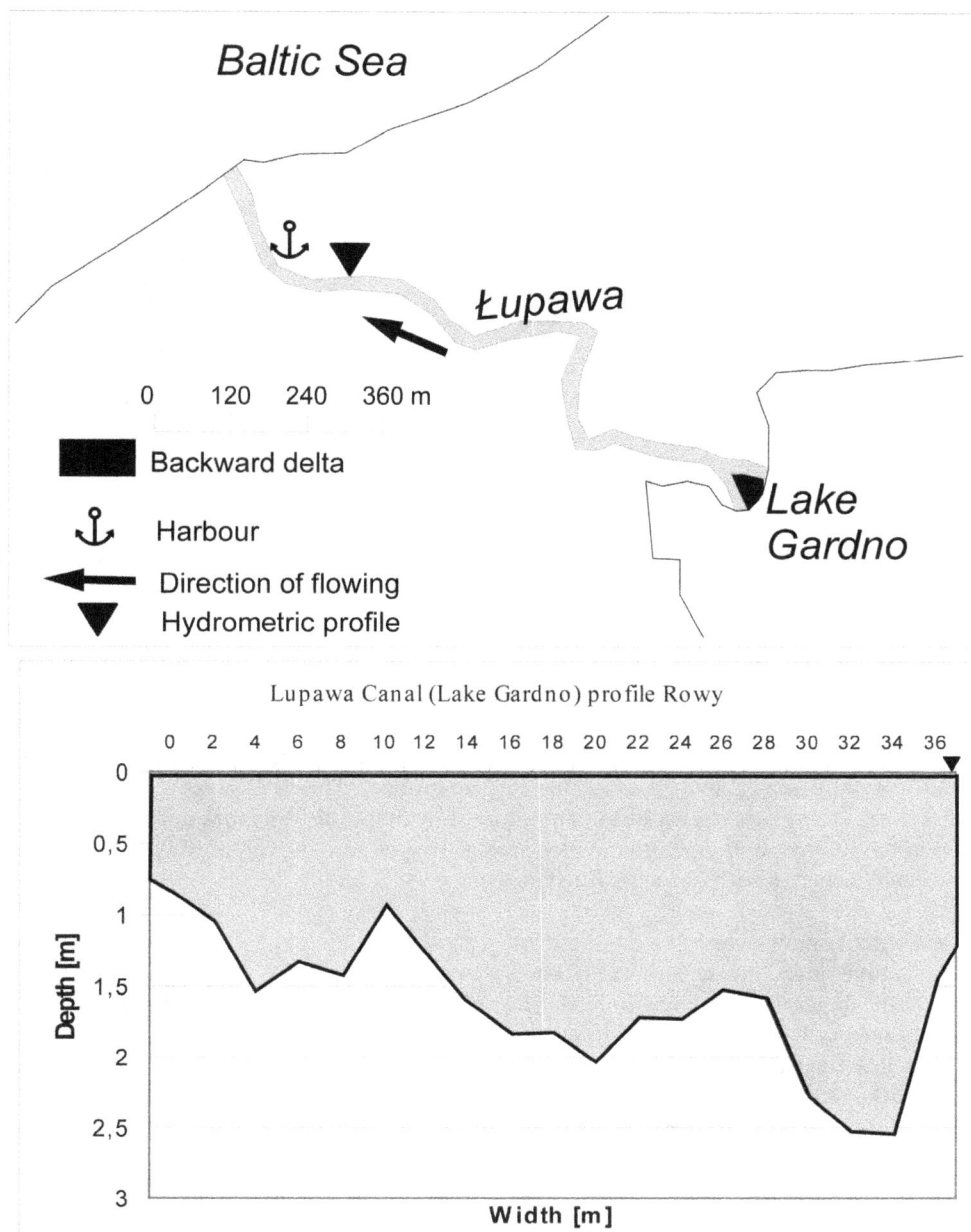

Fig. 7. Łupawa Canal linking Lake Gardno with the Baltic Sea.

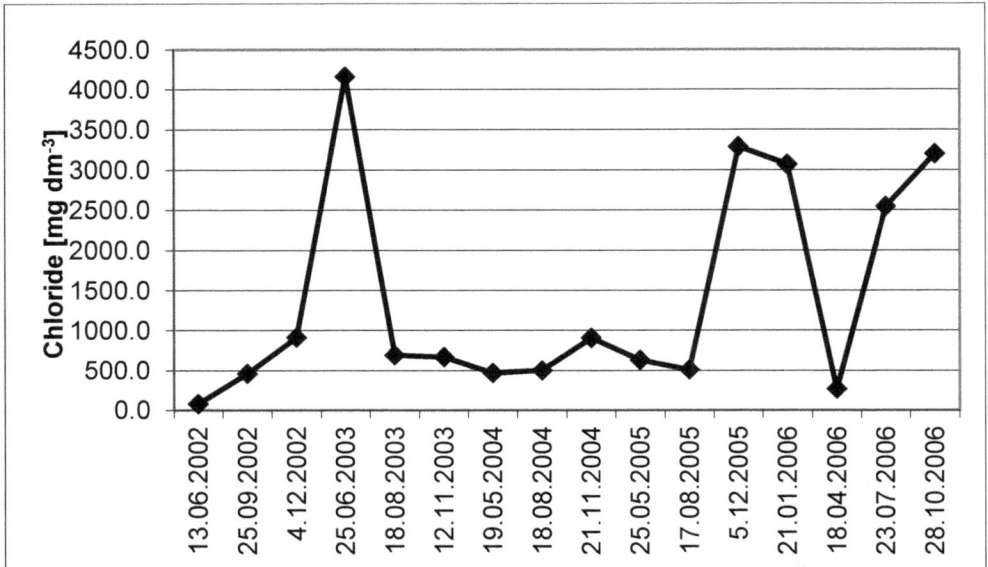

Fig. 8. Chloride concentrations in the canal linking Lake Gardno with the Baltic Sea (2002-06).

Lake Łebsko is linked with the sea by a canal about 2.7 km long and 5-25 m wide (gradient: 0.11‰) (Fig. 9). The rate of water flow from the lake to the sea ranged from 10.2 to 24.3 m^3 s^{-1} during the study period. When the hydrometeorological conditions are right, the direction of flow can become reversed. This occurred on Nov. 10, 2006 with a rate of full channel sea-to-lake water flow of 54 m^3 s^{-1}. The concentration of chloride in the Łeba Canal is persistently high, which indicates that seawater intrusions do not occur (16 observations, 2002-07) regardless of hydrometeorological conditions. The seawater effect is more pronounced during sea storms in the autumn and winter and less pronounced at low sea levels in the summer and during snowmelt season. In addition, the concentration of chloride in the canal increases with decreasing distance to the sea (Fig. 10).

4.4 Mouths of small rivers

The Reda River is 45 km long and has a drainage basin of 485 km^2 and empties into Puck Bay – a part of the larger Gdansk Bay. The Reda river channel is 6-13 m wide and 0.5 – 1.0 m deep. Its discharge is 4.1 – 6.3 m^3 s^{-1}. The duration of temporary increases in chloride concentration in rivers is related to the duration of favorable winds. The concentration of chloride at the mouth of the river was 1,142.7 mg/dm^3 (Fig 11) on May 20, 2004. The presence of water from Puck Bay can be traced to easterly winds with a speed of more than 10 m s^{-1}.

The Płucnica River is 9.2 km long and has a drainage basin of 85.2 km^2. The river's mean discharge is 0.47 m^3 s^{-1}. It empties into Puck Bay. The concentration of chloride did not exceed 100 mg Cl$^-$ dm^{-3} (Fig. 12) during the two-year study period. The concentration of chloride did not fall below 20 mg Cl$^-$ dm^{-3} either. This indicates that the mouth of the river is affected by the sea to some extent but that extent is not significant enough to warrant the conclusion that seawater intrusions are taking place.

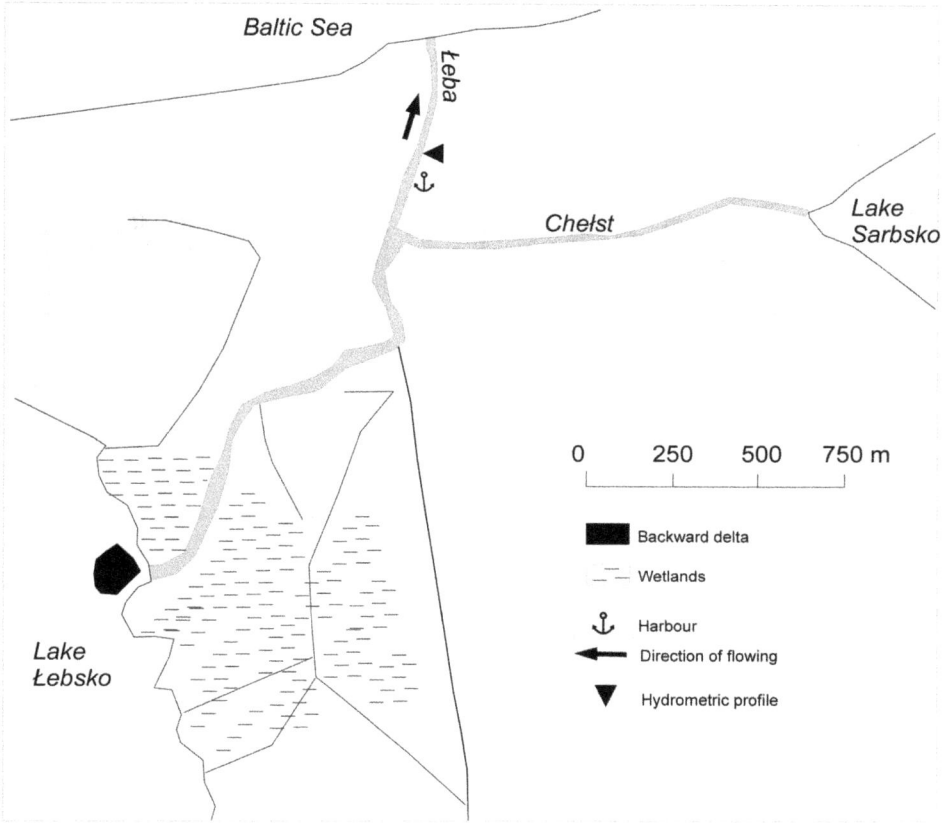

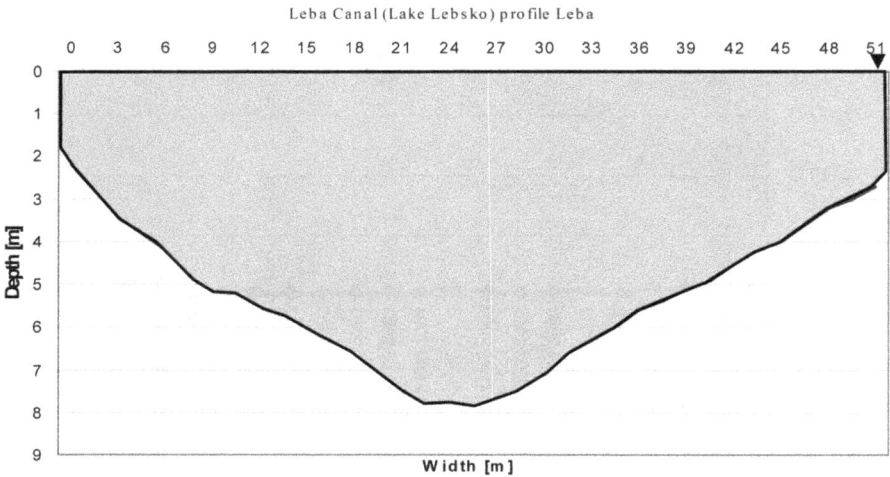

Fig. 9. Łeba Canal linking Lake Łebsko with the Baltic Sea.

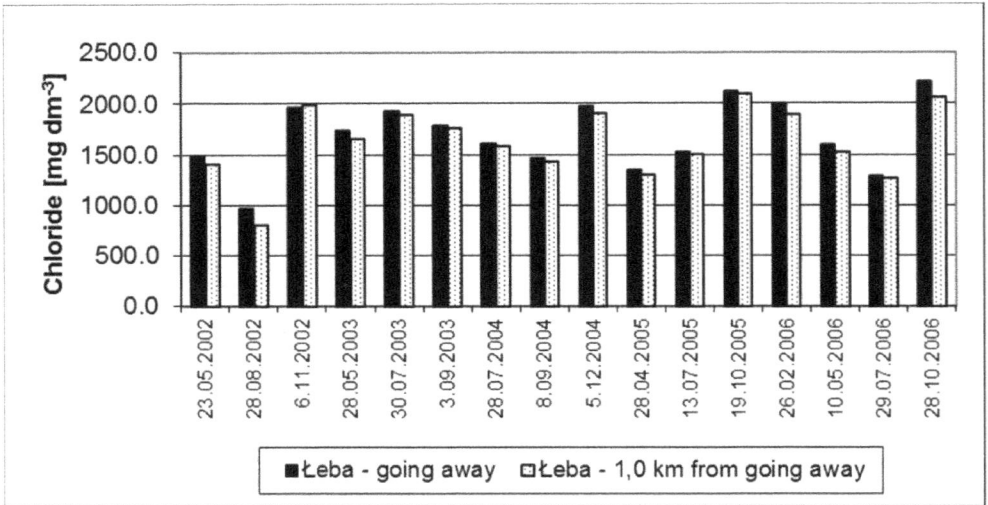

Fig. 10. Concentration of chloride in the Łeba Canal in the period 2002-07.

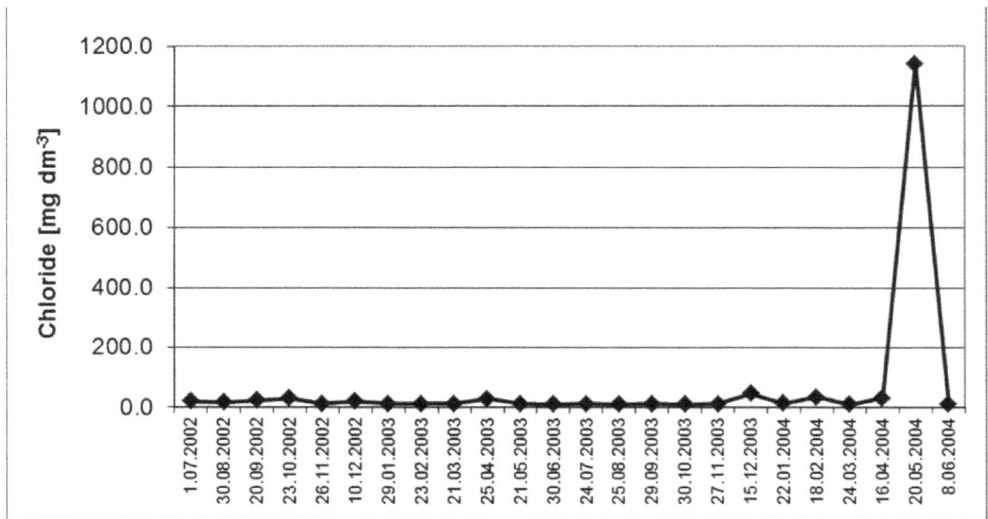

Fig. 11. Concentration of chloride at the mouth of the Reda River in the period 2002-04.

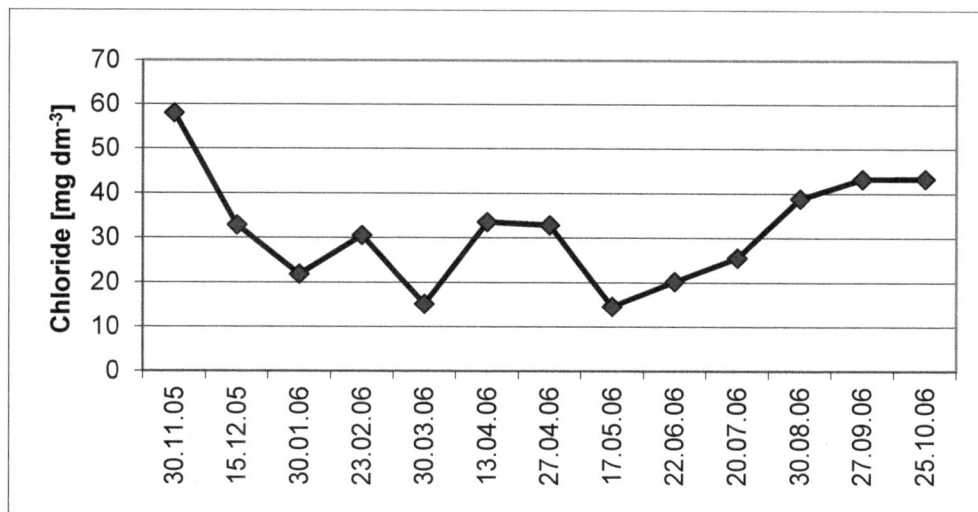

Fig. 12. Chloride concentration in the Płutnica River versus time.

4.5 Wetlands

Wetlands found along the southern coast of the Baltic Sea are unique in that they serve as focal points for seawater intrusions taking place in a variety of bodies of water. Beka Preserve is one such wetland area. The effects of seawater intrusions can be observed here at the mouth sections of the Reda and Zagórska Struga rivers as well as in a dense network of drainage ditches and swamplands. The water balance in Beka Preserve is affected by direct seawater intrusions from Puck Bay, the impediment of surface runoff resulting from temporary increases in base water levels, the flow of seawater over coastal embankments and the influx of brackish groundwater.

This diverse array of water effects divides Beka Preserve into several parts (Fig. 13). The northern part is adjacent to an upland and receives its surface runoff as well as precipitation. Its waters are not very saline. The central part includes a lot of saline puddles and saline groundwater. The western part is affected by freshwater flowing from the interior as well as by brackish water flowing over an embankment from time to time.

Persistently high concentrations of chloride (1,500 to almost 4,500 mg/dm³) were detected in the Beka Canal, the Unnamed Canal and the Jan Canal during the period 2002-04. Chloride concentrations remained under 20 mg*dm³ (Fig. 14) in other bodies of water. All of the investigated bodies of water with high chloride concentrations feature stagnant water and do not possess a direct link to the sea. The wetlands also feature puddles of marine water as a result of embankment overflow and wind-carried spray (Fig. 15).

Research data indicates that groundwater is responsible for hydrological differences in Beka Preserve. This is confirmed for the northern part of the preserve (piezometer A) by chloride concentrations ranging from 50 to 80 mg*dm³ Cl⁻. This concentration is the result of groundwater flowing down from the upland. The remaining piezometers (B - F) recorded high concentrations of chloride (1,000 - 3,500 mg/dm³ Cl⁻), which is a lot higher than the chloride concentrations detected in surface bodies of water. In addition, the fluctuations in chloride concentration in groundwater and the level of groundwater are consistent with fluctuations in Puck Bay. This suggests that seawater from Puck Bay may be encroaching via underground pathways.

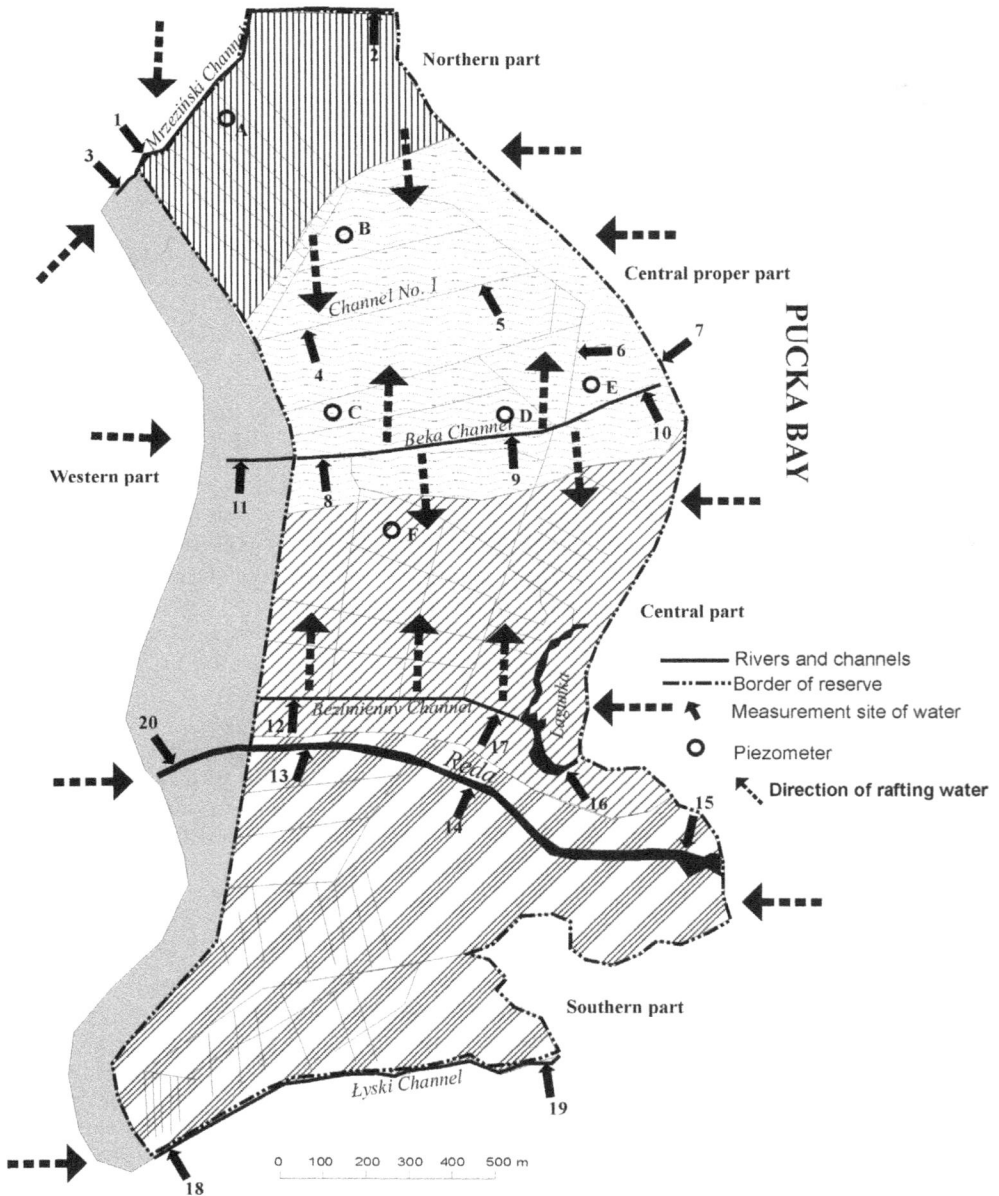

Fig. 13. Differences in water chemistry across Beka Reserve.

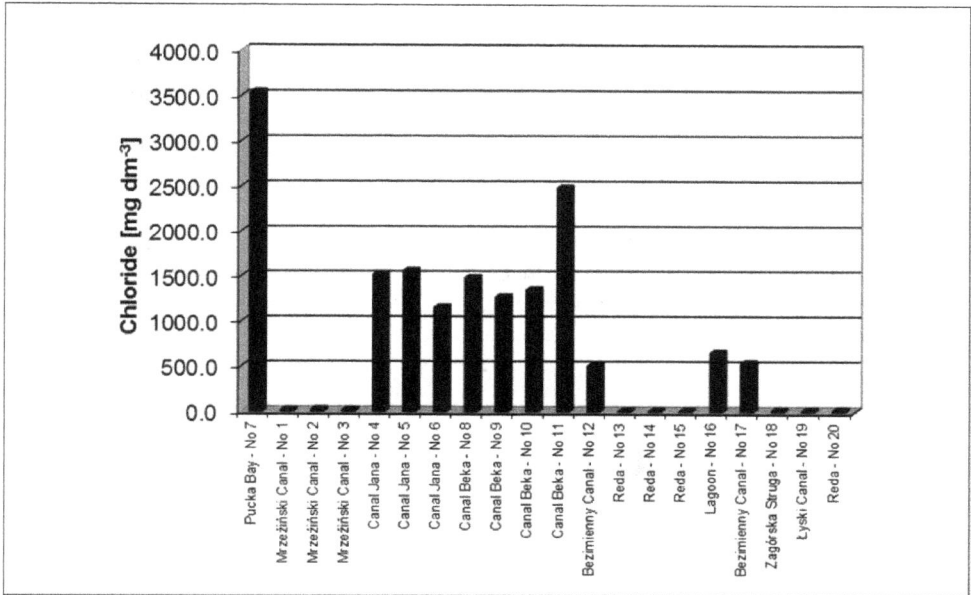

Fig. 14. Concentration of chloride in selected bodies of water on July 15, 2003.

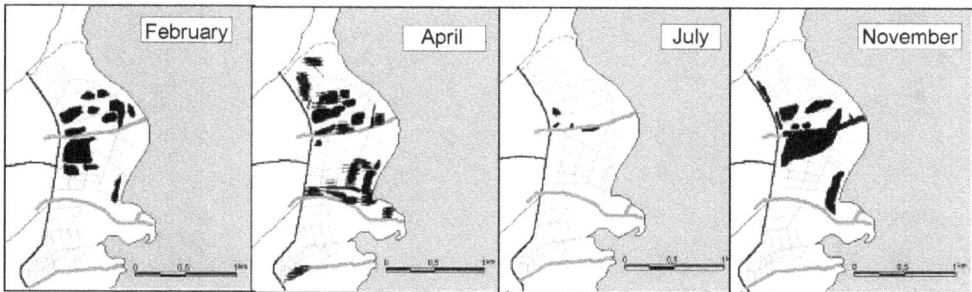

Fig. 15. Changes in wetland surface area on a seasonal basis.

Hence, it may be inferred that this wetland area is affected primarily by periodic seawater intrusions as well as by marine water flowing over coastal embankments and being carried by the wind.

4.6 Hydrographic systems

The extent of seawater impact along the southern Baltic coast is larger in places where bodies of water form hydrographic systems.

The Martwa Wisła River and the Śmiała Wisła River as well as the mouth of the Wisła River (via a sluice) form this type of system. The Martwa Wisła (Fig. 16) is an old arm of the Wisła River and today remains separated from the mouth section of the main river channel. The

Śmiała Wisła is an old breach in a sandbar that occurred in 1840. The Śmiała Wisła is about 2.5 km long and 150-200 m wide.
Today this section of river also connects the Martwa Wisła with the Baltic Sea. The eastern section of the Martwa Wisła remains a flooding arm, whereas the western section serves as a port channel. Discharge in the Martwa Wisła varies substantially as does the direction of flow (Jasińska, 1997). This and the river's direct link with the Bay of Gdansk cause mean annual salinity to remain high where the river meets the sea (6.13‰), somewhat lower in the central part (4.71‰) and the lowest in the western part (3.48‰) (Jasińska, 1997).
This general pattern is interrupted by the influx of river water (salinity in the Motława > 1‰) and seawater via the Śmiała Wisła. Seawater intrusions have been shown to penetrate the river in the period 2002-03 in a manner characteristic of estuaries – a wedge along the bottom (Cameron, Pritchard, 1963; Davidson et al., 1991).

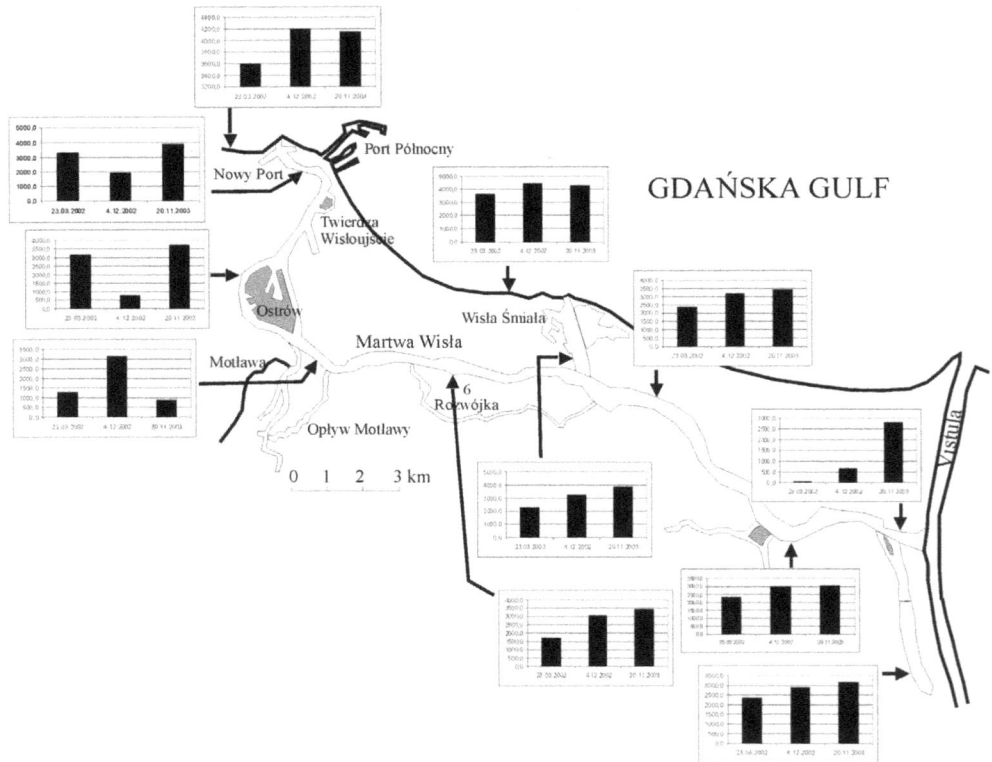

Fig. 16. Chloride concentration in the Martwa Wisła and the Śmiała Wisła in 2002-03.

Elbląg Bay, the Elbląg River and Lake Druzno form yet another hydrograhic system in the eastern Vistula Delta. Lake Druzno is fed by a land drainage basin. Water exits the lake via the Elbląg River, which empties into Elbląg Bay. Seawater periodically enters the lake via this same water route (22 – 33 % of annual discharge).

Mean values of both sea-based and land-based data indicate that the entire system experiences the chemical impact of Vistula Bay. Seawater intrusions were detected in 16 of 35 cases analyzed in 1997-99. Major intrusions were noted on Sept. 24, 1997, Aug. 26, 1998, Feb. 25, 1999 and Oct. 6, 1999. Minor intrusions were noted on July 9, 1997, Oct. 30, 1997 and Jan. 22, 1998.

Mean chloride concentrations were always detected closer to the sea (Fig. 17). It has also been shown that the chloride concentration in the middle part of Lake Druzno is always higher than in the lake's tributaries. This shows that water chemistry in Lake Druzno is substantially affected by seawater intrusions from Vistula Bay.

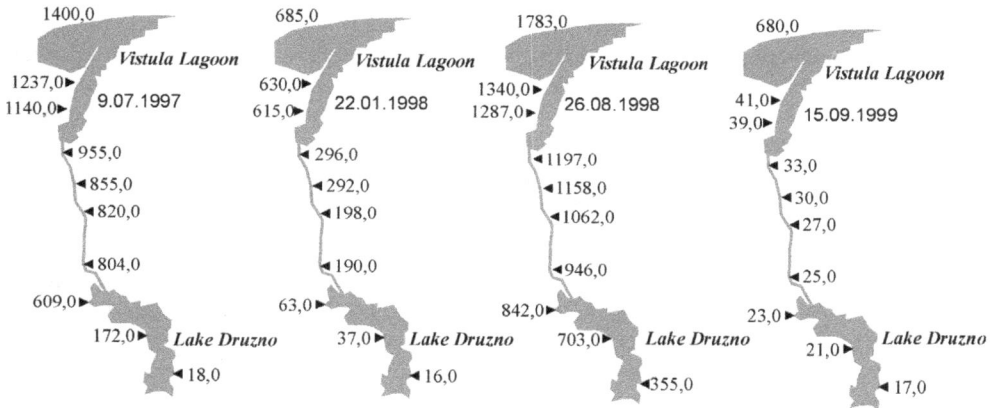

Fig. 17. Chloride concentration in the Elbląg Bay – Elbląg River – Lake Druzno hydrographic system for selected periods.

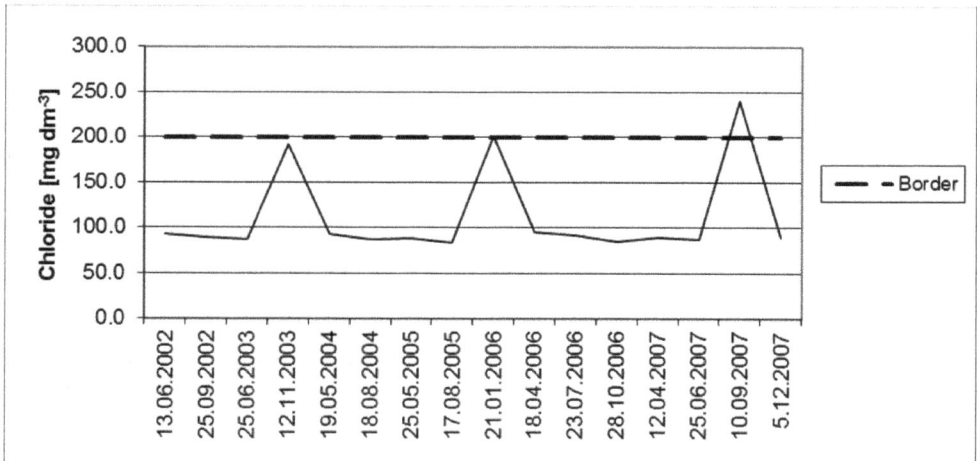

Fig. 18. Chloride concentration in Lake Smołdzińskie in the period 2002-07.

In the period 1996-98, the concentration of chloride ranged from 26 to 1,345 mg/dm³ in Elbląg Bay, from 15 to 1,197 mg/dm³ in the Elbląg River and from 12 to 824 mg/dm³ in the southern part of Lake Druzno.

Another example of a hydrographic system experiencing periodic marine impact is the following system: canal linking Lake Gardno to the sea, Lake Gardno, river linking Lake Gardno with Lake Smołdzińskie, Lake Smołdzińskie. While the concentration of chloride normally does not exceed 200 mg Cl⁻ dm⁻³ in Lake Smołdzińskie, minor exceptions do exist (Fig. 18).

5. Conclusions

The water balance along the southern Baltic coast is affected by land-based water drainage, seawater intrusions produced by fluctuating sea levels driven by wind surges, hydrographic entity types and the water balance between these types. The stated hypothesis has been shown to be true – the hydrochemical effect of seawater intrusions depends on the type of hydrographic entity and whether that entity is stand-alone or part of a hydrographic system.

Bodies of water such as Vistula Bay experience seawater effects throughout, although their intensity varies spatially. In this case, the coastal zone consists of the surface of Vistula Bay and other bodies of water linked to it.

In bodies of standing freshwater, the hydrochemical effects of seawater vary substantially and depend on the size of the given body of water as well as its location relative to the sea. Large lakes such as Koprowo, Resko Przymorskie, Bukowo and Łebsko constantly exerience seawater intrusions, which makes them similar to lagoons (Drwal, Cieśliński, 2007). Lake Gardno experiences seawater intrusions most of the year but also experiences periods of lower salinity. Gradient appears to play a major role in this case. Nevertheless, all of the studied lakes are affected by seawater intrusions.

Lake Jamno is primarily a freshwater lake but it does experience some seawater intrusions, which cause a sudden rise in chloride concentration in some parts of the lake. Small lakes close to the sea such as Ptasi Raj and Karaś constantly experience seawater intrusions. The extent of the coastal zone is the same for bodies of standing water and lagoons – area of the given body of water plus that of other bodies of water linked to it.

Canals linking the lakes of interest with the sea are the only routes where seawater can travel towards the interior of the coastal zone. Freshwater and seawater can travel either inland or towards the sea via the full cross section of each canal – a classic example of an estuary.

The hydrochemical effect of the sea in the mouths of small rivers is small and not strong enough to be labeled a seawater intrusion. The small cross section of small rivers may help explain this inference. In this case, the coastal zone is virtually limited to the mouth of the given river.

The occasionally elevated concentration of chloride in wetlands may be the result of occasional underground seawater intrusions as well as seawater being carried inland by the wind or simply splashing over embankments. Underground penetration is more likely since surface flow is impeded by coastal infrastructure.

Hence, it may be inferred that wetlands experience periodic seawater intrusions primarily via underground pathways as well as via splashing over embankments and seawater carried by the wind.

Seawater penetrates inland in a classic estuary-type wedge fashion along the bottom in hydrographic systems such as the one formed by the Martwa Wisła and Śmiała Wisła and the mouth of the Wisła River (via a sluice). This system is made up of old flooding arms of the Wisła River and a breach in a sandbar. In other systems – such as the Elbląg Bay, Elbląg River, Lake Druzno system – the coastal zone expands. This happens in areas with lagoons or large bodies of standing freshwater.

Effects associated with marine impact on coastal inland water systems along the southern Baltic Sea – a virtually land-locked sea with negligible tides and a young glacial Pleistocene sediment base – are reminiscent of those taking place along open seas.

6. References

Appelo C.A.J., Willemsen A., 1987, Geochemical calculations and observations on salt water intrusions, a combined geochemical/mixing cell model, J. Hydrology, 94, 313-330.

Bear J., Cheng A.H.-D., Sorek S., Ouazar D., Herrera I. (eds.), 1999, Seawater Intrusion in Coastal Aquifers - Concepts, Methods, and Practices, Kluwer Academic Publishers, Dordrecht/Boston/London, ss. 266.

Beletsky D., Schwab D., McCormick M., Miller G., Saylor J., Roebber J., 1999, Hydrodynamic modeling for the 1998 lake Michigan coastal turbidity plume ewent, [in:] M. Spaulding and H.L. Butler (eds), Estuarine and coastal modeling. American Society of Civil Emgineers, Virginia, 597-613.

Bowden K.F., 1967, Circulation and diffusion, [in:] G.H. Lauff (ed.), Estuaries, AAAS Publ., No. 83, Washington, 15–36.

Cameron W. M., Pritchard D. W., 1963 , Estuaries. The Sea , New York-London, vol. 2.

Cieśliński R., 2004, Application of arithmetic formulae in determining volume of sea waters inflow into Elbląska Bay – river Elbląg – lake Druzno hydrological system, Acta Geophysica Polonica, vol. 52, No. 4, 521 – 539.

Cieśliński R., Drwal J., 2005, Quasi - estuary processes and consequences for human activity, South Baltic, Estuarine, Coastal and Shelf Science, vol. 62, 477 – 485.

Davidson N. C., d'Alaffoley D., Doody J. P., Way L. S., Gordon J., Key R., Pieńkowski M. W., Mitchell R., Duff K. L., 1991, Nature conservation and estuaries in Great Britain, Estuaries Review Chief Scientist Directorate Nature Conservancy Council Northmister House, Peterborough.

Davies S., DeWiest R., 1966, Hydrogeology, Wiley, New York, ss. 245.

Drwal J., 1995, Impact of Baltic Sea on Ground Water and Surface Water in Żuławy Wiślane (Vistula Delta), [in:] Polish Coast: Past, Prezent and Future ed. by Karol Rotnicki, Journal of Coastal Research, Special Issue No. 22 CERF, p. 166-171.

Drwal J., Cieśliński R., 2007, Coastal lakes and marine intrusions on the southern Baltic coast, Oceanological and Hydrobiological Studies, Vol. XXXVI, No. 2 2007, 61 – 75.

Dziadziuszko Z., 1994, Sea-level Fluctuations [in:] Atlas of the Baltic Sea, A. Majewski and Z. Lauer (eds), IMGW, Warszawa

Flynn K. M., McKee K. L., Mendelssohn I. A., 1995, Recovery of freshwater marsh vegetation after a saltwater intrusion event, Journal of Oceanology, vo. 103, No. 1, 63–72.

Folk R.L., 1974, The natural history of crystalline calcium zarbonate: effect of magnesium content and salinity, Journal of Sedimentary Petrology, 44, 1, 40-53.

Foster I.D., 1980, Chemical yields in runoff, and denudation in a small arable catchment, East Devon, England, Journal of Hydrology, 47, 349–368.

Giambastiani B.M.S., Antonellini M., Gualbert H.P., Essink O., Stuurman R.J., 2007, Saltwater intrusion in the unconfined coastal aquifer of Ravenna (Italy): A numerical model, Journal of Hydrology, Vol. 340, Issues 1-2, 91–104.

Glover R.E., 1959, The pattern of fresh water flow in coastal aquifer, J. Geoph. Research, vol. 64, No. 4, 439–475.

Godo T., Kato K., Kamiya H., Yshitobi Y., 2001, Observation of wind-induced two-layer dynamice in lake Nakaumi, a coastal lagoon in Japan, Limnology, vol. 2, No 2, 137–143.

Halbfass W., 1901, Beiträge zur Kenntnis der Pommerschen Seen, Pett. Mitt. Erg. Heft., 36, p. 131.

Halbfass W., 1904, Weitere beiträge sur kerntnis der Pommerschen Seen, Pet. Mitt, p. 154.

Haslett S.K., 2008, Coastal system, Taylor & Francis, ss. 256.

Hem J.D., 1989, Study and interpretation of the characteristics of natural waters, U.S. Geol. Survey, Water Supply Paper, vol. 2254, ss. 263.

Hsing-Juh, L., Xiao-Xun, D., Kwang-Tsao, S., Huei-Meei, S., Wen-Tseng, L., Hwey-Lian, H., Lee-Shing, F., Jia-Jang, H., 2006, Trophic structure and functioning in a eutrophic and poorly flushed lagoon in southwestern Taiwan, Marine environmental research, 62 (1), 61-82.

Ishitobi Y., Kamiya H., Yokoyama K., Kumagai M., Okuda S., 1999, Physical conditions of saline water intrusion into a coastal lagoon, lake Shinji, Japan, Japanese Journal of Limnology, vol. 60, No 4, 439–452.

Jasińska E., 1997, Hydrodynamic and dynamics of salt water In the Martwa Vistula, Hydrotechnical Transactions, Polish Academy of Sciences, No. 61, 31–41.

Kunisch E., 1913, Der Gardensee und Gr. Dolgensee. Mit einem anhang: Ein Beitrag zur Kenntnis des Lebasees, XII Jahresb. Ges. Greiswald, p. 44.

Łomniewski K., Drwal J., Gołębiewski R., Pelczar M., Pietrucień Cz., Szeliga J., Ziółkowski J., 1972, Zasolenie wód w delcie Wisły, Rozprawy Wydz. III, z. 9, GTN, Gdańsk, 263-276.

Macdonald B.C.T., Smith J., Keene A.F., Tunks M., Kinsela A., White I., 2006, Impacts of runoff from sulfuric soils on sediment chemistry in an estuarine lake, Science of Total Environment, vol. 329, no. 1-3, 115–130.

Majewski A., 1972, Charakterystyka hydrologiczna estuariowych wód u polskiego wybrzeża, Prace PIHM , zeszyt 105 , 3-37.

Mikulski Z., Bojanowicz M., Ciszewski R.., 1969, Bilans wodny jeziora Druzno, Prace PIHM 96, 73-88.

Mosquera I., Cosoli S., Gačić M., Mazzoldi A., 2005, Wind forcing and notidal flow In the inlet sof the Venice lagoon, Geophysical Research, vol. 7, European Geosciences Union, 267–278.

Murray E., 2002, Determining the environmental status of coastal lakes: science for estuary management, Coast to Coast 2002, 315–317.

Piasecki M., Sanders B, 1999, Control of estuarine salinity using the adjoint metod, [in:] M. Spaulding, H.L. Butler (eds), Estuarine and coastal modeling, American Society of Civil Emgineers, Virginia, 1–16.

Pulido-Leboeuf, P, 2004, Seawater intrusion and associated processes in a small coastal complex aquifer (Castell de Ferro, Spain), Appl. Geochem., 19, 1517–1527.

Sanderson B.G., Baginska B., 2007, Calculating flow into coastal lakes from water level measurements, Environmental Modelling & Software, vol. 22, no 6, 774–786.

Saeijs H.L.F., Stortelder P.B.M., 1982, Converting an estuary to Lake Grevelingen: Environmental review of a coastal engineering project, Environmental Management, Vol. 6, No. 5, 377–405.

Spagnoli F., Specchiulli A., Sirocco T., Carapella G., Villani P., Casolino G., Schiavone P., Franchi M., 2002, The Lago di Varano hydrologic characteristic and sediment composition, Marine Ecology, 23, supplement 1, 384–394.

Tanaka, H., Takasaki M., Lee H. S., Yamaji H., 2005, Field observation of salinity intrusion into Nagatsura Lagoon, Taylor & Francis, 871–875.

Tiruneh, N. D., Motz, L. H. 2003, Three-Dimensional Modeling of Saltwater Intrusion in a Coastal Aquifer Coupled with the Impact of Climate Change, World Water & Environmental Resources Congress 2003, American Society of Civil Engineers, Philadelphia, PA, June 23-26, 1079–1087.

Tomczak K., 1995, Geological structure and Holocene Evolution of the Polish Coastal Zone [in:] Polish Coast: Past, Prezent and Future ed. by Karol Rotnicki, Journal of Coastal Research, Special Issue No. 22 CERF,p.13-31

Rotnicki K., 1995, The Coastal Zone - Prezent, Past and Future [in:] Polish Coast: Past, Prezent and Future ed. by Karol Rotnicki, Journal of Coastal Research, Special Issue No. 22 CERF, p.3-13

Sztobryn M., Stigge H.J., Wielbińska D., Weidig B., Stanisławczyk I., Kańska A., Krzysztofiak K, Kowalska B., Letkiewicz B., Mykita M., 2005, Storm Surges in the Southern Baltic Sea (Western and Central Parts), Berichte des Bundesamtes für Seeschifffahrt und Hydrographie, Nr. 39

Szopowski Z., 1962, Wybrane zagadnienia związane z wymianą wód pomiędzy jeziorem Łebsko a morzem, Materiały do monografii polskiego brzegu morskiego, z. 3, IBW PAN w Gdańsku, PWN, Poznań, ss. 122.

Van der Thuin H. (ed.), 1990, Guidelines on the study of seawater intrusion into rivers, International Hydrological Programme, UNESCO, Paris, ss. 139.

Zeidler R., Wroblewski A., Miętus M., 1995, Wind, Wale, and Storm Surges Regime at the Polish Balic Coast [in:] Polish Coast: Past, Prezent and Future ed. by Karol Rotnicki, Journal of Coastal Research, Special Issue No. 22 CERF, p. 33-56

Whittecar G.R., Nowroozi A.A., Hall J.R., 2005, Delineation of saltwater intrusion through a coastal borrow pit by resistivity survey, Environmental and Emgineering Geoscience, vol. 11, No 3, 209–219.

7

Prediction of Wave Height Based on the Monitoring of Surface Wind

Tsukasa Hokimoto
Graduate School of Mathematical Sciences, The University of Tokyo
Japan

1. Introduction

The ocean wave is one of the physical factors which cause serious sea disasters, and its prediction provides the information available for various human activity related to the sea. More than a half-century has passed since original theories for wave hindcasting techniques have been proposed in the pioneering papers such as Sverdrup and Munk (1947) and Pierson, Neumann, and James (1960) and so on, the method and the technique for wave prediction problem have progressed a great deal, against the background of recent progresses in the technologies of measurement and computation. However, even at the present, the prediction of the wave phenomena is still a difficult problem, and the technology for wave prediction is going on further development. There are several reasons why the prediction of the phenomena related to the sea state is a difficult problem even now. One of the reasons is the complexity of the physical mechanism on the wave development. When the sea is getting rough by wind forcing, the sea surface movement is affected by the interactions among the meteorological factors, such as wind motion and atmospheric pressure, and the topographical influence which varies by region. It means that the theoretical description of the sea surface movement, taking into account of the dynamic relationship among these factors, is very complicated. And another reason is the difficulty of the field measurement at sea. It is often the case that we can not carry out constant monitoring on the necessary meteorological factors, due to the lack of measurement facilities, sudden malfunction of a measurement instrument, and so on.

In the traditional research on the wave prediction problem, various statistical methods for the prediction of the sea state data have been proposed until now. However, most of such methods have been considered based on the measured data obtained by buoys or ships. In Japan, the Japan Meteorological Agency has set up about 1300 regional stations for the ground-based meteorological monitoring, which is called Automated Meteorological Data Acquisition System (AMeDAS), throughout of this country, and over 80 sensors for ultrasonic wave height meters in the coastal areas. They provide measured data on wave height and various meteorological factors constantly, which are available via Internet. It is thought that the physical factors which make influence on the sea condition, such as the wind speed and wind direction, change with spatial and temporal correlations. So, as an approach to the above wave prediction problem, we develop a statistical model for predicting the change of wave height from the change of surface wind, obtained by constant ground-based observation.

In this chapter, we provide two topics on the statistical modeling for the prediction of wave height. The first topic is a modeling for predicting the change of wave height based on ocean wind, by applying the method proposed in our previous paper (Hokimoto and Shimizu (2008)). And the second topic is the development of a statistical model for predicting wave height, based on the change of surface wind, obtained by ground-based observation. Also, the effectiveness in prediction using the proposed models is examined by means of the numerical experiment.

The sections below are organized as follows. In the next section, we outline traditional researches on the statistical models for the sea state data. In section 3, we present a method for the wave height prediction based on the measurement of ocean wind. In section 4, we develop a model for predicting wave height based on the measurement of surface wind, obtained by ground-based observation. And section 5 provides a summary of the result and discussion of further research on this topic.

2. Time series models for the sea state analysis

It is well-known that the wave motion under low wind speed can be approximated by Gaussian process. For the measured data in this aspect, the linear stationary time series models proposed by Box and Jenkins (1976), such as autoregressive (AR) model or autoregressive moving average (ARMA) model, have been widely used to construct a predictor. In fact, various applications to wave height data (e.g., Cunha and Guedes (1999), Yim et. al. (2002)) and the wind data (e.g., Brown et. al. (1984), Daniel and Chen (1991)) have been reported by many authors.

However, as for the wave motion during the wave development process, the above stationary models do not give reasonable predictions. There are also many models which are applicable to the measured data in the transitional aspect. One is standard linear nonstationary time series models. For example, autoregressive integrated moving average (ARIMA) model (Box and Jenkins (1976)), the autoregressive model with time varying coefficients (Kitagawa and Gersch (1985)), and generalized autoregressive conditional heteroskedasticity (GARCH) model (Bollerslev (1986)) are used widely to the nonstationary time series data. Also, if the speed in changing statistical structure can be regarded to be slow, we can apply a stationary AR model to the time series data in the local time interval which can be regarded to be stationary. For example, a model for predicting the change of nonstationary spectral density function of the sea surface movement during the wave development process was developed based on this concept (Hokimoto et. al. (2003)). There are also nonstationary time series models based on the decomposition of the trend and the other components (e.g., Athanassoulis et. al. (1995), Stefanakos et. al. (2002), Walton and Borgman (1990)).

One of interests when we treat the measured data on the sea state is how we treat the directional time series data on the wind direction. This problem is serious when we consider a statistical model to the multivariate time series data including wave height, wind speed and wind direction, because directional data have a unique property that they take the values on the circle. In the framework of directional statistics, various methodologies for statistical inferences on the directional data have been proposed (for example, Mardia and Jupp (2000)). Among them, multivariate regression models, including circular and linear variables, have been often proposed in environment studies. Johnson and Wehrly (1978) considered the theoretical background of the linear parametric regression, which has the linear variable and the angular variable. And the extension of their model has given in Fisher and Lee (1992), SenGupta (2004), SenGupta and Ugwuowo (2006), and so forth. However, there have been only limited attempts to model multivariate angular-linear data. In Hokimoto and Shimizu (2008), we developed an angular-linear time series model to express the dynamic structure among wave height, wind speed and wind direction, by extending the multiple regression model by Johnson and Wehrly (1978), and showed the effectiveness on wave height prediction between the change in ocean wind and the change in wave height. Our interest here is that the model whose structure is similar to the above model may be effective for the description of the dynamic relationship between wave height and surface wind.

3. Wave height prediction based on the change of ocean wind

In this section, we present a statistical method for predicting the change of wave height from the motion of ocean wind, based on Hokimoto and Shimizu (2008). The development of this method was motivated by the measured data obtained from ocean surveys using a research ship, in Hunka-bay, Hokkaido, Japan.

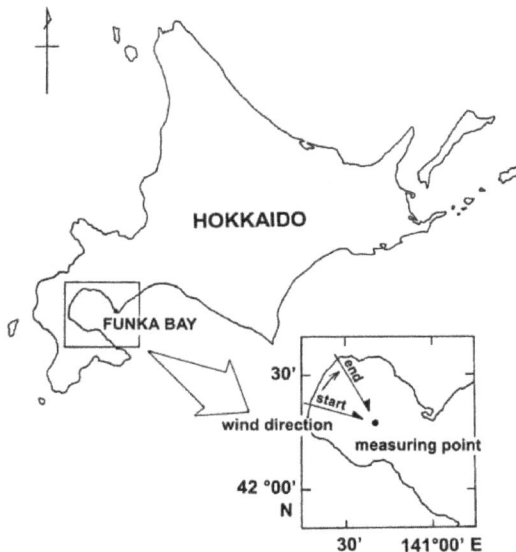

Fig. 1. A map around the measuring point

3.1 In-situ monitoring on wave height and ocean wind

Figure 1 shows a map around the measuring point (42°17′N, 140°40′E). We have measured the changes of relative sea surface level, wind speed and wind direction in Hunka-bay. For the relative sea surface movement, we measured relative displacement from the mean of the sea surface movement over 10 minutes by using an ultrasonic wave height meter of the research ship. Also, the changes in wind speed and wind direction at about 15 meters height from the sea surface were measured by using an ultrasonic wind meter. After the measurement, we obtained the time series data on 1/3 significant wave height, mean wind speed and mean wind direction for every 1 minute, based on the measured records.

Figure 2 displays the time series data obtained in the above, which were measured on Dec. 2, 1999. From the top, the significant wave height (m), the wind speed (m/s), and the wind direction (rad.) are shown, where each sample size is 90. It is noted that the origin of the wind direction data is defined to be north and the positive value means the clockwise direction. According to the weather maps of the sampling day, as well as the days before and after, the location of atmospheric pressure formed typical pattern of winter in Japan. In other words, the high pressure area is extending over the west of Japan Islands and the low pressure area is extending over the east. Under the background of this location, the above observation showed the tendency that the wind direction changed slowly from north-west to north, and the wind speed rapidly increased approximately from 6m/s to 13m/s in 40~50 minutes, and then changed slowly in the range approximately from 12m/s to 15m/s. On the other hand, 1/3 wave height gradually grew up to about 3.5 meters under the background that the wind speed increased and the wind direction changed slowly. The above data can be regarded to be the measurement of the wave development process, because wave height increases under the situation that the wind speed becomes faster and the wind direction does not change so much.

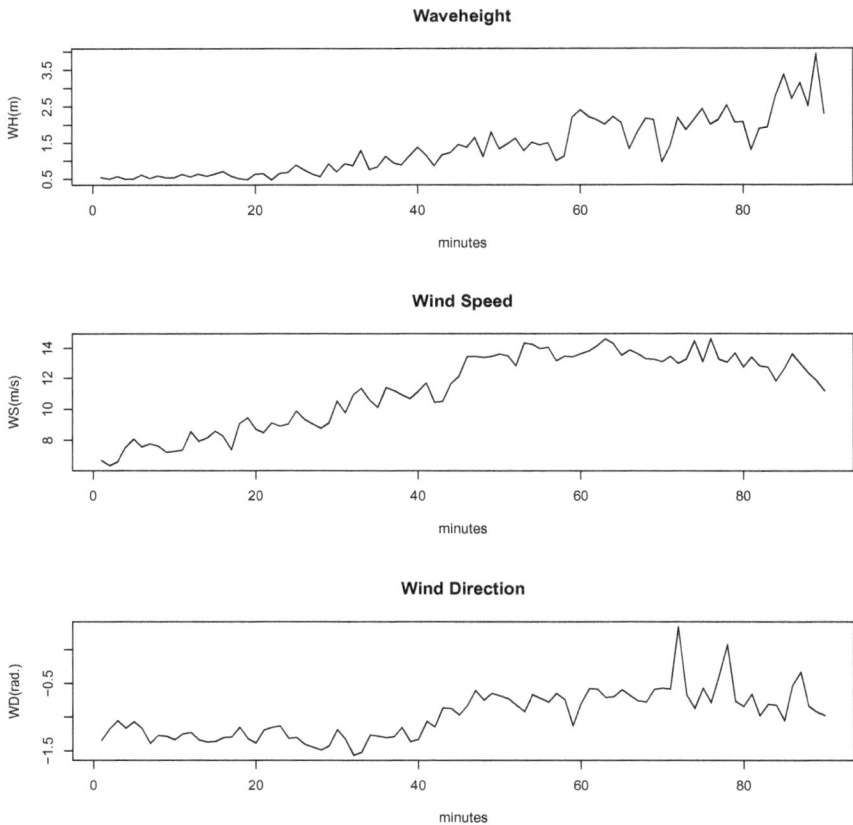

Fig. 2. Measured data on wave and wind (from the top, 1/3 significant wave height (m), wind speed (m/s), wind direction (rad.))

3.2 Some characteristics on the correlation structure

In the following, we make some preliminary analyses on the correlation structure of the measured data, in order to investigate what class of model is suitable for expressing the change of the measured data. In the following, let $\{WH_t\}$, $\{WS_t\}$ and $\{WD_t\}$ $(t = 1, \ldots, N)$ be sets of measurements of significant wave height, wind speed and wind direction, respectively, where t is the time point and N is the sample size.

3.2.1 Circular autocorrelation of the wind direction data

First, we invstigate the correlation structure of the directional time series data of wind direction. As a basic concept of exploratory circular data analysis, we refer to a book by Fisher (1993, Chapter 2) and use the following two transformations of WD_t

$$x_t = \cos(WD_t), \quad y_t = \sin(WD_t) \tag{1}$$

In order to explore the possibility of detecting changes of direction, we use two statistics; one is the cumulative sum (CUSUM) plot displayed by the points

$$C_t = \sum_{i=1}^{t} x_i, \quad S_t = \sum_{i=1}^{t} y_i \tag{2}$$

and the other is the cumulative mean direction plot $\{\Theta_t^c; t = 1, \ldots\}$, such that

$$\cos(\Theta_t^c) = C_t / \sqrt{C_t^2 + S_t^2}, \quad \sin(\Theta_t^c) = S_t / \sqrt{C_t^2 + S_t^2} \tag{3}$$

are satisfied simultaneously. CUSUM plot is displayed in the left of Figure 3, where the horizontal axis denotes C_t and the vertical axis denotes S_t. Also, the cumulative mean directional plot is displayed in the right of Figure 3, where the horizontal axis denotes the time point t and the vertical axis denotes Θ_t^c. It is noted that the change in statistical structure of the directional time series data is admitted, when the trend of CUSUM plot is clearly different from the straight line whose slope is one, and when the value of the cumulative mean directional plot is clearly different from the constant value. The cumulative mean directional plot suggests the possibility that the directional time series data have a change point of statistical structure at $t = 40$ roughly, and in this case the time series exhibits nonstationarity. We also checked the statistical test of change in mean direction by using CircStats (Chapter 11 of Jammalamadaka and SenGupta (2001)). The result showed that there exists a change point at the time point $t = 42$, which suggested that the data exhibit nonstationarity.

Now we are interested in whether there is clear difference in the correlation structures, between the case when we regard the wind direction data to be circular time series data and the case when we regard the data to be linear time series data. For the estimation of correlation, it is necessary to subtract the trend of the data. Therefore, we estimate the trend from the following two standpoints. One is the esimation by regarding the data to be circular time series data. In this case, for estimating trend, we obtain the smoothed series of $\{x_t\}$ and $\{y_t\}$ by using the locally weighted regression (LOWESS). And then, based on the smoothed series, say $\{x_t^*\}$ and $\{y_t^*\}$, we obtain the smoothed trend of wind direction T_t^*, such that

$$\frac{x_t^*}{\sqrt{(x_t^*)^2 + (y_t^*)^2}} = \cos(T_t^*), \quad \frac{y_t^*}{\sqrt{(x_t^*)^2 + (y_t^*)^2}} = \sin(T_t^*) \tag{4}$$

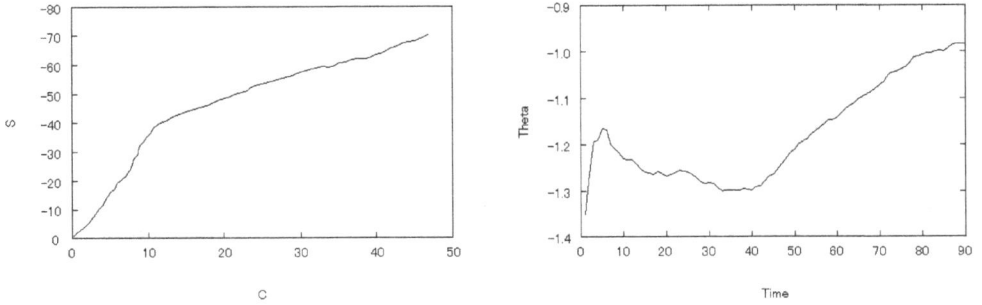

Fig. 3. Cumulative sum plot (left) and Cumulative mean directional plot (right)

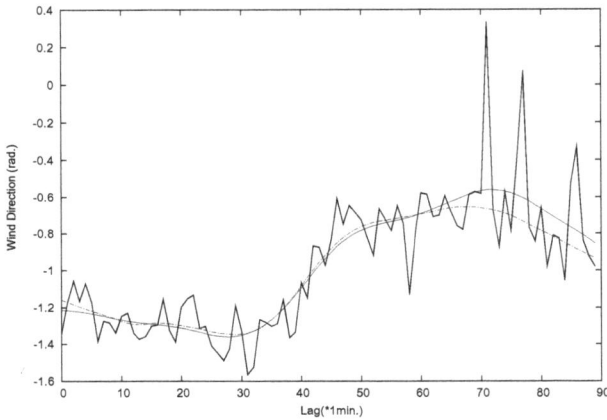

Fig. 4. Trend estimation on wind direction based on $\{T_t^*\}$ (dotted curve) and $\{T_t^{**}\}$ (solid curve)

are satisfied simultaneously. Another is the trend estimation based on linear time series data, which is given in Kitagawa and Gersch (1985). The trend of $\{WD_t\}$ can be obtained by applying the trend model,

$$WD_t = T_t^{**} + \zeta_t, \qquad \zeta_t \sim N(0, \sigma_\zeta^2) \tag{5}$$

and

$$T_t^{**} - T_{t-1}^{**} = v_t, \qquad v_t \sim N(0, \sigma_v^2) \tag{6}$$

where T_t^{**} is the random variable to express the trend, σ_ζ^2 and σ_v^2 are unknown variances of ζ_t and v_t, respectively. Figure 4 shows the trend estimation based on T_t^* and T_t^{**}, where the dotted curve means $\{T_t^*\}$ and the sold curve means $\{T_t^{**}\}$. It looks that there is no clear difference between $\{T_t^*\}$ and $\{T_t^{**}\}$. So we estimate circular autocorrelation coefficient based on the subtracted series, $WD_t^* \equiv WD_t - T_t^*$. Based on circular-circular association (Fisher

(1993, Chapter 6)), the sample circular autocorrelation coefficient is given by

$$\hat{\rho}^*(\tau) = \frac{4(A_\tau B_\tau - C_\tau D_\tau)}{\left[(N^2 - E_\tau^2 - F_\tau^2)(N^2 - G_\tau^2 - H_\tau^2)\right]^{1/2}}, \quad \tau = 0, 1, \ldots \tag{7}$$

where τ is the time lag, and

$$A_\tau = \sum_{t=1}^{N-\tau} \cos WD_t^* \cos WD_{t+\tau}^*, \quad B_\tau = \sum_{t=1}^{N-\tau} \sin WD_t^* \sin WD_{t+\tau}^*, \quad C_\tau = \sum_{t=1}^{N-\tau} \cos WD_t^* \sin WD_{t+\tau}^*,$$

$$D_\tau = \sum_{t=1}^{N-\tau} \sin WD_t^* \cos WD_{t+\tau}^*, \quad E_\tau = \sum_{t=1}^{N-\tau} \cos(2WD_t^*), \quad F_\tau = \sum_{t=1}^{N-\tau} \sin(2WD_t^*), \quad G_\tau = \sum_{t=1}^{N-\tau} \cos(2WD_{t+\tau}^*),$$

$$H_\tau = \sum_{t=1}^{N-\tau} \sin(2WD_{t+\tau}^*) \tag{8}$$

On the other hand, the sample autocorrelation function of the time series data WD_t^* is given by

$$\hat{\rho}^{**}(\tau) = \frac{\sum_{t=1}^{N-\tau}(WD_{t+\tau}^* - \overline{WD^*})(WD_t^* - \overline{WD^*})}{\sum_{t=1}^{N}(WD_t^* - \overline{WD^*})^2},$$

$$\overline{WD^*} = \frac{1}{N}\sum_{t=1}^{N} WD_t^* \tag{9}$$

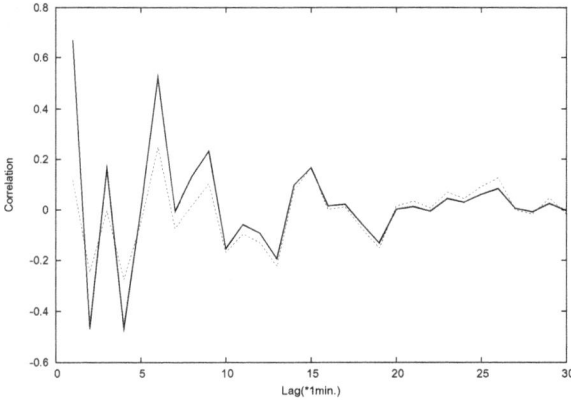

Fig. 5. Comparion between $\{\hat{\rho}^*(\tau)\}$ (bold line) and $\{\hat{\rho}^{**}(\tau)\}$ (dotted line)

Figure 5 displays the estimates of $\hat{\rho}^*(\tau)$ and $\hat{\rho}^{**}(\tau)$ $(0 \leq \tau \leq 30)$, where the vertical axis denotes the correlation, the horizontal axis denotes τ in minutes, and the bold and dotted lines correspond to $\hat{\rho}^*(\tau)$ and $\hat{\rho}^{**}(\tau)$, respectively. We observe that they change similarly with the same tendency, although $|\hat{\rho}^*(\tau)|$ takes slightly larger values than $|\hat{\rho}^{**}(\tau)|$ when τ is small. It is evaluated from this result that the sample circular autocorrelation coefficient can be approximated by the linear correlation to some extent. Also, it suggests the possibility that it is sufficient to express the dynamic structure of the measured data by using the linear time series model.

Lag	(i)	(ii)	Lag	(i)	(ii)
0	0.038	0.084	11	0.174	0.173
1	-0.096	-0.085	12	0.156	0.184
2	0.102	0.093	13	0.183	0.228
3	0.006	-0.014	14	0.176	0.188
4	-0.227	-0.214	15	0.155	0.171
5	-0.139	-0.133	16	0.025	0.032
6	0.067	0.087	17	0.300	0.335
7	0.076	0.119	18	0.046	0.016
8	0.040	0.050	19	-0.080	-0.058
9	-0.139	-0.147	20	-0.070	-0.042
10	-0.039	-0.050			

Table 1. Cross correlation functions ((i) $\{WD_t\}$ and $\{WH_t\}$, (ii) $\{\sin(WD_t)\}$ and $\{WH_t\}$)

3.2.2 Cross correlation among wind speed, wind direction and wave height

Next, we focus on the cross correlations among $\{WS_t\}$, $\{WD_t\}$ and $\{WH_t\}$. We estimated the cross correlation function between $\{WH_t\}$ and the variables $\{WS_t\}$, $\{WD_t\}$ and $\{\sin(WD_t)\}$, by using the time series data after subtraction of their trends estimated by LOWESS method. Figure 6 shows an estimated result of the cross correlation function between $\{WS_t\}$ and $\{WH_t\}$ by using the sample cross correlation function $\gamma(\tau)$ ($\tau = 0, \pm 1, \ldots$),

$$\gamma(\tau) = \frac{\sum_{t=1}^{N-\tau}(WH_t - \overline{WH})(WS_{t+\tau} - \overline{WS})}{\sqrt{\sum_{t=1}^{N}(WH_t - \overline{WH})^2}\sqrt{\sum_{t=1}^{N}(WS_t - \overline{WS})^2}}, \quad \overline{WH} = \frac{1}{N}\sum_{t=1}^{N}WH_t, \quad \overline{WS} = \frac{1}{N}\sum_{t=1}^{N}WS_t \quad (10)$$

where the horizontal means the time lag in minutes and two parallel lines denote Bartlett's bounds (i.e., $\pm 1.96N^{-1/2}$). It suggests the possibility that the change in wind speed affects the one of wave height after 10~20 minutes.

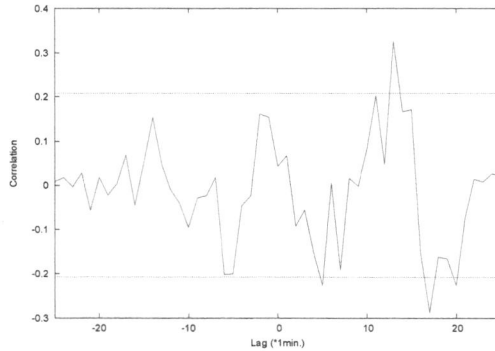

Fig. 6. Cross correlation function between $\{WS_t\}$ and $\{WH_t\}$ with Bartlett's bounds

For estimation of the correlation between $\{WD_t\}$ and $\{WH_t\}$, it is of interest how we treat the directional variable WD_t. Table 1 gives estimated values of the cross correlation functions in the two cases, (i) $\{WD_t\}$ and $\{WH_t\}$ and (ii) $\{\sin(WD_t)\}$ and $\{WH_t\}$. It is observed that the absolute value of (ii) tends to take larger values than the one of (i). This result suggests the possibility that it is expected to improve the prediction accuracy by adopting the variables $\sin(WD_t)$ (and $\cos(WD_t)$) as the explanatory variables, instead of using WD_t.

3.3 A statistical modeling on the change of wave height by wind forcing

Suppose that we predict the future values of wave height $\{WH_{N+l}; l = 1, \ldots, L\}$, based on the historical data $\{WH_t, WS_t, WD_t\}$ ($t = 1, \ldots, N$). We start our consideration by assuming that the time series $\{WH_t\}$, $\{WS_t\}$ and $\{WD_t\}$ are stationary, after applying a proper transformation (the detail is described later in this section). We write the change of $\{WH_t\}$ as

$$WH_t = m_L + \sum_{i=1}^{p} \beta_i^{(1)} WH_{t-i} + \sum_{i=1}^{p}\sum_{k=1}^{K} \beta_{i,k}^{(3)} \cos(k \cdot WD_{t-i}) + \sum_{i=1}^{p}\sum_{k=1}^{K} \beta_{i,k}^{(4)} \sin(k \cdot WD_{t-i})$$

$$+ \sum_{i=1}^{p} \beta_i^{(2)} WS_{t-i} + \varepsilon_t^{(1)}, \qquad \varepsilon_t^{(1)} \sim WN(0, \sigma_{WH}^2) \qquad (11)$$

where p and K are orders, m_L is the unknown mean, β's are unknown weights, and $\varepsilon_t^{(1)}$ is the random variable which follows a white noise process with $E(\varepsilon_t^{(1)})=0$ and $V(\varepsilon_t^{(1)}) = \sigma_{WH}^2$. Similarly, we write

$$WS_t = m_S + \sum_{i=1}^{p} \gamma_i^{(1)} WH_{t-i} + \sum_{i=1}^{p}\sum_{k=1}^{K} \gamma_{i,k}^{(3)} \cos(k \cdot WD_{t-i}) + \sum_{i=1}^{p}\sum_{k=1}^{K} \gamma_{i,k}^{(4)} \sin(k \cdot WD_{t-i})$$

$$+ \sum_{i=1}^{p} \gamma_i^{(2)} WS_{t-i} + \varepsilon_t^{(2)}, \qquad \varepsilon_t^{(2)} \sim WN(0, \sigma_{WN}^2) \qquad (12)$$

and $\sin(h \cdot WD_t)$ and $\cos(h \cdot WD_t)$ ($h = 1, \ldots, K$) as

$$\sin(h \cdot WD_t) = m_h + \sum_{i=1}^{p} \delta_i^{(1)} WH_{t-i} + \sum_{i=1}^{p}\sum_{k=1}^{K} \delta_{i,k}^{(3)} \cos(k \cdot WD_{t-i}) + \sum_{i=1}^{p}\sum_{k=1}^{K} \delta_{i,k}^{(4)} \sin(k \cdot WD_{t-i})$$

$$+ \sum_{i=1}^{p} \delta_i^{(2)} WS_{t-i} + \delta_t^{(h)}, \qquad \delta_t^{(h)} \sim WN(0, \sigma_h^2) \qquad (13)$$

and so forth, where m_s, m_h, γ's and δ's are unknown weights. Put the state vector at time t by

$$y_t^{(K)} \equiv (WH_t, WS_t, \cos(WD_t), \sin(WD_t), \ldots, \cos(K \cdot WD_t), \sin(K \cdot WD_t))' \qquad (14)$$

Then we can write

$$y_t^{(K)} = m^{(K)} + A_1^{(K)} y_{t-1}^{(K)} + \cdots + A_p^{(K)} y_{t-p}^{(K)} + \delta_t^{(K)} \quad \delta_t^{(K)} \sim WN(0, \Sigma^{(K)}) \qquad (15)$$

where $m^{(K)}$ is the unknown mean vector, $A_i^{(K)}$ ($i = 1, \ldots, p$) is the unknown coefficient matrix, and $\delta_t^{(K)}$ follows the multivariate white noise process with mean 0 and the dispersion matrix $\Sigma^{(K)}$. This is a multivariate vector autoregressive model of the pth order, and therefore, the estimates for elements of unknown matrices $A_i^{(K)}$ can be obtained by using the least squares method (e.g., Brockwell and Davis (1996)). Thus, we can construct an l-step ($l = 1, \ldots, L$) ahead predictor based on (15) by

$$\hat{y}_{N+l}^{(K)} = \hat{m}^{(K)} + \hat{A}_1^{(K)} z_{N+l-1}^{(K)} + \hat{A}_2^{(K)} z_{N+l-2}^{(K)} + \cdots + \hat{A}_p^{(K)} z_{N+l-p}^{(K)} \qquad (16)$$

and $z_{N+l-m}^{(K)} = y_{N+l-p}^{(K)}$ $(l \leq p)$, $z_{N+l-m}^{(K)} = \hat{y}_{N+l-p}^{(K)}$ $(l > p)$, where $\hat{A}_i$ is the least squares estimator of A_i, The predicted values of WH_{N+l} $(l = 1, \ldots, L)$ can be obtained from the prediction of $\hat{y}_{N+l}^{(K)}$.

However, the model (15) with the state vector (14) has a drawback in computational aspect. It is probable that the accuracy of the estimates of parameter becomes worse when both K and p become large, because (15) has $(2 + 2K) + p(2 + 2K)^2$ unknown parameters to be estimated. For improving the prediction accuracies, the dimension of the state vector $\hat{y}_t^{(K)}$ should be small. In order to taking account of the multiple directional information with the small numbers of variables, we focus on the following linear sum

$$\widetilde{WD}_t^{(K)} \equiv \omega_1 \cos(WD_t) + \omega_2 \sin(WD_t) + \cdots + \omega_{2K-1} \cos(K \cdot WD_t) + \omega_{2K} \sin(K \cdot WD_t) \quad (17)$$

where ω_i $(i = 1, \ldots, 2K)$ are unknown weights. And we propose to use the model (15) with the state vector

$$\tilde{y}_t^{(K)} \equiv (WH_t, WS_t, \widetilde{WD}_t^{(K)})' \quad (18)$$

Here, it is necessary to determine the optimum order K and the value of ω_i. For determining ω_i, we introduce the concept of principal component analysis. $\widetilde{WD}_t^{(K)}$ can be written as

$$\widetilde{WD}_t^{(K)} = \Omega_K' D_t^{(K)} \quad (19)$$

where $\Omega_K = (\omega_1, \ldots, \omega_{2K})'$ and $D_t^{(K)} = (\cos(WD_t), \sin(WD_t), \ldots, \cos(K \cdot WD_t), \sin(K \cdot WD_t))'$.

We select the values of Ω_K so that

$$V(\widetilde{WD}_t^{(K)}) = \Omega_K' \Sigma_t^{(K)} \Omega_K \quad (20)$$

is maximized under the constraints $\Omega_K' \Omega_K = 1$, where $\Sigma_t^{(K)}$ is the dispersion matrix of $D_t^{(K)}$. Ω_K can be obtained as the eigenvector $b^{(K)}$ of the eigen equation,

$$\Sigma_t^{(K)} b^{(K)} = \lambda b^{(K)} \quad (21)$$

Let $\lambda_1 \geq \cdots \geq \lambda_{2K}$ be $2K$ eigenvalues of the eigen equation. We choose the eigenvector which corresponds to λ_1 with unit norm, say $\tilde{b}_M^{(K)}$, with K fixed. We estimate $\widetilde{WD}_t^{(K)}$ by

$$\widehat{\widetilde{WD}}_t^{(K)} = \tilde{b}_M^{(K)} D_t^{(K)} \quad (22)$$

As for the selection of the order K, we choose the value of K such that the squared sum of the prediction errors,

$$S_l(K) = \frac{1}{N - l - N^* + 1} \sum_{t=N^*}^{N-l} (WH_{t+l} - \widehat{WH}_{t+l}^{(K)})^2 \quad (23)$$

is minimized for every l, where $\widehat{WH}_{t+l}^{(K)}$ is the predicted value by (16) and N^* is a prefixed value. For selection of p in (15), we use AIC (Akaike Information Criterion), under the value of K is fixed.

As observed in Figure 2, the time series data of WH_t, WS_t and WD_t during the wave development process exhibit nonstationarity. We follow the method of ARIMA model by Box and Jenkins (1976) and focus on the differenced time series. In other words, we regard the differenced series to be stationary and then fit

$$x_t^{(K)} = B_1^{(K)} x_{t-1}^{(K)} + \cdots + B_p^{(K)} x_{t-p}^{(K)} + \epsilon_t^{(K)}, \qquad \epsilon_t^{(K)} \sim WN(0, \Sigma_{\epsilon^{(K)}}) \tag{24}$$

and

$$x_t^{(K)} \equiv (\nabla WH_t, \nabla WS_t, \widehat{\nabla WD_t}^{(K)})' \tag{25}$$

where ∇ is the back-shift operator such that $\nabla WH_t = WH_t - WH_{t-1}$.

3.4 The effect of angular-linear structure on the prediction of wave height

In the following, we examine the availability of the proposed method through the evaluation of the prediction accuracy on wave height. For this purpose, we carried out the numerical experiments on prediction accuracy by using the measured data shown in Figure 2. The procedure of the prediction experiment is as follows. First, we fit the model (24) to the multivariate time series data $\{WH_t, WS_t, WD_t; t = 1\ldots,50\}$ and then obtain the prediction values of WH_t up to 5 steps ahead (1 step corresponds to 1 minute). Next, we fit the model to the time series data from $t=2$ to $t=51$ and obtain the predicted values in the same way. After repeating this procedure, the prediction accuracy is evaluated based on the predicted values and realizations. As criteria for evaluation, we define the mean absolute error (MAE) and the correlation coefficient (COR) by

$$MAE(l) \equiv \frac{1}{M} \sum_{i=1}^{M} |WH_{N+l}^{(i)} - \widehat{WH}_{N+l}^{(i)}| \tag{26}$$

$$COR(l) \equiv \frac{\sum_{i=1}^{M} (WH_{N+l}^{(i)} - \overline{WH}^{(i)}(l))(\widehat{WH}_{N+l}^{(i)} - \overline{\widehat{WH}}^{(i)}(l))}{\sqrt{\sum_{i=1}^{M}(WH_{N+l}^{(i)} - \overline{WH}^{(i)}(l))^2} \sqrt{\sum_{i=1}^{M}(\widehat{WH}_{N+l}^{(i)} - \overline{\widehat{WH}}^{(i)}(l))^2}},$$

$$\overline{WH}(l) = \frac{1}{M} \sum_{i=1}^{M} WH_{N+l}^{(i)}, \quad \overline{\widehat{WH}}(l) = \frac{1}{M} \sum_{i=1}^{M} \widehat{WH}_{N+l}^{(i)} \tag{27}$$

where l is the prediction step ($l = 1,\ldots,5$), $WH_t^{(i)}$ is the realization of WH_t at the ith experiment, $\widehat{WH}_t^{(i)}$ is the predicted value of WH_t at the ith experiment, and M is the number of repetitions of the experiment. MAE gives better evaluation as the predicted value gets closer to the observation. COR is defined as the sample correlation between the observations and predicted values, in order to evaluate the degree of accordance to their trends.

We first investigate whether the angular-linear structure of the proposed model give the positive effect on the prediction accuracy of wave height. For this purpose, we analyze whether it is possible to improve the prediction accuracy by taking into account the variables $\{\sin(k \cdot WD_t), \cos(k \cdot WD_t)\}$ ($k = 1,\ldots, K$), instead of using the variable WD_t directly. In this experiment, we compare the prediction accuracy by using the model (24), under assuming the following three state vectors. The first is the vector consisted from the difference of WH_t, WS_t and WD_t,

$$x_t \equiv (\nabla WH_t, \nabla WS_t, \nabla WD_t)' \tag{28}$$

	MAE					COR				
p	$L=1$	$L=2$	$L=3$	$L=4$	$L=5$	$L=1$	$L=2$	$L=3$	$L=4$	$L=5$
1	0.385	0.446	0.517	0.490	0.585	0.501	0.335	0.200	0.360	0.205
2	0.394	0.454	0.464	0.490	0.582	0.450	0.292	0.258	0.366	0.201
3	0.415	0.500	0.482	0.474	0.601	0.346	0.195	0.159	0.403	0.155
4	0.428	0.518	0.505	0.487	0.613	0.322	0.123	0.081	0.357	0.152
5	0.421	0.528	0.494	0.493	0.604	0.333	0.093	0.063	0.320	0.156

Table 2. Prediction accuracy by using the model (24) with the state vector (28) ($M=35$)

	MAE					COR				
p	$L=1$	$L=2$	$L=3$	$L=4$	$L=5$	$L=1$	$L=2$	$L=3$	$L=4$	$L=5$
1	0.360	0.480	0.506	0.552	0.589	0.492	0.291	0.208	0.389	0.221
2	0.357	0.473	0.502	0.544	0.573	0.483	0.281	0.188	0.333	0.154
3	0.362	0.477	0.508	0.550	0.598	0.465	0.255	0.145	0.258	-0.078
4	0.356	0.474	0.502	0.540	0.586	0.479	0.260	0.146	0.312	-0.007
5	0.372	0.492	0.506	0.539	0.590	0.442	0.207	0.130	0.285	0.015

Table 3. Prediction accuracy by using the model (24) with the state vector (29) ($M=35$)

	MAE					COR				
p	$L=1$	$L=2$	$L=3$	$L=4$	$L=5$	$L=1$	$L=2$	$L=3$	$L=4$	$L=5$
1	0.378	0.445	0.490	0.485	0.600	0.481	0.338	0.234	0.370	0.163
2	0.353	0.435	0.456	0.502	0.574	0.433	0.296	0.259	0.360	0.180
3	0.367	0.440	0.478	0.493	0.575	0.380	0.256	0.174	0.381	0.146
4	0.367	0.446	0.473	0.480	0.571	0.384	0.258	0.178	0.396	0.173
5	0.365	0.439	0.475	0.488	0.576	0.381	0.260	0.171	0.387	0.147

Table 4. Prediction accuracy by using the model (24) with the state vector (25) ($M=35, K=25$)

The second is

$$x_t \equiv \left(\nabla WH_t, \nabla WS_t, \nabla \cos\left(WD_t\right), \nabla \sin\left(WD_t\right) \right)' \tag{29}$$

And the third is (25), the proposed method.

Tables 2, 3 and 4 show MAE's and COR's in the above three cases, respectively. It is noted that each experiment was carried out under the condition that the order p was fixed in the range from 1 to 5. Overall, the result by using (29) tends to give smaller MAE's than the one by using (28). It suggests the possibility that taking into account the angular-linear structure is effective for improving the prediction accuracies by the predictor based on (28). The result of COR also shows the similar tendency. It is noted that, as the order p and the prediction step L are larger, the prediction accuracy based on (29) becomes worse to take negative correlations. The prediction based on the model with the state vector (25) tends to give the best prediction accuracy among the three models. This suggests that the principal component structure of (25) worked effectively, which contributed to the improvement of the prediction accuracy.

4. Predicting the change of wave height from surface wind

In the previous section, we developed a statistical model for explaining the dynamic relationship between ocean wind and wave height. Now we consider the prediction problem on wave height based on the motion of the surface wind, observed by ground-based observation. For this purpose, we develop a new method by applying the model presented in the previous section. Also, in order to evaluate the availability of the developed model, we compare the prediction accuracies between the proposed model and traditional time series models.

Fig. 7. Locations of the sensor for ultrasonic wave height meter (inverted triangle) and major AMeDAS stations around the sensor (black circles)

4.1 Ground-based observation on surface wind and measurement of wave height

In Japan, as described in Introduction, many stations of AMeDAS and the sensors for ultrasonic wave height meters have been located in various regions and coastal areas of this country by the Japan Meteorological Agency. In the following, we consider a case study on prediction of the wave height in Matsumae-oki, the sea area in the southwest of Hokkaido. Figure 7 shows a map of the locations of the sensor of a wave height meter in Matsumae-oki (42°24'38"N, 140°05'50"E) and major AMeDAS stations located around the sensor. The monitoring of the changes of wind speed and wind direction, and the measurement of wave height are carrying out constantly, and the measured data are available via Internet. For the following analysis, we obtained the dataset of wave height measured in Matsumae-oki and the datasets of wind speed and wind direction monitored at the AMeDAS station in Matsumae-cho, which is located roughly 5 km away from the measuring point of wave height. Figure 8 displays the changes in the significant wave height (m) measured in Matsumae-oki, and wind speed (m/s) and wind direction (rad.) monitored in Matsumae-cho, which were measured every hour on the hour. They are the records for every four seasons in the period from April 2010 to February 2011. As the datasets for four seasons, we obtained the measured data in the period from Apr. 1 to May 31 for spring, Jul. 1 to Aug. 31 for summer, Oct. 1 to Nov. 31 for autumn and Jan. 1 to Feb. 28 of 2011 for winter. The measured data on wind speed and wind direction are provided as the mean value over the past 10 minutes, and the measured data on 1/3 significant wave height is calculated based the sea surface data over

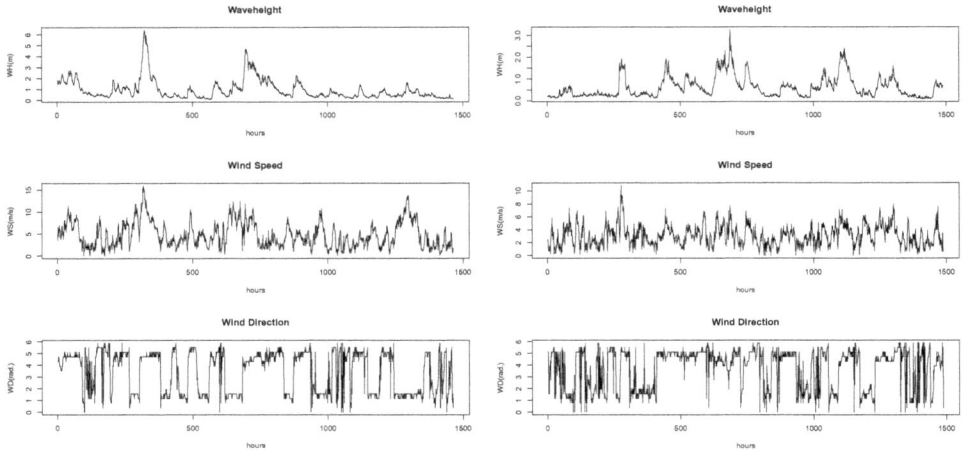

(a) Spring (Apr. 1- May 31)

(b) Summer (Jul. 1- Aug. 31)

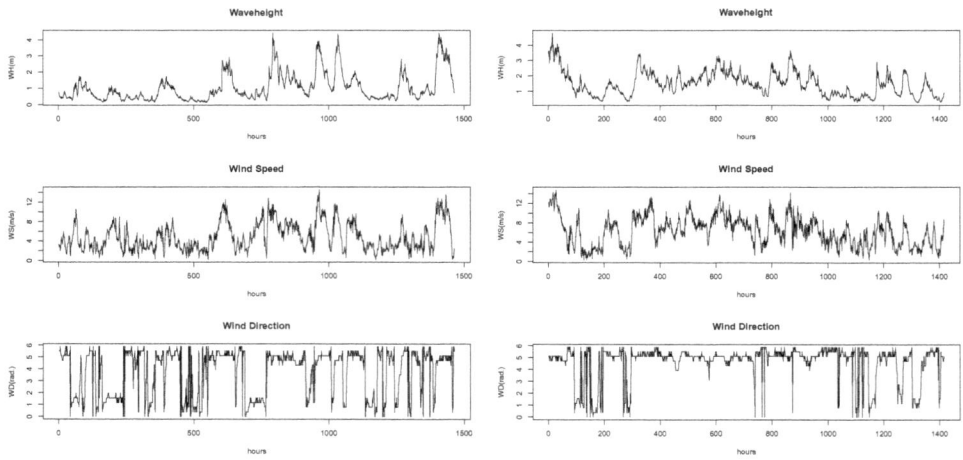

(c) Autumn (Oct. 1- Nov. 31)

(d) Winter (Jan. 1- Feb. 28)

Fig. 8. The changes of 1/3 significant wave height (m) (top), wind speed (m/s) (middle) and wind direction (rad.) (bottom) for four seasons (Apr. 2010 - Feb. 2011)

the past 25 minutes. It is noted that the origin of the wind direction data is defined to be north and the positive value means the clockwise direction.

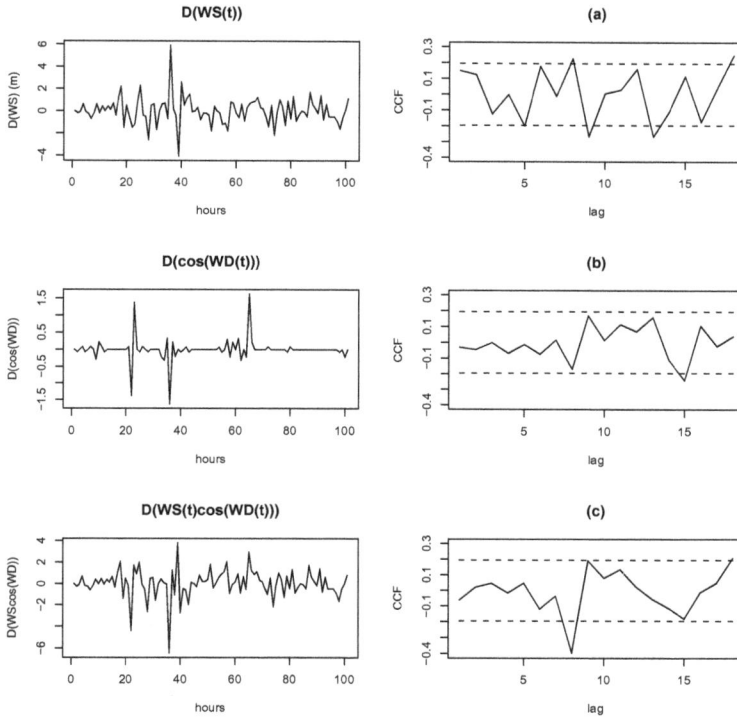

Fig. 9. Time series plots of $\{\nabla WS_t^*\}$, $\{\nabla \cos(WD^*_t)\}$, $\{\nabla(WS_t^* \cos(WD_t^*))\}$ (left column) and the cross correlation functions in the cases (a)-(c) (right column)

4.2 Cross correlation among the measured data

First, we consider how we treat the measured data of the wind direction. In the following, let $\{WS_t^*\}$ and $\{WD_t^*\}$ be the measured time series data on wind speed and wind direction of the surface wind.

Figure 9 shows the time series plots of $\{\nabla WS_t^*\}$, $\{\nabla \cos(WD_t^*)\}$ and $\{\nabla(WS_t^* \cos(WD_t^*))\}$, and the cross correlation functions in the 3 cases, (a) $\{\nabla WS_t^*\}$ and $\{\nabla WH_t\}$, (b) $\{\nabla \cos(WD_t^*)\}$ and $\{\nabla WH_t\}$, and (c) $\{\nabla(WS_t^* \cos(WD_t^*))\}$ and $\{\nabla WH_t\}$, which were estimated by (10), where the dotted lines mean the Bartlett's bounds. We observe that the case (c) gives larger cross correlation than the cases of (a) and (b).

4.3 Modeling the change of wave height by taking into account the change of surface wind

We consider a statistical model to express the change in wave height based on the change in surface wind, monitored at an AMeDAS station. Following the result in the previous section, we build a nonstationary time series model focusing on the change of $\nabla(WS_t^* \cos(WD_t^*))$. We write ∇WH_t as

$$\nabla WH_t = \sum_{i=1}^{p} \alpha_i \nabla WH_{t-i} + \sum_{i=1}^{p} \sum_{k=1}^{K} \beta_{i,k} \nabla (WS^*_{t-i} \cos(kWD^*_{t-i}))$$

$$+ \sum_{i=1}^{p} \sum_{k=1}^{K} \gamma_{i,k} \nabla (WS^*_{t-i} \sin(kWD^*_{t-i})) + \varepsilon_{1,t}, \quad \varepsilon_{1,t} \sim WN(0, \sigma^2_{WH}) \tag{30}$$

where p and K are orders, (α, β, γ) are unknown coefficients and $\varepsilon_{1,t}$ is the random variable which follows a white noise process with $E(\varepsilon_{1,t})=0$ and $V(\varepsilon_{1,t}) = \sigma^2_{WH}$. Similarly, we write

$$\nabla (WS^*_t \sin(hWD^*_t)) = \sum_{i=1}^{p} \alpha_i^{(h)} \nabla WH_{t-i} + \sum_{i=1}^{p} \sum_{k=1}^{K} \beta_{i,k}^{(h)} \nabla (WS^*_{t-i} \cos(kWD^*_{t-i}))$$

$$+ \sum_{i=1}^{p} \sum_{k=1}^{K} \gamma_{i,k}^{(h)} \nabla (WS^*_{t-i} \sin(kWD^*_{t-i})) + \varepsilon_{2,t}^{(h)} \tag{31}$$

$$\nabla (WS^*_t \cos(hWD^*_t)) = \sum_{i=1}^{p} \alpha_i^{(h)} \nabla WH_{t-i} + \sum_{i=1}^{p} \sum_{k=1}^{K} \beta_{i,k}^{(h)} \nabla (WS^*_{t-i} \cos(kWD^*_{t-i}))$$

$$+ \sum_{i=1}^{p} \sum_{k=1}^{K} \gamma_{i,k}^{(h)} \nabla (WS^*_{t-i} \sin(kWD^*_{t-i})) + \varepsilon_{3,t}^{(h)} \tag{32}$$

for $h = 1, \ldots, K$, where $\varepsilon_{2,t}^{(h)} \sim WN(0, \sigma^2_{2,h})$ and $\varepsilon_{3,t}^{(h)} \sim WN(0, \sigma^2_{3,h})$.
Put the state vector at time point t by the $(2K+1)$ dimensional vector

$$\boldsymbol{y}_t^{(K)} \equiv (\nabla WH_t, \nabla WC_1, \nabla WS_1, \ldots, \nabla WC_K, \nabla WS_K)' \tag{33}$$

where $WC_h = WS^*_t \cos(hWD^*_t)$ and $WS_h = WS^*_t \sin(hWD^*_t)$ $(h = 1, \ldots, K)$. Then the above models can be rewritten by a multivariate AR model,

$$\boldsymbol{y}_t^{(K)} = A_1^{(K)} \boldsymbol{y}_{t-1}^{(K)} + \cdots + A_p^{(K)} \boldsymbol{y}_{t-p}^{(K)} + \boldsymbol{\delta}_t^{(K)}, \qquad \boldsymbol{\delta}_t^{(K)} \sim WN(\boldsymbol{0}, \Sigma^{(K)}) \tag{34}$$

where $A_i^{(K)}$ $(i = 1, \ldots, p)$ are unknown coefficient matrices and $\boldsymbol{\delta}_t^{(K)}$ follows the multivariate white noise process with mean $\boldsymbol{0}$ and the dispersion matrix $\Sigma^{(K)}$. An l-step ahead predictor can be constructed by

$$\hat{\boldsymbol{y}}_{N+l}^{(K)} = \hat{A}_1^{(K)} \boldsymbol{z}_{N+l-1}^{(K)} + \hat{A}_2^{(K)} \boldsymbol{z}_{N+l-2}^{(K)} + \cdots + \hat{A}_p^{(K)} \boldsymbol{z}_{N+l-p}^{(K)},$$

$$\boldsymbol{z}_{N+l-m}^{(K)} = \boldsymbol{y}_{N+l-p}^{(K)} (l \leq p), \qquad \hat{\boldsymbol{y}}_{N+l-p}^{(K)} (l > p) \tag{35}$$

where $\hat{A}_i^{(K)}$ is the least squares estimator of $A_i^{(K)}$. Thus the l-step ahead predicted values, WH_{N+l} $(l = 1, \ldots, L)$, can be obtained by the predictor $\hat{\boldsymbol{y}}_{N+l}^{(K)}$.

Model	MAE					COR				
	$L=1$	$L=2$	$L=3$	$L=4$	$L=5$	$L=1$	$L=2$	$L=3$	$L=4$	$L=5$
(i)	0.114	0.167	0.222	0.242	0.267	0.982	0.955	0.927	0.886	0.862
(ii)	0.096	0.142	0.176	0.219	0.246	0.985	0.962	0.946	0.914	0.903
(iii)	0.103	0.143	0.178	0.215	0.239	0.984	0.960	0.944	0.912	0.903
(iv)	0.097	0.143	0.178	0.217	0.241	0.985	0.961	0.945	0.913	0.902
(v)	0.098	0.138	0.176	0.213	0.235	0.986	0.965	0.949	0.919	0.907

Table 5. MAE's and COR's based on spring data

4.4 Evaluation of the prediction accuracy

In the following, we evaluate the effectiveness of the proposed method by means of the prediction experiment which is similar to the one given in the subsection 3.4.

The procedure for the experiment is as follows. We select the time point to start prediction randomly in the range of the dataset. And then fit the proposed model to the measured time series data for 100 hours (i.e., sample size is 100), and obtain the predicted values up to 5 steps ahead (1 step corresponds to 1 hour). After repeating the procedures, we evaluate the prediction accuracy by MAE and COR. For evaluation of the prediction accuracy, we also obtain the predicted values when we used traditional time series models. The models introduced for comparison are defined as follows;

(i) $WH_t = \sum_{i=1}^{p} \alpha_i WH_{t-i} + \delta_{1,t}, \quad \delta_{1,t} \sim WN(0,\sigma_1^2)$

(ii) $\nabla WH_t = \sum_{i=1}^{p} \beta_i \nabla WH_{t-i} + \delta_{2,t}, \quad \delta_{2,t} \sim WN(0,\sigma_2^2)$

(iii) $y_t = A_1 y_{t-1} + \cdots + A_p y_{t-p} + \delta_{3,t}, \quad \delta_t \sim WN(0,\Sigma_{3,t}), \quad y_t = (\nabla WH_t, \nabla WS_t^*)'$

(iv) $y_t = B_1 y_{t-1} + \cdots + B_p y_{t-p} + \delta_{4,t}, \quad \delta_t \sim WN(0,\Sigma_{4,t}), \quad y_t = (\nabla WH_t, \nabla WS_t^* \cdot \nabla \cos(WD_t^*))'$

(v) $y_t = C_1 y_{t-1} + \cdots + C_p y_{t-p} + \delta_{5,t}, \quad \delta_t \sim WN(0,\Sigma_{5,t}), \quad y_t = (\nabla WH_t, \nabla(WS_t^* \cos(WD_t^*)))'$

where $\{\alpha_i, \beta_i, A_i, B_i, C_i\}$ are unknown parameters. (i) and (ii) are univariate time series models based on wave height. The former is a stationary AR(p) model and the latter is a nonstationary ARIMA(p,1,0) model. (iii) is a multivariate AR model taking into account the wind speed as a covariate, and (iv) and (v) are multivariate AR models taking into account wind speed and wind direction as covariates. It is noted that if the changes of wind speed and wind direction are dependent, the prediction accuracy of (v) becomes better than that of (iv). Table 5 shows MAE's and COR's of the above five models, based on the measured data in spring. The number of repetitions is 130. It is noted that we selected the order of the model by Akaike Information Criterion (AIC). By the comparison between (i) and (ii), we confirm that the nonstationary ARIMA model gives better prediction performance than the stationary AR model. Also, the comparisons between (ii) and (iii), (ii) and (iv), and (ii) and (v) show the tendency that the model taking into account the change of wind motion as covariate improves the prediction accuracy when we used the univariate time series model on wave height. Furthermore, the comparison between (iv) and (v) shows the tendency that the prediction accuracy by using (v) becomes better, which suggests that there exists the dependency between wind speed and wind direction.

(A) Spring

Model	MAE					COR				
	$L=1$	$L=2$	$L=3$	$L=4$	$L=5$	$L=1$	$L=2$	$L=3$	$L=4$	$L=5$
(i)	0.114	0.167	0.222	0.242	0.267	0.982	0.955	0.927	0.886	0.862
(ii)	0.096	0.142	0.176	0.219	0.246	0.985	0.962	0.946	0.914	0.903
(iii)	0.103	0.143	0.178	0.215	0.239	0.984	0.960	0.944	0.912	0.903
(iv)	0.097	0.143	0.178	0.217	0.241	0.985	0.961	0.945	0.913	0.902
(v)	0.098	0.138	0.176	0.213	0.235	0.986	0.965	0.949	0.919	0.907

(B) Summer

Model	MAE					COR				
	$L=1$	$L=2$	$L=3$	$L=4$	$L=5$	$L=1$	$L=2$	$L=3$	$L=4$	$L=5$
(i)	0.077	0.089	0.120	0.153	0.191	0.970	0.949	0.930	0.895	0.852
(ii)	0.074	0.085	0.106	0.136	0.155	0.980	0.974	0.959	0.929	0.896
(iii)	0.073	0.085	0.108	0.137	0.155	0.980	0.976	0.957	0.927	0.893
(iv)	0.072	0.081	0.107	0.136	0.156	0.981	0.977	0.958	0.927	0.892
(v)	0.072	0.083	0.105	0.132	0.149	0.981	0.975	0.959	0.930	0.897

(C) Autumn

Model	MAE					COR				
	$L=1$	$L=2$	$L=3$	$L=4$	$L=5$	$L=1$	$L=2$	$L=3$	$L=4$	$L=5$
(i)	0.094	0.169	0.235	0.293	0.327	0.984	0.937	0.903	0.837	0.789
(ii)	0.086	0.145	0.187	0.236	0.268	0.989	0.962	0.938	0.898	0.872
(iii)	0.083	0.144	0.188	0.231	0.259	0.990	0.964	0.940	0.901	0.875
(iv)	0.088	0.149	0.191	0.238	0.264	0.989	0.962	0.939	0.900	0.873
(v)	0.085	0.146	0.188	0.233	0.259	0.989	0.963	0.940	0.899	0.874

(D) Winter

Model	MAE					COR				
	$L=1$	$L=2$	$L=3$	$L=4$	$L=5$	$L=1$	$L=2$	$L=3$	$L=4$	$L=5$
(i)	0.107	0.168	0.234	0.266	0.301	0.979	0.948	0.905	0.873	0.828
(ii)	0.103	0.163	0.208	0.245	0.273	0.979	0.951	0.916	0.891	0.854
(iii)	0.103	0.156	0.202	0.243	0.271	0.979	0.952	0.916	0.890	0.852
(iv)	0.106	0.163	0.208	0.245	0.273	0.976	0.947	0.909	0.886	0.848
(v)	0.100	0.156	0.203	0.243	0.272	0.980	0.952	0.917	0.889	0.851

Table 6. Comparisons of MAE and COR for every season (Apr. 2010 - Feb. 2011)

4.5 Robustness on the predictability of wave height for every season

It is also of interest whether or not the effectiveness in prediction using the developed model is robust throughout a year. In Japan, there exists unique characteristics on the pressure pattern for every season. Therefore, it is necessary to investigate whether the model has the ability to improve the prediction accuracies by traditional models, for all seasons of a year.

Table 6 shows MAE's and COR's obtained by using the measured time series data for four seasons in the period from April 2010 to February 2011. Overall, the result has the tendency that the proposed model (v) has the ability to give the best prediction accuracy among the five models, although the degree on improvement of the accuracy are different for every season.

5. Conclusion

Our goal in this chapter is the development of a statistical approach for predicting the change of wave height, based on the measured data of the surface wind obtained by ground-based observation. In section 3, we presented a method for predicting the change of wave height based on ocean wind, which was proposed by Hokimoto and Shimizu (2008). And in section 4, we developed a model for predicting the wave height from the change of surface wind, by applying the model given in the previous section. The evaluation on the prediction accuracy suggested the possibility that the method proposed in section 4 improves the prediction accuracies by using the predictors based on traditional time series models. As described at the beginning, the physical factors which impacts on the change in the sea state will change with correlations on space and time. At the present, the models presented in this chapter do not have spatial structure. For example, the development of the model, taking into account the directional change of the wind direction observed at multiple AMeDAS stations, will be available for deeper understandings on the dynamic interaction between the motions of wind and wave.

6. References

Athanassoulis, G.A., Stefanakos, C.N. (1995). A nonstationary stochastic model for long-term time series of significant wave height, *Journal of Geophysical Research*, 100(C8), 16149-16162.

Bollerslev, T.(1986). Generalized Autoregressive Conditional Heteroskedasticity, *Journal of Econometrics*, 31, 307-327.

Box, G.E.P., Jenkins, G.M. (1976). *Time Series Analysis, Forecasting and Control* (revised edition), Holden-Day, San Francisco.

Brockwell, P.J., Davis, R.A. (1996). *Introduction to Time Series and Forecasting*, Springer-Verlag, New York.

Brown, B.G., Katz, R.W., Murphy A.H. (1984). Time series models to simulate and forecast wind speed and wind power, *Journal of Climate and Applied Meteorology*, 23, 1184-1195.

Cunha C, Guedes S.C. (1999). On the choice of data transformation for modelling time series of significant wave height, *Ocean Eng*, 26, 489-506.

Daniel, A.R., Chen, A.A. (1991). Stochastic simulation and forecasting of hourly average wind speed sequences in Jamaica, *Sol Energy*, 46(1), 1-11.

Fisher, N.I. (1993). *Statistical Analysis of Circular Data*, Cambridge University Press, Cambridge.

Fisher, N.I., Lee, A. J. (1992). Regression models for an angular response, *Biometrics*, 48, 665-677.

Johnson, R.A., Wehrly, T.E. (1978). Some Angular-Linear Distributions and Related Regression Models, *Journal of the American Statistical Association*, 73, 602-606.

Hokimoto, T., Kimura, N., Iwamori, T., Amagai, K., Huzii, M. (2003). The effects of wind forcing on the dynamic spectrum in wave development: A statistical approach using a parametric model, *Journal of Geophysical Research*, 108(C10), 5-1-5-12.

Hokimoto, T., Shimizu, K. (2008). An angular-linear time series model for wave height prediction, *Ann Inst Stat Math*, 60, 781-800.

Kitagawa, G., Gersch, W. (1985). A Smoothness Priors Time-Varying AR Coefficient Modeling of Nonstationary Covariance Time Series, *IEEE Transactions on Automatic Control*, 30, 48-56.

Mardia, K.V., Jupp, P.E. (2000). *Directional Statistics*, John Wiley, New York.

O' Carroll (1984). Weather modelling for offshore operations, *The Statistician*, 33, 161-169.

Pierson, W.J., Neumann, G, and James, R.W. (1960). Practical Methods for Observing and Forecasting Ocean Waves By Means of Wave Spectra and Statistics, *U.S. Navy Hydrographic Office*, Reprint edition.

SenGupta, A. (2004). On the constructions of probability distributions for directional data, *Bulletin of the Calcutta Mathematical Society*, 96(2), 139-154.

SenGupta, A., Ugwuowo, F.I. (2006). Asymmetric circular-linear mutilvariate regression models with application to environmental data, *Environmental and Ecological Statistics*, 13, 299-309.

Stefanakos, C.N., Athanassoulis, G.A., Barstow, S.F. (2002). Mutivariate time series modelling of significant wave height, *Proceedings of International Society of Offshore and Polar Engineers Conference*, III, 66-73.

Sverdrup, H.U., Munk, W.H. (1947). Wind sea and swell: Theory of relation for forecasting, *U.S. Navy Hydrographic Office, Washington, D.C.*, No.601.

Walton, T. L., Borgman, L.E. (1990). Simulation of non-stationary, non-gaussian water levels on the great lakes, *Journal of the ASCE*, Waterway Port, Coastal, and Ocean Engineering Division, 116(6), 664-685.

Yim, J.Z., Chou, C., Ho, P. (2002). A study on simulating the time series of siginificant wave height near the keelung harbor, *Proceedings of International Society of Offshore and Polar Engineers Conference*, III, 92-96.

8

Variability of Internal Solitary Waves in the Northwest South China Sea

Zhenhua Xu and Baoshu Yin
Institute of Oceanology, Chinese Academy of Sciences,
Key Laboratory of Ocean Circulation and Waves (KLOCAW),
Chinese Academy of Sciences,
China

1. Introduction

Internal waves are waves that travel within the interior of the water column. Its existence is owned to the stratified density structure between two or continuous layers of fluids (Apel et al., 1987). Internal solitary waves (ISWs) are nonlinear internal waves, which are frequently observed all over the world oceans, where strong tides and stratification occur over varying topography features (Apel et al., 1985; Colosi et al., 2001; Osborne & Burch, 1980). They typically occur in packets at tidal intervals, suggesting that they mainly originate from the tide-topography interactions over variable topography (Gerkema et al., 1995). Depending on the different environmental conditions, there are two main mechanisms for the generation of ISWs: lee-wave mechanism (Maxworthy, 1979) and nonlinear internal tide mechanism (Lee & Beardsley, 1974).

The lee-wave mechanism states that the lee-wave is formed by the ebb tide and released when the tide changes from ebb tide to flood tide, and evolves into a rank-ordered internal solitary wave (ISW) packet. By the nonlinear internal tide mechanism, internal tides spawn ISWs in three steps: initial generation of a front due to topographic blocking, nonlinear steepening of the front, and formation of a rank-order ISW packet under effects of nonlinearity and dispersion (Helfrich & Melville, 2006; Zhao & Alford, 2006).

Internal solitary waves are important for many practical reasons. As they are commonly observed wherever strong tides and stratification occur next to the irregular topography, thus they are often prominent features seen in optical and radar satellite imagery of coastal waters. They can propagate over several hundred kilometers and transport both mass and momentum. They can also induce considerable velocity shears that can impose unexpectedly large stresses on offshore oil-drilling rigs and lead to turbulence and mixing. In addition, the mixing often introduces bottom nutrients into the water column, thereby fertilizing the local region and modifying the biology system therein (Jackson, 2004).

ISWs in the South China Sea (SCS) have been observed at a variety of locations from Luzon Strait to the continental shelf (Cai et al., 2002; Lien et al., 2005; Liu et al., 2004; Ramp et al., 2004; Xu et al., 2010; Zhao et al., 2004). Until recently, considerable effort has been focused on the study of ISWs in the northeastern SCS (Farmer et al., 2009; Moore et al., 2007; Ramp et al., 2004). In contrast, Due to the shortage of high-quality data sets, studies of the ISWs in the northwestern SCS are quite limited and nonlinear internal waves in this region have been observed mainly by satellite remote sensing (Liu et al., 1998; Jackson, 2004; Li et al, 2008). Thus few attempts have been done on the statistical analysis of the ISWs and detailed study of the features of ISWs in this area based on long-term observations is rarely reported. In order to investigate the typical characteristics of ISWs in Wenchang area of the northwestern SCS, the Wenchang Internal Wave Experiment (WCIWE) was conducted in 2005. A complete set of current and temperature data of mooring measurements were acquired and numerous ISWs were observed in Wenchang area during this experiment. Based on the newly-acquired observation data, a preliminary study of the ISWs has been undertaken in this area (Xu et al., 2010). This chapter intends to present a summary of the interesting observations of the ISWs in the northwestern region, investigate the statistical characteristics of the ISWs, make a theoretical analyse using the KdV models and examine the possible effects of the ISWs on the platforms and marine biological system in the SCS.

2. Theoretical description of internal solitary waves

The earliest recognition of internal solitary wave phenomena was reported by Scott Russell (1838, 1844) in the 19th century. Later Korteweg and devries (1895) derived some of the interesting mathematical properties of such wave and produced the now-famous analytical soliton solutions. Despite the fact that the oceanic observations show the presence of mode-one internal waves that are often highly nonlinear, weakly nonlinear KdV-type theories have played the primary role in elucidating the essential features of the observations, if not always the precise quantitative details. They have the advantage of permitting modeling of unsteady wave evolution under various conditions with reduced wave equations (Helfrich and Melville, 2006; Koop and Butler, 1981; Small and Hornby, 2005).

The Korteweg and devries (KdV) equation arises from an assumption that nonlinearity, scaled by $\alpha = a/H$, and nonhydrostatic dispersion, $\beta = (l/H)^2$, are comparable and small: $\beta = O(\alpha) \ll 1$. Here a is a measure of the wave amplitude, H is an intrinsic vertical scale, and l is a measure of the wavelength (Holloway et al , 1997; Liu et al., 1998).

$$\frac{\partial \eta}{\partial t} + c\frac{\partial \eta}{\partial x} + \alpha\eta\frac{\partial \eta}{\partial x} + \beta\frac{\partial^3 \eta}{\partial x^3} = 0 \tag{1}$$

For a two-layer system with a rigid lid and no mean flow, in the Boussinesq approximation,

$$\alpha = \frac{3}{2}\frac{H_1 - H_2}{H_1 H_2}C_0 \tag{2}$$

$$\beta = \frac{1}{6}H_1 H_2 C_0 \tag{3}$$

$$C_0 = \left[\frac{\Delta\rho g H_1 H_2}{\rho(H_1 + H_2)}\right]^{\frac{1}{2}} \tag{4}$$

Here g is the gravitational acceleration, $\Delta\rho = \rho_2 - \rho_1$ is the layer density difference, ρ_1 (ρ_2) is the density of the upper (lower) layer, and H_1 and H_2 are the mean upper and lower layer depths, respectively. This equation has the solution:

$$\eta = \eta_0 \operatorname{sech}^2\left(\frac{x - Ct}{\lambda}\right) \tag{5}$$

The nonlinear velocity C and the characteristic width λ of the soliton being related to the linear speed C_0 and the amplitude of the displacement η_0 by

$$C = C_0 + \frac{\alpha \eta_0}{3}, \quad \lambda^2 = \frac{12\beta}{\alpha \eta_0}. \tag{6}$$

When the interface is close to the middle of the water layer, $H_1 \approx H_2$, the nonlinear coefficient α is small or even equal to zero. In this case, high-order nonlinear coefficient must be included in the KdV equation, which results into the extended Korteweg-de Vries (eKdV) equation.

$$\frac{\partial \eta}{\partial t} + (c + \alpha \eta + \alpha_1 \eta^2)\frac{\partial \eta}{\partial x} + \beta \frac{\partial^3 \eta}{\partial x^3} = 0 \tag{7}$$

Where for the case of two-layer fluid the second nonlinear coefficient is

$$\alpha_1 = \frac{3c}{h_1^2 h_2^2}\left[\frac{7}{8}\left(\frac{\rho_2 h_1^2 - \rho_1 h_2^2}{\rho_2 h_1 - \rho_1 h_2}\right) - \frac{\rho_2 h_1^3 - \rho_1 h_2^3}{\rho_2 h_1 + \rho_1 h_2}\right] \approx -\frac{3}{8}\frac{(h_1 + h_2)^2 + 4h_1 h_2}{(H_1 H_2)^2} C_0 \tag{8}$$

With $\alpha_1 = 0$ this reduces to the well-known KdV equation. This equation has a solitary wave solution of the form

$$\eta(x, t) = \eta_0 \frac{\operatorname{sech}^2[\kappa(x - C_m t)]}{1 - \mu \tanh^2[\kappa(x - C_m t)]} \tag{9}$$

Where t is time, and x is the spatial variable in the direction of wave propagation. η_0 is the maximum interface elevation at x = 0, and κ and μ are parameters.
The characteristic wave-width λ predicted by the EKdV model is related to the mass of the wave by computing

$$\lambda = \frac{1}{2\eta_0} \int_{-\infty}^{\infty} \eta(x) dx \tag{10}$$

As follows from Eq. 8, within the framework of the two-layer model, α_1 is always negative. However, in the general case the coefficient α may be either negative or positive. In the latter case, solitons of both positive and negative polarities may exist. In addition, nonstationary solitons, called breathers, are also possible. The evolution of initial pulse-type perturbations may be fairly complex. If the pycnocline is located just at the critical level so that the parameter is exactly zero, Eq.7 reduces to the well-known modified Korteweg–de Vries (mKdV) equation.

$$\frac{\partial \eta}{\partial t} + (c + \alpha_1 \eta^2)\frac{\partial \eta}{\partial x} + \beta \frac{\partial^3 \eta}{\partial x^3} = 0 \tag{11}$$

The mKdV equation also has soliton-type solutions, but only those propagating on a constant nonzero pedestal (Apel et al., 2006).

3. Internal solitary waves in the northeast SCS

Based on their different characteristics, Ramp et al. (2004) denoted the two types of ISW packets on the slope of the northeastern SCS as a-wave and b-wave, respectively. The a-wave arrives with remarkable regularity at the same time each day, 24 hours apart; the b-wave arrives one hour later each day. The ISWs in the a-wave packets have greater amplitude than those in the b-wave packets. For the a-wave, the largest ISW is always in the lead with smaller ISWs behind, sometimes rank-ordered and sometimes not. The b-wave generally consists of a single ISW growing out of the center of the packet (Klymak et al., 2006; Zhao & Alford, 2006).

ISWs in this area appear to originate near the Luzon Strait and propagate westward across the SCS basin until they encounter the shelf near China's coast (Zhao & Alford, 2006). Liu et al. (1998) have studied the elevation ISWs east of Hainan Island by SAR images on the basis of the assumption of a semidiurnal tidal origin. They inferred that the wave packets were transformed from the depression ISWs generated near the Luzon Strait.

Large amplitude ISWs in the northeastern South China Sea (SCS) have been frequently reported since the Asian Seas International Acoustics Experiment. Hence, there has been an increasing interest in investigating the highly nonlinear effects on the ISWs in different models (Ramp et al., 2004; Farmer et al., 2009). As pointed by Ramp et al. (2004), simulating large amplitude ISWs in the northeastern SCS is limited in the EKdV model, but KdV theory is more suitable for the simulation of the solitons in this area.

4. Internal solitary waves in the northwest SCS

4.1 Observational methods

Field observations were made from late Spring to early Autumn 2005 at Wenchang Station (112°E, 19°35'N) on the northwestern shelf of the SCS. The water depth at the station is 117 m.

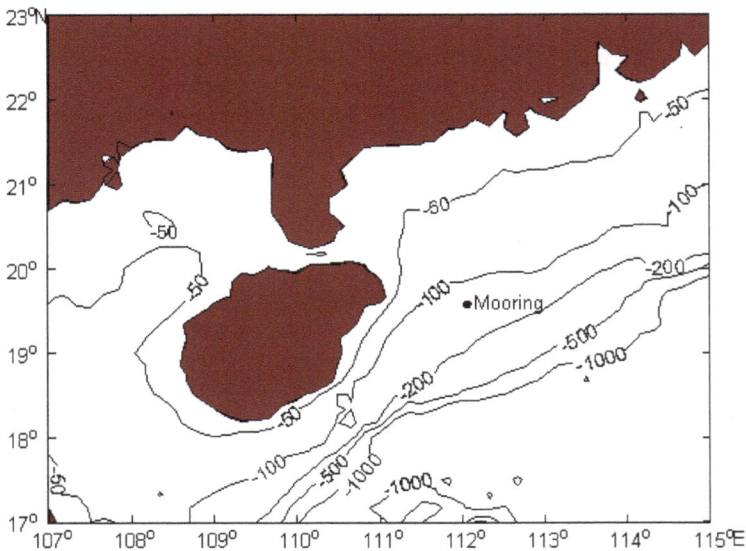

Fig. 1. Map of the study area with isobaths showing the seafloor topography Contours mark isobaths in meters. Symbol * indicates the mooring position.

The study area and mooring position are indicated in Fig.1. An array of temperature and salinity sensors, and an acoustic Doppler current profiler (ADCP) were deployed at Wenchang Station to examine the thermal and hydrodynamic structure on the Wenchang shelf. A 190 kHz down-looking ADCP was positioned at a depth of 8 m. The depth of the available current data measured by ADCP ranged from 10 to 114 m, with a vertical interval of 2 m. Current measurements were recorded with a precision of $1\times10\text{-}4$ m/s at a time interval of 10 min. The temperature sensor information with a precision of 0.01°C and a time interval of 1 min were collected at 23 layers. Most of the temperature sensors were placed between 4 and 40 m below the sea surface with a vertical separation of less than 4 m, whereas the bottom two sensors were located at depths of 50 and 75 m. Salinity measurements taken at a interval of 1 min were acquired at five layers by sensors placed at 8, 20, 30, 40, and 75 m (Xu et al., 2011).

4.2 Characteristics of the ISWs
4.2.1 General description of the ISW packets
In order to investigate the properties of the ISWs, the temperature data were converted to vertical displacements of isotherms using linear interpolation method. The semidiurnal tide dominated at the beginning of the observed record while the diurnal tide gradually became more pronounced on Sep 9-13, followed by semidiurnal tide again on Sep 14 and 15 (Figure 2a). In general, the ISWs occurred during the entire observation period, while the largest waves were observed once every diurnal tidal cycle from Sep 9 to 13 and the smaller waves were found to occur irregularly at roughly semidiurnal period. The dominate diurnal ISWs in our study area, with the possible local generation mechanism,are quite different from the principal solitons observed every semidiurnal tidal cycle in the northeastern SCS (Ramp et al., 2004).

Figure 2b shows three abrupt deepening of the upper stratified layer followed by developed packets of high-frequency ISWs. The regular arrival of internal wave packets riding on these depression bores of strong tides at roughly the diurnal tidal period strongly suggests that the main energy source of the waves is the diurnal tide. The internal tides on 10 and 11 September are obviously stronger than that of 9 September.

The seasonal thermocline, generally between 20 m and 50 m depth during this period, was deepened more than twenty meters by the strong internal tides and ISWs (Figure 2c). The strongest stratification of the water column appears around 05:00 on 10 September when the temperature gradient reaches as large as 0.55 ℃/m. The most noticeable difference is that the thermocline on 9 September is relatively weaker than that of 10 September through the whole observed water column. Therefore, the ISWs display distinct characteristics during these two days, which will be discussed in the next section.

4.2.2 Depression internal solitary waves
A closer inspection of the time series of temperature data for 9 September is shown to get a better view of the internal wave structure. Figure 3a shows an internal tide steepening trailed by a packet of ISWs. The first soliton emerged on the front face of the internal bore. The spike-like fluctuation of temperature first increased and then decreased. This indicates that the solitons are depression waves. The temperature fluctuates in phase vertically for the entire water column, which suggests that these solitons all behave as the first mode depression waves. The largest soliton appears around 12:15 on 9 September with the amplitude and period of order 25 m and 20 min, respectively (Figure 3b). There are several small solitons appearing before and after the main soliton.

Before the ISWs arrived, the average temperature in the upper mixed layer was about 28.5 ℃ and the mixed layer thickness was roughly 25 m in the water of 117 m depth (Figure 3c). Below the upper mixed layer, there was a strong thermocline with the temperature gradient of 0.35 ℃/m. When the main soliton arrived around 12:15 on 9 September, the temperature structure changed dramatically with the mixed layer deepened to 50 m depth. This result again confirms our previous description that the amplitude of the main soliton is approximately 25 m.

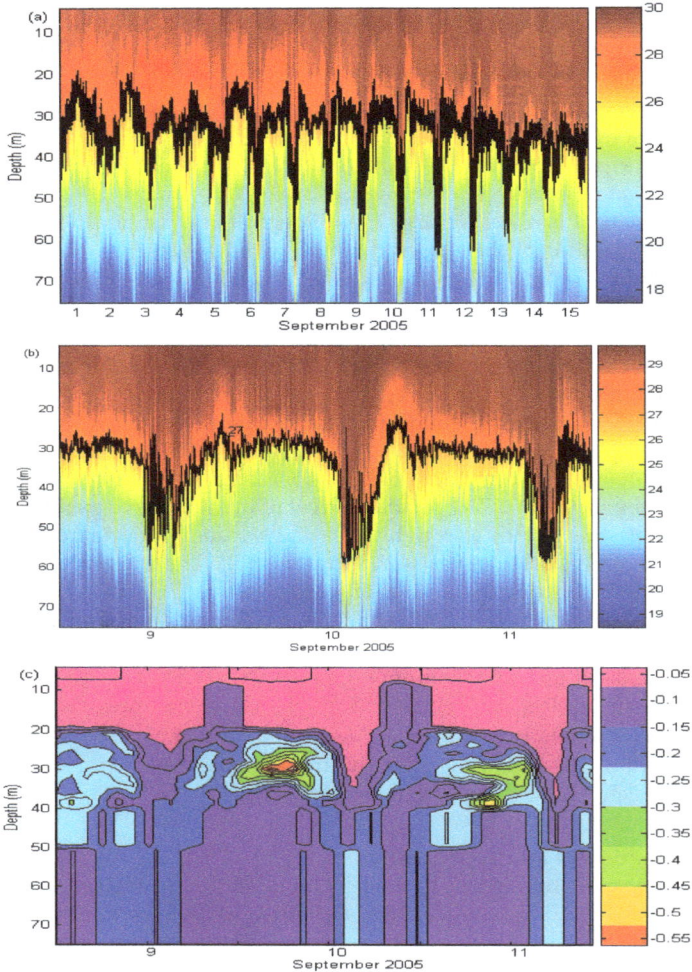

Fig. 2. (a) 15-day time series of isotherm depths from the temperature data from September 1 to September 15, 2005. The isotherm of 27 ℃ is noted, (b) 3-day time series of isotherm depths from the temperature data from September 9 to September 11, 2005. The isotherm of 27 ℃ is noted, (c) 3-day time series of temperature gradient depths from September 9 to September 11, 2005. After Xu et al. (2010).

The largest soliton arriving around 12:15 on 9 September was classic mode-1 wave, with opposing velocities in the upper and lower layers (Figure 3d). The point of the zero crossing of the velocity was close to 55 m, which was associated with the strong thermocline during the arrival of the main soliton indicated by figure 3c. The direction of the horizontal velocity changed with depth, roughly northwestward in the upper layer and southeastward in the lower layer. According to the KDV theory of the ISW, the direction of the depression ISW is the same as that of the currents in the upper layer water column. These results suggest that this soliton behaves as the first mode depression wave, propagating nearly northwestward.

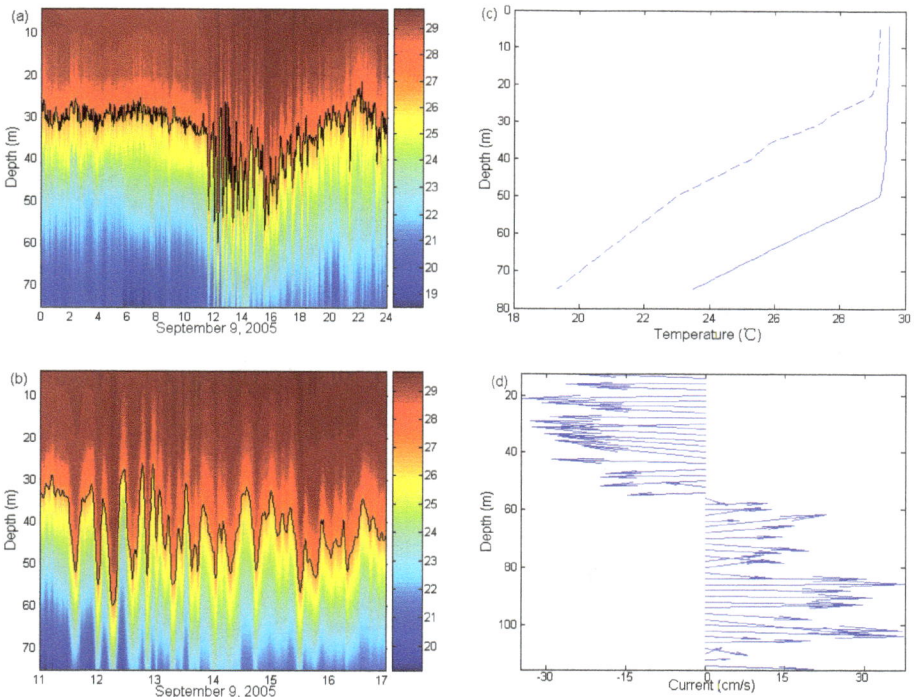

Fig. 3. 24-h (a) and 6-h (b) time series of isotherm depths from the temperature data on September 9, 2005. The isotherm of 27 °C represents the thermocline, (c) Temperature profiles measured before and during the passing of largest soliton which appears around 12:15 on 9 September. The dashed line indicates the profiles obtained by averaging the temperature data from 10:20 to 10:40 on 9 September, while the solid line indicates the profiles during the passing of this soliton, (d) ISW current speed around 12:15 on 9 September obtained by high-pass filtering the raw current data with a 3-h cutoff period. The arrows denote the vectors of current speed. Upward direction of the arrow is due the North. After Xu et al., (2010).

4.2.3 Elevation internal solitary waves

The ISWs on 10 September have similar characteristics to the September 9 solitons, but there are also some remarkable and interesting differences. Figure 4a shows an abrupt deepening of the upper stratified layer followed by several rank-ordered solitons. The temperature also fluctuated in phase vertically for the whole observed water column. The waves are rank-ordered in amplitude with the largest soliton in the lead. The remarkable feature to notice is the occurrence of elevation waves riding on top of a large depression bore. The spike-like fluctuation of temperature shows they first decrease and then increase, which indicates that the solitons are elevation waves. The largest wave arriving around 14:50 on 10 September with amplitude and period of order 35 m and 15 min was consistent with the elevation wave of first mode (Figure 4b).

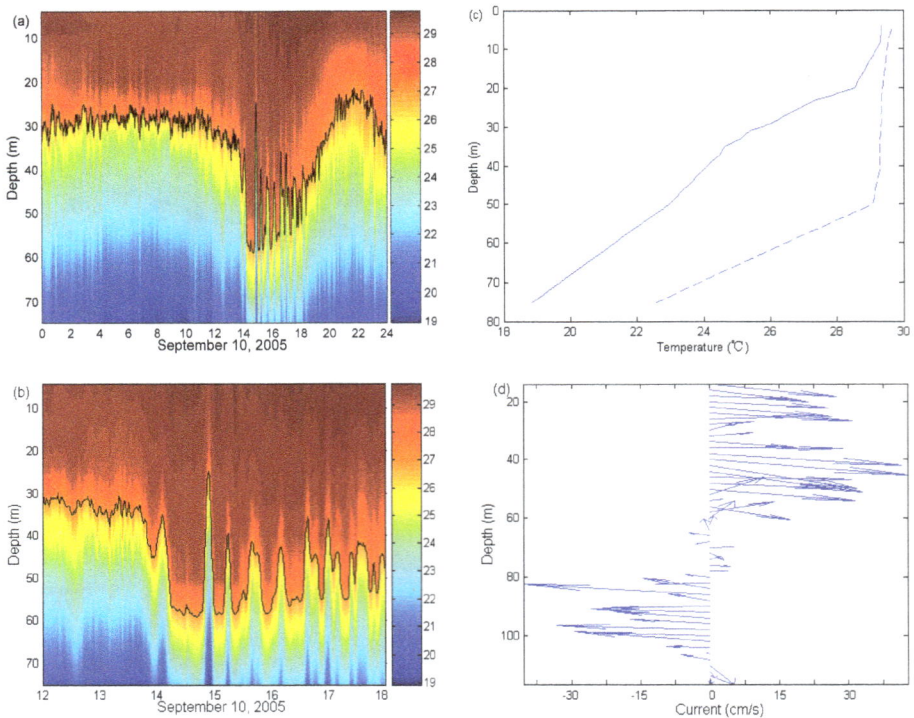

Fig. 4. 24-h (a) and 6-h (b) time series of isotherm depths from the temperature data on September 10, 2005. The isotherm of 27 °C represents the thermocline, (c) Temperature profiles measured before and during the largest soliton which appears around 14:50 on 10 September. The dashed line indicates the profiles obtained by averaging the temperature data from 14:20 to 14:40 on 10 September; while the solid line indicates the profiles during the passing of this soliton, (d) ISW current around 14:50 on 10 September obtained by high-pass filtering the raw current data with a 3-h cutoff period. After Xu et al., (2010).

Before the ISWs arrived, the average temperature in the upper mixed layer was about 29 ℃ and the mixed layer thickness was about 60 m in the water of 117 m depth due to the passing of a large depression bore. Below the upper mixed layer, there was a strong thermocline with the temperature gradient of 0.25 ℃/m. When the main soliton arrived around 14:50 on 10 September, the mixed layer was raised to 20 m (Figure 4c). This again confirms our previous observation that the amplitude of the main soliton is more than 35 m. The largest soliton arriving around 14:50 on 10 September was also classic mode-1 wave, with opposing velocities in the upper and lower layers (Figure 4d). The vertical location of the zero crossing of the velocity was between 60 m and 80 m. This soliton has opposite polarity to the ISWs appearing on 9 September. The current directions in the upper and lower layers were eastward with slightly southward and westward with slightly northward respectively. According to the KDV theory of the ISW, the elevation ISWs propagate in the same direction as that of the currents in the lower layer water. These results indicate that the September 10 soliton also propagate mainly westward, but deflect slightly to the north. However, on the contrary to the solitons on 9 September, the solitons on 10 September behave as mode-1 elevation waves due to the passing of a large depression of tidal bore.

This type of elevation ISWs observed on 10 September during WCIWE is not unique. However, it's exciting and interesting that both elevation and depression ISWs were observed at the same mooring location in different days. There were several similar wave packets propagating roughly northwestward from 9 September to 13 September, 2005. These wave packets were observed around the time of the spring tides in the study area. The seasonal thermocline generally locates at a constant depth during a short period, and the upper mixed layer of 30 m during the autumn is usually thinner than the bottom layer on the shelf of the northwestern SCS. However, the study area is dominated by the diurnal internal tides with the amplitude of 30 m much larger than that caused by the semidiurnal semidiurnal tides. There exists the possibility of a location at which the thermocline of these tidal bores with depression of 30 m is closer to the bottom than to the surface. Prior to the arrival of the ISWs during 10 September, the thermocline lowered progressively owning to the strong diurnal internal tide. So, the upper layer thickness of 60 m almost exceeded the bottom layer thickness of 57 m in the wake of this sudden depression, and the elevation waves were expected to be evolved from the internal tides due to the changing sign of nonlinear effects at this critical depth according to the theory of ISWs. On the contrary, the diurnal tide on 9 September is a little weaker than that on 10 September, the thermocline was slightly above mid-depth even during the trough of the internal tides and so the usual depression waves formed. The largest elevation soliton on 10 September with the amplitude of 35 m was very strong in related to the total depth of 117 m. The high energy of this type of elevation ISW must be important for sediment resuspension (Bogucki et al., 1997; Klymak et al., 2003) and in the energy budget. Hsu et al. (2000) have demonstrated using KdV-type numerical model that both depression and elevation waves can be generated in the upwelling area dependent on the stratification and the initial tidal mixing condition near the generation area. In the next section, we will also use the KdV equations to investigate the nonlinear effects, and then demonstrate the significant role of the ISWs in affecting the platform and biological system in the ocean interior.

4.3 Theoretical analysis of the highly ISWs
4.3.1 General description of the highly ISW packets

During the WCIWE experiment, as described in the former section, extremely strong ISWs were observed frequently. The maximum amplitude of the ISWs during the observational period reached as much as 45 m, which are the largest internal waves that have been observed in this area, which suggests that the highly nonlinear ISWs in the northwestern SCS are an indispensable part of the energy budget of the internal waves in the northern SCS. A detail study on the 3-h time series of temperature for May 5 (Figure 5) provides a good view of the internal wave structure. The prominent feature to note for the ISWs on May 5 is that the large soliton extending down from the leading edge of the low frequency wave displacement by up to 45 m. In addition, several small solitons appeared before and after the main soliton. Furthermore, it is apparently to see that larger waves have longer wave periods, to which we will discuss in detail in the next section.

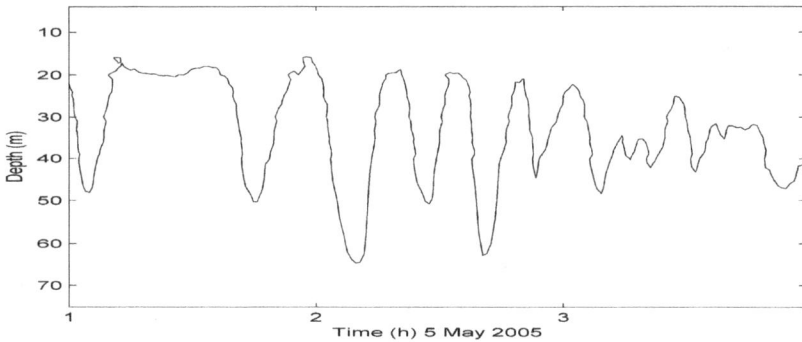

Fig. 5. 3-hour time series of 25°C isotherm depths from the temperature data for 5 May 2005.

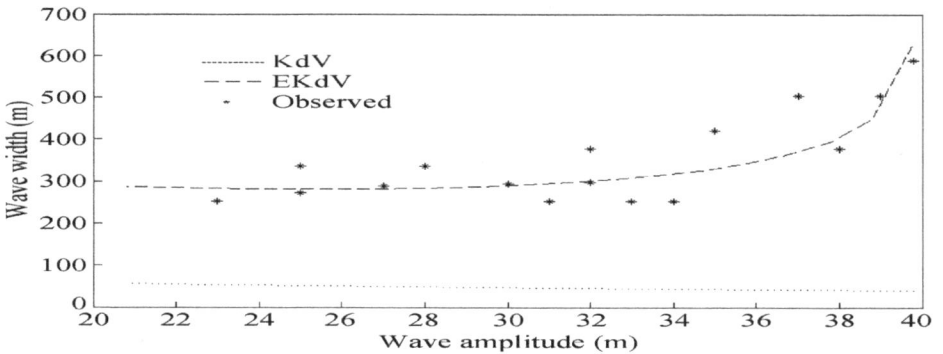

Fig. 6. Wave widths versus wave amplitude for observed ISWs. The prediction from the EKdV and KdV theory are plotted for reference. After Xu et al., (2010).

4.3.2 Theoretical analysis

Michiallet et al. (1998) showed in their laboratory experiment that large amplitude waves can be better approximated by the second-order EKdV, (formed by including the cubic non-linear term) than by the KdV theory. Based on the observation in the northwestern SCS, we investigated high-order nonlinear effects on the depression solitary waves by comparing the solutions of EKdV model in first-order KdV theory.

The relationship between the wave widths and amplitude is an important characteristic for an ISW and represents the most fundamental difference between various theories and the observation (Koop et al., 1980). Thus, the classical KdV and EKdV solutions were compared to the observed width of ISWs with moderate to large amplitudes. The widths of these waves are plotted against wave amplitude (Figure 6). The environmental parameters used to derive the theoretical values include upper layer thickness of 30 m, lower layer thickness of 90 m, and density contrast of 3 kg m^{-3}. The wave widths were computed from the time durations measured between the points where the amplitude decayed to 0.42 times of the maximum amplitude, and converted wave widths were assumed at a constant linear phase speed of 0.7 m s^{-1} estimated from the KdV and EKdV theory (Colosi et al., 2001). As expected, it was found that the EKdV solutions produced a satisfactory fit to the scatter data. On the other hand, using classical KdV theory produced much narrower solitons than those by observation and by the EKdV computation. For example, waves in 117 m water depth between 20 and 40 m in amplitude and between 250 and 500 m in width are clustered around the analytical curve of the EKdV equation, while for KdV waves of 20 to 40 m in amplitude, the wave width is less than 50 m, suggesting that the classical KdV equation is incapable of simulating highly nonlinear solitons. In particular, wave width is expected to decrease linearly as wave amplitude grows based on the KdV equation. The EKdV equation, on the contrary, suggests that wave width increases with wave amplitude, which conforms well to the features of large amplitude ISWs observed during our experiment. These finding were similar to those of Michiallet et al. (1998) and Chen et al. (2007, 2008, 2009); they compared theoretical solutions with experimental data.

To provide some estimate of the variance between two profiles, two representative individual waveform at different amplitudes are compared and presented. By comparing observed wave profiles with those predicted by internal wave theory (Fig.7), the EKdV equation is shown to satisfactorily represent the ISW at moderate to large amplitudes. However, KdV theory predicted a much shallower width than what was observed. A remarkable feature to be investigated is that the width of the observed wave grows with the amplitude, which fits well with the wave width predicted by EKdV. In contrast, the predicted width by KdV decreases with increasing amplitude and does not capture the broadening of observed waves, i.e., the KdV solution deviates more from the observation as the wave grows. This phenomenon may be due to the fact that in this shallow water situation, the wave pushes the thermocline down to a region where the nonlinear coefficients in KdV approach zero, under which the KdV theory does not predict the presence of ISWs. As mentioned previously, the EKdV theory with an extra cubic nonlinear term is particularly appropriate for this situation. Therefore, we suggest that the discrepancy between the observations and the KdV theory wave shape is likely due to high-order cubic nonlinear effects. The cubic nonlinear terms become more important than the quadratic nonlinear term and should not be neglected in simulation of large amplitude ISWs on continental shelves.

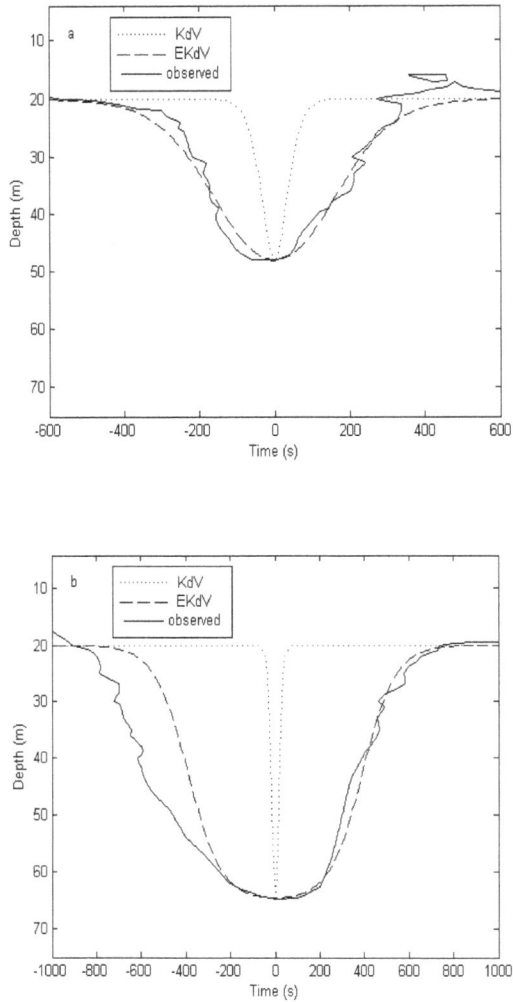

Fig. 7. Observed wave shapes in different amplitudes from May 1 at 28 m (a) and 5 May at 45 m (b)The prediction from the EKdV and KdV theory are plotted for reference. After Xu et al., (2010).

However, as pointed by Ramp et al. (2004), simulating large amplitude ISWs in the northeastern SCS is limited in the EKdV model. This is because large amplitude waves in deep water (>500 m) do not push the thermocline deep enough in a water column, and therefore the effect of the cubic nonlinear term on the soliton is less important than that of the quadratic nonlinear term. Hence, the EKdV model is less appropriate in this situation (Fan et al., 2008). However, as we described above, for highly nonlinear ISWs in shallow water, such as on the shelf of the northwestern SCS where the ratio of the cubic nonlinear term to the quadratic nonlinear term becomes much greater than those in deep water, the EKdV model is more suitable.

4.4 Effects on marine biological variations

Internal waves might play an important role in affecting the nutrient regeneration in a tropical reef ecosystem. Unfortunately, we are unable to acquire the chlorophyll and dissolved oxygen data during our experiment. However, in the northeastern SCS, previous studies have highlighted the significant effects of the internal waves on the reef ecosystem, based on the field measurements at Dongsha Atoll in the northern SCS (Wang et al., 2007). They found that mixing and advection by internal waves could generate upward flues of nutrients and suggested that the episodic upwelling of deep water due to internal waves might account for the high coral cover on the shallow reef slopes of Dongsha Atoll, in contrast to the extreme high mortality of corals inside the lagoon that occurred during the 1997-98 EI Eino warming event. Further studies based on more field measurement data are needed to quantify the temporal and spatial variations of nutrients and microbial abundance to demonstrate the ecological effects of internal waves in the coral reef ecosystem.

4.5 Wave-induced force on a pile

Based on first-order KdV theory, Cai et al. (2003) introduced Morison's empirical formula to estimate the forces and torques exerted by ISWs on cylindrical piles near Dongsha Islands in the northeastern SCS (Cai et al., 2008; Morison et al., 1950). However, as we have described above, classical KdV equation was completely inapplicable for the simulation of highly nonlinear solitons in the northwestern SCS, whereas EKdV equation could satisfactorily represent the ISW in moderate to large amplitude in this area. As a result, in this section, we use EKdV theory to estimate the induced current by the ISWs based on the Morison formula (See Cai et al., 2003 for details), and then calculate the wave force on a cylindrical pile in the northwestern SCS.

The results show that the horizontal current induced by the soliton reverse the direction in the upper and lower layer, and the maximum current speed in the lower layer is smaller than that in the upper layer. Similarly, the wave forces acting on the pile also reverse their directions due to the variation of the horizontal velocity and the maximum wave forces are 300 kN and 100 kN for the upper layer and the lower layer, respectively.

5. Conclusion

Based on the observation data of ADCP and thermistor chain, an unprecedented detailed study of the ISWs on the continental shelf of the northwestern South China Sea has been presented. Both elevation and depression ISWs are observed and behave as the first mode nonlinear wave in the mooring area. The ISWs occur at the interval of diurnal frequencies, which are distinctly different from the ISW in the northeastern area, where the solitons mainly appear with the semi-diurnal frequency, suggesting that the ISWs in the northwestern SCS might be generated by the local tide-topography interactions. In addition, a highly nonlinear wave was observed to displace the thermocline down to 45 m, making it the largest wave that was ever observed in this area. Furthermore, the comparison between the results of theoretical predictions and observations shows that the high nonlinearity of the depression waves is represented by second-order EKdV theory better than first-order KdV theory. Maximum internal wave forces on a cylindrical Pile on the northwestern shelf of SCS are estimated to be 300 kN and 100 kN for the upper layer and the lower layer, respectively. Thus, the ISWs in our study area can cause serious threat to ocean engineering structures. In addition, mixing and advection by internal waves were found to generate upward flues of nutrients and affect the nutrient regeneration in a tropical reef ecosystem. Further measurements and numerical simulations are needed to study the generation mechanism of the solitons and to demonstrate the ecological effects of internal waves in the coral reef ecosystem.

6. Acknowledgment

This work is supported by National Natural Science Foundation of China (No. 41106017 and No. 41030855), Key program of Knowledge Innovation Project of Chinese Academy of Sciences (No.KZCX1-YW-12), Natural Science Foundation of Jiangsu Province of China (BK2011396), and the National 863 program (No. 2008AA09A401).

7. References

Apel, J.R. (1987). *Principles of ocean physics*, Academic Press, Ltd., London, 634pp.

Apel, J.R.; Holbrook, J. R.; Liu, A. K. & Tsai J. J. (1985). The Sulu Sea internal soliton experiment. *J. Phys. Oceanogr.*, 15: 1625– 1651.

Apel, J.R.; Ostrovsky, L.A.; Stepanyants, Y.; & Lynch, J.F. (2007). Internal solitons in the ocean and their effect on underwater sound. *Journal of the Acoustical Society of America*, 121(2): 695-722.

Baines, P.G. (1982). On internal tide generation models. *Deep Sea Res.*, 29: 307–338.

Bogucki, D.J.; Redekopp, L. & Dickey, T. (1997). Sediment resuspension and mixing by resonantly generated internal solitary waves. *J. Phys. Ocean.*, 27: 1181-1196.

Cai, S.; Gan, Z. & Long, X. (2002). Some characteristics and evolution of the internal soliton in the northern South China Sea. Chin Sci Bull, 47(1): 21-26.

Cai, S.Q.; Long, X.M. & Gan, Z.J. (2003). A method to estimate the forces exerted by internal solitons on cylindrical Piles. Ocean Engineering, 30: 673-689.

Cai, S.Q.; Long, X.M. & Wang, S.A. (2008). Forces and torques exerted by internal solitons in shear flows on cylindrical piles. Applied Ocean Research, 30: 72–77.

Chen, C.Y. (2009) Amplitude decay and energy dissipation due to the interaction of internal solitary waves with a triangular obstacle in a two-layer fluid system: the blockage parameter. Journal of Marine Science and Technology, 14(4): 499-512.

Chen, C.Y.; Hsu, J.R.C.; Cheng, M.H. & Chen, C.W. (2008). Experiments on mixing and dissipation in internal solitary waves over two triangular obstacles. Environmental Fluid Mechanics, 8(3): 199-214.

Chen, C.Y.; Hsu, J.RC. & Cheng, M.H. 2007. An investigation on internal solitary waves in a two-layer fluid: Propagation and reflection from steep slopes. *Ocean Engineering*, 34(1): 171-184.

Colosi, J.A.; Beardsley, R.C.; Lynch, J.F.; Gawarkiewicz, G.; Chiu, C.S. & Scotti, A. (2001). Observations of nonlinear internal waves on the outer New England continental shelf during the summer Shelf break Primer study. *J. Geophys. Res.*, 106: 9587– 9601.

Duda,T.F.; Lynch, J.F.; Irish, J.D.; Beardsley, R.C.; Ramp, S.R.; Chiu, C.S.; Tang, T.Y. & Yang, Y.J. (2004). Internal tide and nonlinear internal wave behavior at the continental slope in the northern South China Sea. *IEEE J. Oceanic Eng.*, 29: 1105–1130.

Fan, Z.S.; Zhang, Y.L. & Song, M. (2008). A study of SAR remote sensing of internal solitary waves in the north of the South China Sea: I. Simulation of internal tide transformation. Acta Oceanologica Sinica. 27(4): 39-56.

Fan, Z.S.; Zhang, Y.L. & Song, M. (2008). A study of SAR remote sensing of internal solitary waves in the north of the South China Sea: II. Simulation of SAR signatures of internal solitary waves. Acta Oceanologica Sinica. 27(5): 36-48.

Farmer, D.; Li, Q. & Park, J.H. (2009). Internal Wave Observations in the South China Sea: The Role of Rotation and Non-Linearity. *Atmosphere-Ocean*, 47(4): 267-280.

Gerkema, T., & Zimmerman, J. T. F. (1995). Generation of nonlinear internal tide and solitary waves. *J. Phys. Oceanogr.*, 25: 1081– 1094.

Grimshaw, R.; Pelinovsky, E. & Talipova, T. 2002. Higher-order Korteweg-de Vries models for internal solitary waves in a stratified shear flow with a free surface. *Nonlin. Processes Geophys.*, 9: 221-235.

Helfrich, K.R. & Melville, W.K. (2006). Long nonlinear internal waves. *Annu. Rev. Fluid Mech.*, 38, 395–425.

Henyey, F.S. & Hoering, A. (1997). Energetics of borelike internal waves. *J. Geophys. Res.*, 102: 3323– 3330.

Holloway, P.E.; Pelinovsky, E.; Talipova, T. & Barnes, B. (1997). A nonlinear model of internal tide transformation on the Australian north west shelf. J. Phys. Oceanogr., 1997,27:871-896.

Hsu, M.K.; Liu, A. K. & Liu, C. (2000). A study of internal waves in the China Seas and Yellow Sea using SAR. *Continental Shelf Res.*, 20: 389-410.

Jackson, C.R. (2004). An Atlas of Internal Solitary-Like Waves and Their Properties, 2nd ed., 560 pp., Global Ocean Assoc., Alexandria, Va. (Available at http://www.internalwaveatlas.com).

Klymak, J.M. & Moum, J. N. (2003). Internal solitary waves of elevation advancing on a shoaling shelf,. *Geophys. Res. Lett.*, 30(20), 2045, doi:10.1029 /2003GL017706.

Klymak, J.M.; Pinkel, R.; Liu, C.T.; Liu, A.K. & David L. (2006). Prototypical solitons in the South China Sea. Geophys. Res. Letters, 33, L11607, doi:10.1029/ 2006GL025932.

Koop, C.G. & Butler, G.. (1981). An investigation of internal solitary waves in a two-fluid system. J. Fluid Mech., 112: 225-251.

Lee, C.Y. & Beardsley, R.C. (1974). The generation of long nonlinear internal waves in a weakly stratified shear flow. *J. Geophys. Res.*, 79: 453– 462.

Li, X.; Zhao, Z. & Pichel, W.G. (2008). Internal solitary waves in the northwestern South China Sea inferred from satellite images. *Geophy. Res. Let.*, 35, L13605, doi:10.1029/2008GL034272.

Lien, R.C.; Tang, T.Y.; Chang, M.H. & D'Asaro, E.A. (2005). Energy of nonlinear internal waves in the South China Sea, *Geophys. Res. Lett.*, 32, L05615, doi: 10.1029/2004GL022012.

Liu, A.K., & Hsu, M.K. (2004). Internal wave study in the South China Sea using synthetic aperture radar (SAR). *Int. J. Remote Sens.*, 25: 1261– 1264.

Liu, A.K., Chang, Y.S.; Hsu, M.K. & Liang, N.K. (1998). Evolution of nonlinear internal waves in the East and South China Seas. *J Geophys Res.*, 103(C4): 7995-8008.

Maxworthy, T. (1979). A note on the internal solitary waves produced by tidal flow over a three-dimensional ridge. *J. Geophys. Res.*, 84: 338– 346.

Michallet, H.; Barthelemy, E. (1998). Experimental study of interfacial solitary waves. *Journal of Fluid Mechanics*, 366: 159-177.

Moore, S.E. & Lien, R.C. (2007). Pilot whales follow internal solitary waves in the South China Sea. *Marine Mammal Science*, 23(1): 193-196.

Morison, J.R.; O'Brien, M.P.; Johnson, J.W. & Schaaf, S.A. (1950). Forces exerted by surface waves on piles. AIME Petroleum Transactions. 189: 149-154.

Osborne, A. R., & Burch, T. L. (1980). Internal solitons in the Andaman Sea. *Science*, 208: 451– 460.

Ramp, S.R.; Tang, T.Y.; Duda, T.F.; Lynch, J.F.; Liu, A.K.; Chiu, C.S.; Bahr, F.L.; Kim, H.R. & Yang, Y.J. (2004). Internal solitons in the northeastern South China Sea, Part I: Sources and deep water propagation, *IEEE J. Oceanic Eng.*, 29: 1157– 1181.

Russell, J. S. (1838). Report of the Committee on Waves. Rep. Meet. British Assoc. Adv. Sci 7th, Liverpool. John Murray, London, 417-496.

Russell, J. S. (1844). Report on Waves. 14th Meeting Brit. Assoc. Adv. Sci., 311-390.

Small, R.J. & Hornby, R.P. (2005). A comparison of weakly and fully non-linear models of the shoaling of a solitary internal wave. Ocean Modelling, 8: 395–416.

Xu, Z.H.; Yin, B.S. & Hou, Y.J. (2010). Highly nonlinear internal solitary waves over continental shelf of northwestern South China Sea. *Chin.J.Oceanol.Limonol*, 28(5): 1049-1054.

Xu, Z.H.; Yin, B.S. & Hou, Y.J. (2011). Multimodal structure of the internal tides on the continental shelf of the northwestern South China Sea. Estuarine, Coastal and Shelf Science, doi:10.1016/j.ecss.2011.08.026.

Xu, Z.H.; Yin, B.S.; Hou, Y.J.; Fan, Z.S. & Liu, A.K. (2010). A study of internal solitary waves observed on the continental shelf in the northwestern South China Sea. *Acta Oceanologica Sinica*, 29:18-25.

Zhao, Z. & Alford, M. H. (2006). Source and propagation of internal solitary waves in the northeastern South China Sea. *J. Geophys. Res.*, 111, C11012, doi:10.1029/2006JC003644.

Zhao, Z.; Klemas, V.; Zheng, Q. & Yan, X. (2004). Remote sensing evidence for baroclinic tide origin of internal solitary waves in the northeastern South China Sea. Geophys. Res. Lett., 31. L06302, doi:10.1029/2003GL019077.

Part 3

Chemical Oceanography

9

Organic-Aggregate-Attached Bacteria in Aquatic Ecosystems: Abundance, Diversity, Community Dynamics and Function

Xiangming Tang[1], Jianying Chao[2], Dan Chen[1],
Keqiang Shao[1] and Guang Gao[1]
*[1]State Key Laboratory of Lake Science and Environment,
Nanjing Institute of Geography and Limnology, Chinese Academy of Sciences,
[2]Nanjing Institute of Environmental Science,
Ministry of Environmental Protection,
P.R. China*

1. Introduction

Micro- and macroscopic aggregates are a ubiquitous and abundant component of aquatic ecosystems. The occurrence and ecological importance of macroscopic aggregates (> 500 μm, also known as marine snow) in the pelagic environment has been extensively studied for more than 30 years (Alldredge and Silver 1988, Grossart et al. 2007, Kiørboe and Jackson 2001, Smith et al. 1992, Silver et al. 1978). And several studies on macroscopic aggregates in deep lake and lotic ecosystems, known as lake snow and river snow, have been reported in recent years (Böckelmann et al. 2002, Grossart and Simon 1993, Grossart and Simon 1998). It has also been suggested that aggregates serve as transient microhabitats suitable for kinds of biogeochemical processes (Paerl and Prufert 1987, Azam and Richard 2001). Abundant of bacteria, rich in nutrients and high metabolic activity of attached microorganisms made the aggregates "hotspots" in energy fluxing, biogeochemical cycling and food web dynamics (Azam and Richard 2001, Caron et al. 1982, Paerl 1974, Simon et al. 2002, Grossart and Ploug 2000).

Compared with pelagic systems in which low hydrodynamic stress allows macroaggregates to form, microaggregates (5-500 μm) dominate in shallow, turbid, and eutrophic systems, such as Lake Taihu (Tang et al. 2009). Lake Taihu, the third largest lake in China, has an area of 2,338 km², and a catchment area of about 36,500 km²; its maximum length is 68.5 km and the maximum width is 56 km (Fig. 1). It's a typical shallow lake with the mean depth <2 m.

In contrast to marine and deep lake systems, Taihu is more productive and it has significant horizontal environmental gradients from northwest to southeast. These gradients, in such parameters as trophic status, concentration of suspended particles, and concentration of Chlorophyll *a*, provide almost unique opportunities for revealing mechanisms that control the composition of bacterial communities within the lake (Tang et al. 2010). Moreover, wind-driven sediment resuspension and intensive cyanobacterial blooms are two major ecological features of this lake (Qin 2008). However, while macroaggregates-attached microbial communities in marine, estuary, river, and deep lake systems have been well-characterized, little is known about the composition and diversity of the smaller organic-aggregate-associated bacterial communities (OABC), and the environmental factors which shape their dynamics, in large shallow eutrophic subtropical lakes.

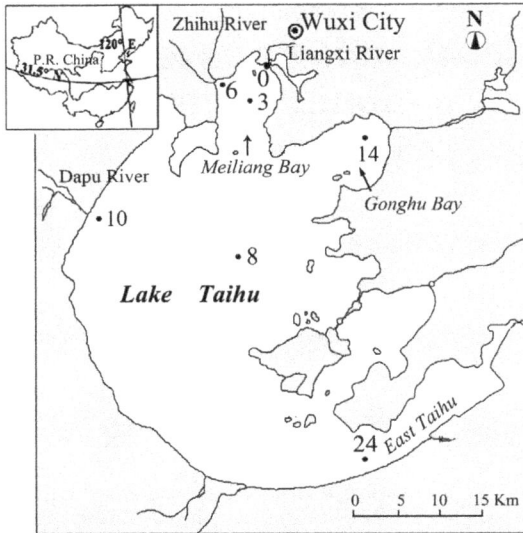

Fig. 1. Map of Lake Taihu and the sampling sites.

Based on terminal restriction fragment length polymorphism (T-RFLP) and 16S ribosomal ribonucleic acid (rRNA) gene clone libraries, we investigated the spatial and temporal heterogeneity of OABC in Lake Taihu during May 2006 to May 2008. Samples were collected at seven stations representing different trophic states and food web structures (Fig. 1.). Site 0, Site 6 and Site 10 are located in highly eutrophic areas near the river mouth of Liangxi River, Zhihu River and Dapu River, respectively; the eutrophication is due to nitrogen and phosphorus from domestic and industrial wastewater discharged from the rivers. Site 3 is located in Meiliang Bay, which experiences intensive blooms of algae, dominated by *Cyanobacteria*, during summer and autumn. Site 8 is located in the open lake, where the water is less enriched with nitrogen and phosphorus but exposed to frequent wind mixing (Wu et al. 2007b). Site 14 is located in Gonghu Bay, a transition area from phytoplankton- to macrophyte- dominated lake habitats. Site 24 is located in East Taihu, which is characterized by submersed macrophyte communities and relatively low phytoplankton concentrations (Tang et al. 2010, Tang et al. 2009).

From the studies on the organic aggregates (OA)-attached bacteria in the large, shallow and eutrophic Lake Taihu, China, some new results which differed from pelagic ecosystems were found. This is a short review of the current understanding of the role of microorganisms attached to organic aggregates in aquatic ecosystems, the gaps in our knowledge and some suggestions of future directions. And this study is the first attempt to summarize the current research achievements on the abundance, diversity, community dynamics and function in different aquatic ecosystems.

2. The abundance of OA-attached bacteria

2.1 The abundance of OA-attached bacteria in different aquatic systems

Bacteria have been found to be colonized on nearly all types of aggregates studied so far, including marine, lacustrine, and riverine macro- and microaggregates (Simon et al. 2002). Macroscopic organic aggregates is densely colonized by microbes, which are not uniformly distributed on aggregates but often form microcolonies (Simon et al. 2002 and references therein). Obviously, the nature of particles and their adsorption capacities determine the abundance of colonized bacteria (Berger et al. 1996). Due to different hydrodynamic conditions and heterogeneity of phytoplankton communities, aggregates in different aquatic systems differed in size and components. Hence, the abundance of attached bacteria and their relative proportion to total bacterial numbers in different systems varies greatly (Table 1).

In marine environments, numbers of attached bacteria on macroscopic organic aggregates (> 500 μm) have been documented since 1986. Densities of bacteria ranged from $1.25 - 1.69 \times 10^6$ cells per aggregate in Southern California Bight to $1.83 - 278 \times 10^6$ cells per aggregate in North Atlantic (Alldredge et al. 1986). On a giant aggregate, the density of bacteria reached to as high as 5.4×10^8 cells (Silver et al. 1998). However, the ratio of macroaggregates-attached bacteria to total bacteria of the surrounding water is relatively low. In most case, they constitute <10% of total bacterial numbers (Table 1). On microaggregates, the percentage of attached bacteria is higher. For example, in a tidally affected coastal ecosystem located in German Wadden Sea, the particle-attached bacteria constituted proportions of 7–47% of total bacteria (Rink et al. 2008). Moreover, the abundance of attached bacteria varied with water depth. In NW Mediterranean Sea, they increased with water depth and reached the most abundant of 3.4×10^5 cells ml^{-1} in 80 m depth, accounting for 14% of the total bacteria (Ghiglione et al. 2007).

Natural lake snow aggregates were densely colonized by microbes (Table 1). In deep lakes, 0.002×10^6 to 23×10^6 bacteria per ml were counted in Lake Constance, a large mesotrophic freshwater lake in Central Europe (Schweitzer et al. 2001, Grossart and Simon 1993), and where transparent exopolymer particles (TEP) and Coomassie-stained particles (CSP) aggregates in Lake Aydat and Lake Pavin could carry similar orders of bacteria (Lemarchand et al. 2006).

The size of riverine aggregates is usually smaller than lake snow or marine snow (Zimmermann-Timm et al. 2002 and references therein). However, the abundance of aggregate -attached bacteria is much higher than that in marine and deep lake systems. In Columbia River estuary, the numbers ranged from 0.1×10^6 to 12.6×10^6 cells ml^{-1}, where enriched aggregates in River Elbe could carry much larger numbers, up to $120-250 \times 10^6$ cells ml^{-1} (Böckelmann et al. 2000). In contrast to marine and deep lake systems, aggregates in riverine and estuarine systems can constitute as much as 90% of total bacterial numbers and production (Simon et al. 2002 and references therein; Table 1). In lab conditions, the abundance of aggregate-attached bacteria can reach to as high as 37.0×10^9 cells (ml agg.)$^{-1}$ (Grossart et al. 2003a).

Based on a 25-month observations of OA and the attached bacteria in the large, shallow, eutrophic Lake Taihu, Tang (2009) revealed that the abundance of OA range from 1.3 to 23.5×10^5 ind. ml^{-1} (average 7.1×10^5 ind. ml^{-1}), and most of them are microaggregates with a size of 10~200 μm. Detrital and algal aggregates dominate in Lake Taihu. Furthermore, there are temporal and spatial dynamics of the origins and compositions of OA. The average abundance of OA-associated bacteria is 15.5×10^6 (range $3.6 - 32.4 \times 10^6$) cells ml^{-1}, accounted for 53% (range 28.7 – 80.4%) of the total bacteria. The mean number of bacteria per particle was much higher in Lake Taihu (25.6 bacteria particle^{-1}) than in River Danube (8 bacteria particle[-1]) (Berger et al. 1996). The abundance of OA-associated bacteria are related in OA, water temperature, total suspended solids (TSS), total phosphorus (TP) and Chl a. Results of scanning electron microscopy (SEM) and 4' 6-diamidino-2-phenylindole dihydrochloride (DAPI) demonstrated that the abundance of OA-associated bacteria is much higher than that of in oceans and deep lakes, which indicates the ecological importance of OA in shallow lakes. Aggregate abundances are higher in lotic systems, such as in rivers and shallow eutrophic lakes, than in deep lakes and in the sea (see above) due to high resuspension rates (Zimmermann-Timm et al. 2002 and references therein).

Bacterial abundance on aggregates is initially driven by a balance between bacterial attachment and detachment (Kiørboe et al. 2002), and subsequently by bacterial growth on the particle surfaces. Previous study (Grossart et al. 2006) showed that percentages of aggregates-attached bacteria greatly changed over time. In addition to bacterial colonization and growth, predation may also affect bacterial abundance and community composition on macroaggregates (Kiørboe et al. 2004, Kiørboe et al. 2003, Jürgens and Sala 2000).

System	Location	Abundance ($\times$ 10[6] cells ml[-1])	Ratio of attached bacteria (%)	Density of colonization (cells particle[-1])	OA type	Source
Marine	German Wadden Sea	1.0–2.9	7 –47	/	DAPI	Rink et al. 2008
	NE Atlantic	0.25–0.43	/	/	Macroaggregates	Turley and Mackie 1994
	North Atlantic	/	0.1–4.4	1.83–278$\times$10[6]	Marine snow	Alldredge et al. 1986
	NW Mediterranean Sea	/	9–32 (3–30m) <10 (>50m) 37.9 (250m) 14.9 (500m)	/	Alcidine orange	Ghiglione et al. 2007
	NW Mediterranean Sea	/	15±5	/	SYBR Green	Mevel et al. 2008
	NW Mediterranean Sea	0.17 (10m) 0.17 (30m) 0.31 (50m) 0.34 (80m) 0.18 (100m) 0.10 (150m)	9.1 (10m) 8.9 (30m) 9.4 (50m) 14.0 (80m) 11.1 (100m) 7.8 (150m)	/	SYBR Green I	Ghiglione et al. 2009
	Isefjord in Denmark	/	< 25	/	SYBR Green I	Tang et al. 2006
	Southern California Bight	/	0.9-3.0	1.25–1.69$\times$10[6]	Marine snow	Alldredge et al. 1986

	Southern California Bight	0.1–1.7	/	/	Marine snow	Simon et al. 1990
	Monterey Bay	/	/	$3.2–5.4 \times 10^8$	Giant aggregates	Silver et al. 1998
Estuaries and Rivers	Columbia River estuary (1995)	2.4 (0.7–5.4)	57.1	/	DAPI	Crump et al. 1998, Crump et al. 1999
	Columbia River estuary (1997)	3.3 (0.1–12.6)	70.2	/	DAPI	Crump et al. 1998, Crump et al. 1999
	River Danube	0.78 ± 0.08	30.34±3.09	/	Alcian Blue, Coomassie Brilliant Blue G-250	Luef et al. 2007
	River Danube	0.09–2.3	51.0	/	SYBR Green I	Peduzzi and Luef 2008
	River Danube	0.1–1.4	9.5	8 ± 3	DAPI	Berger et al. 1996
	Elbe Estuary	5.0–50	/	/	DAPI	Zimmermann-Timm et al. 1998
	Elbe Estuary	/	75.0	$0.3–2.5 \times 10^6$	Alcian Blue	Zimmermann 1997
	River Elbe	120–250	/	/	River snow	Böckelmann et al. 2000
Deep lakes	Lake Constance	0.002–0.016	/	/	Lake snow	Schweitzer et al. 2001
	Lake Constance	/	4.3	$4.04±3.08 \times 10^6$ (6 m) $8.77±7.83 \times 10^6$ (25 m)	Macroaggregates (<3 to 20 mm)	Grossart and Simon 1998
	Lake Constance	1.2–23	/	/	Macroaggregates	Grossart and Simon 1993
	Lake Constance	/	/	$0.5 –2 \times 10^6$	Macroaggregates	Weiss et al. 1996
	Lake Aydat	0.533±0.124	7.4	/	TEP & CSP	Lemarchand et al. 2006
	Lake Pavin	0.099±0.036	2.6	/	TEP & CSP	Lemarchand et al. 2006
Shallow lakes	Lake Taihu	3.6–32.4	53.2	25.6±12.9	DAPI	Tang 2009

Table 1. Abundances of aggregate-attached bacteria in various aquatic environments.

2.2 The abundance of OA-attached bacteria vs. free-living bacteria

Bacteria are often highly enriched on aggregates as compared to the surrounding water (Alldredge et al. 1986, Becquevort et al. 1998). For example, in surface waters of Southern California Bight, the bacterial densities on sinking marine snow aggregates were >2000-fold higher than in the surrounding water (Ploug et al. 1999). In Lake Constance, the bacterial abundance on macroaggregates on a per volume basis was $\sim 10^8$ ml^{-1}, which is $100 \times$ higher than in the bulk water (Grossart and Simon 1993). Though the relative abundance of attached bacteria is lower than that of free-living bacteria in most aquatic systems, in many cases, the aggregates-attached bacteria are bigger and more active than free-living ones.

3. The diversity of OA-attached bacteria

3.1 Morphological and phylogenetic diversity

Organic aggregates are often colonized by bacteria. Bacteria can attached to newly formed aggregates within 5 min, which made bacteria the pioneer colonizers of organic aggregates (Wörner et al. 2000). However, the morphological and phylogenetic diversity of OA-attached bacteria are depend on the composition, age, size of aggregates and the physicochemical characteristics of the surrounding water.

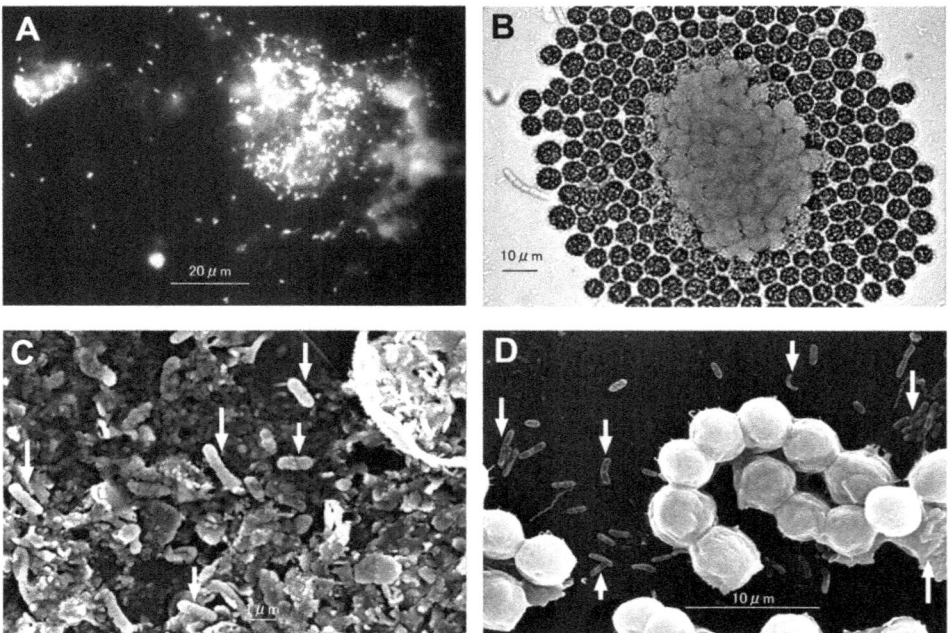

Fig. 2. Typical organic aggregates and the attached bacteria in Lake Taihu. (A) Microphotographs of DAPI-stained OA. The small bright dots represent bacteria. (B) Senescent *Microcystis* spp. aggregates surrounded by abundant bacteria (bright dots). Scanning electron micrographs of detritus-like aggregates (C) and phytodetrital aggregates (D). Small arrows indicate the presence of bacterial microcolonies.

Generally, the composition of aggregates determines the morphological traits of OA and the attached bacteria. Observations of aggregates from coastal macrophyte-derived dissolved organic material by Alber & Valiela (1994) showed that large numbers of rod-shaped bacteria attached to this kind of aggregates. Clear morphological differences can be observed on SEM micrographs for OA and OA-associated bacteria collected from the four sampling stations in Lake Taihu (Tang et al. 2009). Intense colonization of mucilage surrounding cyanobacterial cells was observed by DAPI stain and SEM (Fig. 2).

Newly formed aggregates are often colonized by small rod and coccidial-shaped bacteria, whereas older aggregates are colonized by filamentous bacteria >0.5 μm (Zimmermann-Timm 2002). A 12-day observation of lab-made aggregates incubated in rolling cylinders showed both morphological and phylogenetic successions of the attached bacteria (Grossart and Simon 1998). During the first 2 d, small aggregates (<5 mm Ø) were dominated by cocci (<1 μm). After 2 to 3 d, larger rod bacteria were dominant and formed colonies. Filamentous or flagellated bacteria dominated after 5 to 7 d. In the end, large aggregates (>9 mm Ø) densely colonized by long and thick rods. During the experiment succession of a-, β-, and γ-proteobacteria were documented. And β-proteobacteria was usually dominant, particularly in aged aggregates (Grossart and Simon 1998).

The bacterial communities of macroscopic organic aggregates (≥1 mm in diameter; from the river Weser) incubated in roller tanks were investigated by (Grossart and Ploug 2000). Using in situ hybridization, they found that the percentage of a- and β-proteobacteria decreased from 13% and 33.7% to 2.6% and 9.0%, respectively, whereas those of the filamentous γ-proteobacteria and Cytophaga increased from 31.9% to 50.4% and from 8.5% to 24.9%, respectively, during the 14 d of incubation. The morphological and phylogenetic succession of aggregate-attached bacteria seems to be related to the interactions between bacteria and protozoans. The occurrence of high abundance of aggregate-attached filamentous bacteria, which are known to be a phenotypic response to protozoan grazing, can be the result of intense bacterial grazing by protists during a later stage (Grossart and Ploug 2000, Grossart and Simon 1998, Ploug and Grossart 2000, Simon et al. 2002, Jürgens and Matz 2002).

3.2 The diversity of attached bacteria vs. free-living bacteria

Difference between the community composition of particle-attached and free-living bacteria were found in most types of the aquatic systems, including Santa Barbara Channel (DeLong et al. 1993), Chesapeake Bay (Bidle and Fletcher 1995), Columbia River estuary (Crump et al. 1999), Mecklenburg Lake District (Allgaier and Grossart 2006), and other aquatic ecosystems (Kellogg and Deming 2009, Riemann and Winding 2001, Selje and Simon 2003; also Table 3).

Abundant researches carried out in marine and deep lakes showed that attached bacteria were always less abundant and less diverse but generally more active than free-living bacteria (Ghiglione et al. 2007, Grossart et al. 2003b, Simon et al. 2002). Using 16S ribosomal desoxynucleic acid (rDNA) based clone library analysis, however, we found that in shallow productive eutrophic Lake Taihu, OA harbors diverse bacterial clusters (Tang et al. 2010, Tang et al. 2009). The bacterial diversity index of S_{Chao1} and Shannon index reached to 323.3± 77.5 and 4.75, respectively (Table 2). And there are no significant differences of diversity index between aggregates-attached and free-living bacterial communities (Unpublished data). These different patterns may reflect different shaping effects in different aquatic ecosystems on aggregates-attached bacterial communities.

Use of DNA- and RNA-derived capillary electrophoresis single-strand conformation polymorphism fingerprinting, Ghiglione et al. (2009) examined the total and metabolically active communities of attached and free-living bacteria in the euphotic zone in the NW Mediterranean Sea. They found that 52-69% of the DNA-derived operational taxonomic units (OTUs) were common in both attached and free-living fractions, suggesting an exchange or co-occurrence between them. Even if colonization on and detachment of particles appear to be ubiquitous, most of the particulate organic carbon remineralisation appeared to be mediated by a rather low number of dominant active OTUs specialized in exploiting such specific microenvironment. In Lake Taihu, we also found some clues of exchanging among aggregates-attached, free-living and sediment-derived bacterial communities (Fig. 3). Comparative statistical analyses of the habitats of OA-associated bacteria highlight the potential ecological importance of the exchange between OABC and the surrounding planktonic community, because 41.5%~78.8% of the sequences are related to freshwater habitats. However, we also found 21.2%~58.5% of sequences closely related to ones previously found in soils, sludge, sediments and other habitats, indicating either the potential importance of allochthonous bacteria in OABC, or similar functions of these bacteria in multi-habitats (Rath et al. 1998, Simon et al. 2002).

Sampling time	Clone library	Site	clones	S_{Chao1}	OTUs	H'	RSI	C (%)
October 2006	A	0#	243	214.3 ± 39.5	104	4.53	45.4	74.1
	B	3#	225	**323.3 ± 77.5**	106	4.31	19.5	65.3
	C	8#	264	276.8 ± 47.0	127	**4.75**	43.5	68.6
	D	24#	220	220.0 ± 45.4	96	4.40	36.1	71.4
Feburary 2007	E	3#	250	257.8 ± 53.1	109	4.58	48.3	72.4
	F	8#	270	292.4 ± 58.0	119	4.56	35.8	70.7
May 2007	G	3#	264	323.0 ± 70.1	118	4.52	30.8	69.3
	H	8#	269	268.2 ± 46.0	122	4.57	29.1	69.5
August 2007	I	3#	189	125.1 ± 18.1	81	4.37	46.3	77.8
	J	6#	149	101.3 ± 16.8	65	4.15	35.6	77.9
	K	8#	212	196.9 ± 36.7	98	4.57	48.5	72.6
	L	24#	169	146.8 ± 26.8	79	4.39	48.8	71.6

Table 2. Spatial and temporal dynamics of organic-aggregate-associated bacterial operational taxonomic units (OTUs) diversity in Lake Taihu, by means of the Chao1 richness estimator (S_{Chao1}) and the reciprocal Simpson's dominance index (RSI). H': Shannon index; C: Coverage. Sampling sites as shown in Fig. 1.

4. Dynamics of OA-attached bacterial communities

4.1 Spatial and temporal dynamics of OA-attached bacterial communities
4.1.1 Marine
The structural composition of bacterial communities on marine snow appears to be dominated by the *Cytophaga/Flavobacteria* (i.e. *Bacteroidetes*) cluster and γ-*proteobacteria*. For example, DeLong et al. (1993) found that *Cytophaga*, *Planctomyces*, or γ-*proteobacteria* were the dominant bacterial clusters dwelled in the marine snow collected in the Santa Barbara Channel. In the Gulf of Trieste (northern Adriatic Sea), a high diversity of bacteria was found on marine snow with dominant phylotypes of the *Cytophaga-Flavobacteria-Bacteroides*

lineage (Rath et al. 1998). On the basis of denaturing gradient gel electrophoresis (DGGE) of polymerase chain reaction (PCR)-amplified 16S rDNA fragments, Bidle & Azam (2001) found that the bacterial community on aggregates of marine diatom detritus was dominated by *γ-proteobacteria* and *Sphingobacteria-Flavobacteria* by comprising 65% and 25% of detected phylotypes, respectively. Other investigations on marine snow-attached bacterial communities in the Southern California Bight (Ploug et al. 1999), in the polar frontal zone of the Southern Ocean (Simon et al. 1999), in Schaproder Bodden (Hempel et al. 2008) and in German Wadden Sea (Rink et al. 2008) resulted in similar conclusion (Table 3).

4.1.2 Estuaries and rivers

The structure of aggregate-attached bacterial communities in river system, such as rivers and estuaries, is strikingly different from that on marine snow (Table 3). Using fluorescent in situ hybridization, Böckelmann et al. (2000), in the Elbe River of Germany, found that the bacterial aggregate community varied over the course of the year. During all seasons, *β-proteobacteria* constituted the numerically most important bacterial group constituting ~54% of the total DAPI-stainable cells. They were characterized by short, rod-shaped bacteria (1 μm in length) showed typical polyalkanoate inclusion bodies and occurred as single cells or short chains, or globular microcolonies within the river snow community. In spring the community had been characterized by great bacterial diversity and by a high abundance of *Cytophagae* (up to 36%). And the predominant bacterial morphotypes were long thin filaments (up to 40 μm in length). In contrast, the number of filamentous *Cytophaga-Flavobacteria* decreased significantly in autumn and winter, with small, rod-shaped cells (0.5 × 1 μm). The percentage of *γ-proteobacteria* peaked at 26% in winter, which is significantly higher than that in autumn (14%). Typical morphotypes within the *γ-proteobacteria* were thick, rod-shaped bacteria (1 × 3 μm). The numbers of *a-proteobacteria* was around 24% in spring and 4% in summer, with rod-shaped (0.5 × 2 μm) and irregularly formed coccoid cells (diameter 1 – 1.5 μm). *Planctomycetales* and sulfate-reducing bacteria (SRB) were of lower significance and constituted 2 to 11% and 2 to 17%, respectively.

Differences among the macroaggregates-attached bacterial communities in the limnetic, the brackish and the marine sections of the River Elbe, were reported by Simon et al. (2002). Using group-specific oligonucleotide probes, the authors found that in the limnetic section, *β-proteobacteria* and *Cytophaga/Flavobacteria* accounted for 20 – 40% and 25 – 36% of the DAPI-stainable cells, respectively, in May, and 18 – 45% and 20%, respectively, in October. In the brackish and marine section of the Elbe estuary, however, *γ-proteobacteria* largely dominated whereas *β-proteobacteria* constituted not more than 12%. Salinity gradient seems to be one of the most important factors controlling the variations of the aggregate-attached bacterial communities from freshwater upstream to downstream estuary. With the increase in salinity, the *β-proteobacteria* was gradually replaced by *γ-proteobacteria* because salinity is a strong adaptive barrier for *β-proteobacteria* (Methé et al. 1998, Simon et al. 2002).

A 14-day observation of aggregates incubated in water samples of the river Weser, Northern Germany, in roller tanks revealed that the percentage of filamentous *γ-proteobacteria* and Cytophaga increased from 31.9% to 50.4% and from 8.5% to 24.9%, respectively, whereas those of *a-* and *β-proteobacteria* decreased from 13% and 33.7% to 2.6% and 9.0%, respectively (Grossart and Ploug 2000). The low percentage of *β-proteobacteria* presumably due to relative higher salinity (<1‰) of the incubation water.

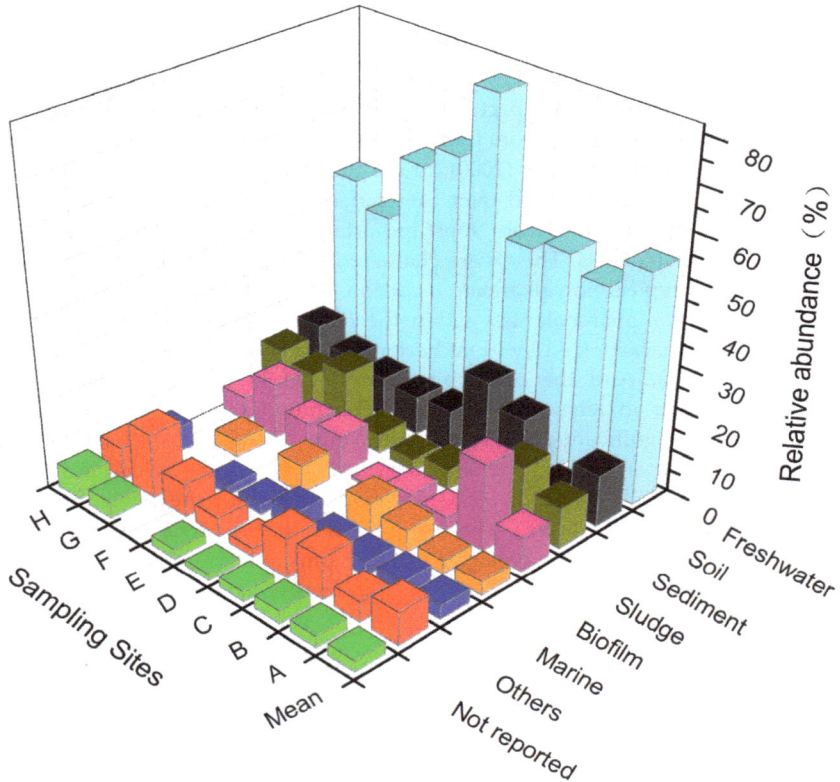

Fig. 3. Proportion of habitat affiliations of the sequences based on comparison of our 16S rRNA clone library sequences with the closest relatives from the GenBank. A~H, represent different OA samples (detailed information is presented in Table 2); the bar labeled "Mean" gives the frequency distribution for all samples. Habitats affiliated with anaerobic swine lagoon, subsurface aquifer, wastewater, rumen fluid, human gut and fuel cell anode as well are included in 'Others'.

System	Location	Bacterial type	Composition of associated bacterial communities (%)					Source
			β-prot.	*a-prot.*	*γ-prot.*	*Bact.*	*Acti.*	
Marine	German Wadden Sea	Attached	/ 15.3±11.4 (2000) 8.2±5.1 (2005)	9.5±3.0 (1999) 30.1±19.9 (2000) 17.8±15.3 (2005)	16.5±2.3 (1999) 27.0±17.3 (2000) 28.8±9.8 (2005)	20.9±2.0 (1999) 27.6±16.4 (2000) 15.0±7.3 (2005)	/	Rink et al. 2008
		Free-living	/ 1.6±1.3 (2000) 1.4±1.2 (2005)	13.8±4.0 (1999) 7.9±7.2 (2000) 7.2±4.1 (2005)	17.1±3.5 (1999) 10.7±7.3 (2000) 9.8±3.5 (2005)	19.2±3.6 (1999) 13.6±7.2 (2000) 11.5±5.2 (2005)	/	Rink et al. 2008
	Northern Adriatic Sea	Attached	/	15.3	15.3	/	/	Rath et al. 1998
	Schaproder Bodden	Attached	/	3-17	/	5-35	/	Hempel et al. 2008
Estuaries and Rivers	Elbe River, Germany	Attached	50 ± 10 ~ 54 ± 6	5 – 25	14 ± 5 ~ 26 ± 8	/		Böckelmann et al. 2000
	Elbe River, Germany	Attached	54	/	/	36	/	Böckelmann 2001
	Weser estuary, Germany	Attached	10.0	/	/	28±8.9	/	Selje and Simon 2003
Deep lakes	Lake Aydat	Attached	30	11.7	/	/	/	Lemarchand et al. 2006
		Free-living	17.4	9.7	/	/	/	Lemarchand et al. 2006
	Lake Constance	Attached	27–42	/	/	/	/	Lemarchand et al. 2006

	Lake Constance	Attached	14.2±10.2 (25m)	10.5±7.9 (25m)	4.2±9.5 (25m)	5.8	/	Schweitzer et al. 2001
			54.0±5.9 (50m)	12.0±3.3 (50m)	1.9±0.7 (50m)			
			41.1±8.4 (110m)	2.7±1.4 (110m)	2.5±0.8 (110m)			
	Lake Constance	Free-living	12.8±4.7 (1996)	4.3±4.6 (1996)	2.6±1.9 (1996)	/	/	Zwisler et al. 2003
			11.2±3.8 (1997)	3.9±2.5 (1997)	1.8±0.7 (1997)			
	Lake Constance	Attached	27 – 42	11 – 25	9 – 33	/	/	Weiss et al. 1996
Shallow lakes	Lake Taihu	Attached	34.4	15.2	9.6	11.2	5.5	Tang et al. 2010, Tang et al. 2009
		Free-living	10.8	5	7.4	2.7	52.1	Wu et al. 2007a
	Mecklenburg Lake District	Attached	3.2	8.1	3.2	24.2	1.6	Allgaier and Grossart 2006
		Free-living	10.4	8.3	0.7	14.6	44.4	Allgaier and Grossart 2006

Table 3. Main phylogenetic composition of aggregate-attached and free-living bacterial communities in various aquatic environments. *β-prot.*, *a-prot.*, *γ-prot.*, *Bact.* and *Acti.* represent *β-proteobacteria, a-proteobacteria, γ-proteobacteria, Bacteroidetes,* and *Actinobacteria,* respectively.

4.1.3 Deep lakes

Lake Constance is a typical deep large mesotrophic lake with maximum depth of 254 m and a surface area of 571 km^2. The bacterial communities on lake snow aggregates in this lake have been intensively studied. Using FISH with rRNA-targeted oligonucleotides, Brachvogel et al. (2001) found that *β-proteobacteria* and *Cytophaga/Flavobacteria* dominated the bacterial community on microaggregates in Lake Constance, constituting 8 to 78% of the DAPI-stainable cells, which equals 14 to 82% of Bacteria. In contrast, *a-proteobacteria* was not detected at all. And *γ-proteobacteria* usually constituted only minor proportions except on zooplankton debris and on phytodetrital aggregates composed of *Dinobryon* spp. (Brachvogel et al. 2001).

Schweitzer et al. (2001) examined the colonization of naturally formed lake snow aggregates and found that they are inhabited by a limited number of β- and a-proteobacteria, which undergo a distinct succession. Detailed studies in Lake Constance, Lake Aydat and Lake Kinneret revealed that the bacterial communities on natural deep lake aggregates were dominated by few species of β-proteobacteria, Cytophaga/Flavobacteria and a-proteobacteria as well (Schweitzer et al. 2001, Grossart and Simon 1998, Simon et al. 2002, Lemarchand et al. 2006). This colonization pattern may reflect the similar adaptation of a specialized bacterial community to the unique environmental conditions on aggregates (Simon et al. 2002). The microbial community on lake snow was dominated by β-proteobacteria, especially during aggregate aging when filamentous and thus grazing resistant bacteria dominated (Grossart and Simon 1998). At earlier stages a-proteobacteria also comprised substantial fractions of the community. This community structure is similar to that of activated sludge flocs, suggesting that lake snow has a function comparable to that of activated sludge flocs in sewage treatment plants (Rath et al. 1998).

4.1.4 Shallow eutrophic lakes
Regional variability of OABC and diversity in Taihu were studied by amplified ribosomal DNA restriction analysis, and comparative analysis of eight large 16S rRNA clone libraries (Fig. 5). Our results demonstrate that OABC were numerically dominated by members of the β-proteobacteria (34.4%), a-proteobacteria (15.2%), Bacteroidetes (11.2%), and Planctomycetes (10.3%) groups. Clones affiliated with γ-proteobacteria, Actinobacteria, Acidobacteria, δ-proteobacteria, Verrucomicrobia, Chloroflexi, Firmicutes, Gemmatimonadetes, Nitrospira, and candidate division OP10 were also found in low frequencies. The dominance of the Bacteroidetes group was related to algae-based aggregates.

The spatial and temporal variations of OABC were also determined by terminal restriction fragment length polymorphism (T-RFLP) analysis. A total of 246 T-RFs were detected from the studied sites, but only about 20 T-RFs were dominant, suggesting that specific microbial populations were adapted to the unique niche provided by the organic aggregates. Analysis of similarity (ANOSIM) revealed significant temporal shifts in OABC, and significant intra-lake heterogeneity.

In Mecklenburg Lake District, Allgaier and Grossart (2006) found that Bacteroidetes (24.2%) is the most abundant phylum attached to particles, while Actinobacteria accounted for 44.4% of the free-living bacterial community. In Lake Taihu, Actinobacteria is the most abundant bacteria, representing 52.1% of the bacterioplankton community (Wu et al. 2007a).

4.2 Environment factors structured the dynamics of OA-attached bacterial communities
To our knowledge, environment factors structured the dynamics of attached bacterial communities have not been extensively studied, especially in shallow lakes. Allgaier and Grossart (2006) found that organic carbon (DOC), phytoplankton biomasses, and primary production have strong correlations with both particle-associated and free-living bacterial communities. In Lake Taihu, significant seasonal and station-dependent variations in OABC were found both in Meiliang Bay and in the lake center. Canonical correspondence analysis (CCA) demonstrated that temperature, dissolved oxygen, total suspended solids (TSS), nutrient levels, and some ions were significantly related to the spatio-temporal dynamics of OABC (Tang et al. 2010, Fig. 4). The findings add new insights into our understanding of the

ecological importance of OABC in large shallow lakes, in particular the intra-lake heterogeneity, temporal evolution, and the interaction between OA and its surrounding water. On the overall, food web structures and water physicochemical conditions shape the aggregates-attached bacterial communities in aquatic systems. Obviously, attached bacterial colonization patterns reflect the different adaptation properties of the members of special bacteria to the given environmental conditions (Simon et al. 2002).

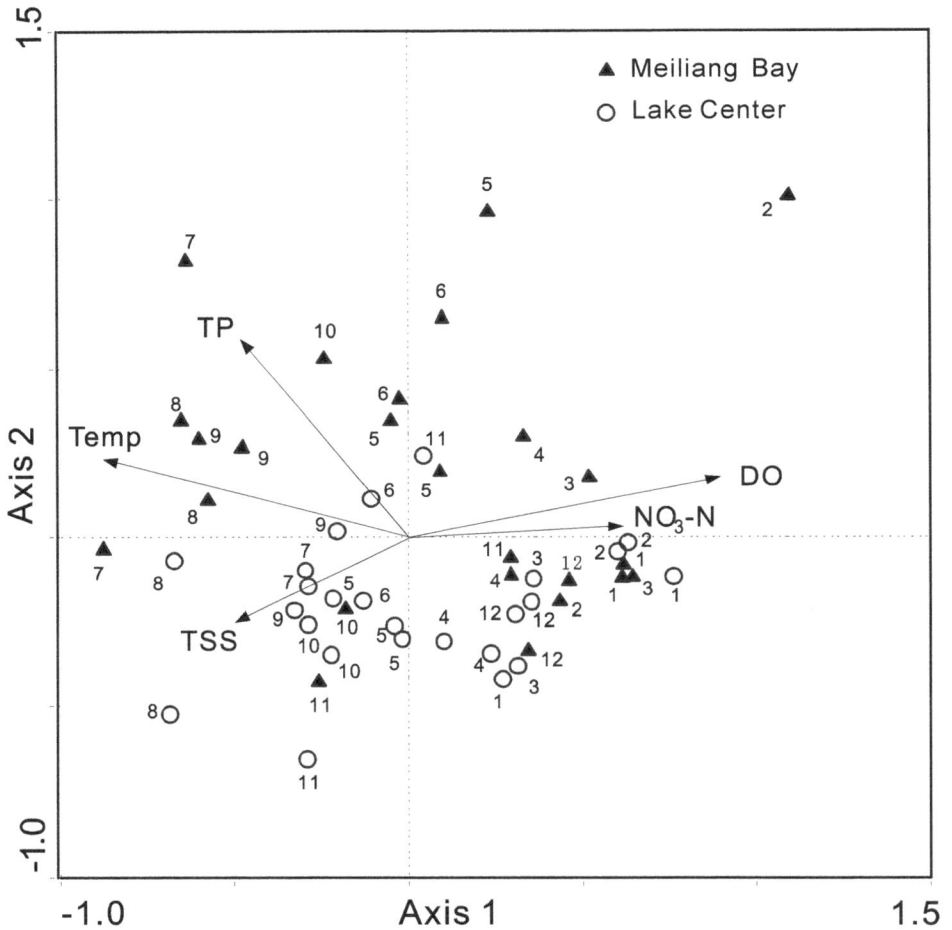

Fig. 4. Canonical Correspondence Analysis (CCA) biplot showed the impact of the statistically significant explanatory environmental variables on OABC in Meilaing Bay and Lake Center from May 2006 to May 2008. The matrices containing presence-absence of T-RFs were used as dependent variable. DO, TP, Temp, TSS, NO_3-N, refer to dissolved oxygen, total phosphorus, water temperature, total suspended solids and Nitrate, respectively. Months are represented by the numbers 1 (January) to 12 (December).

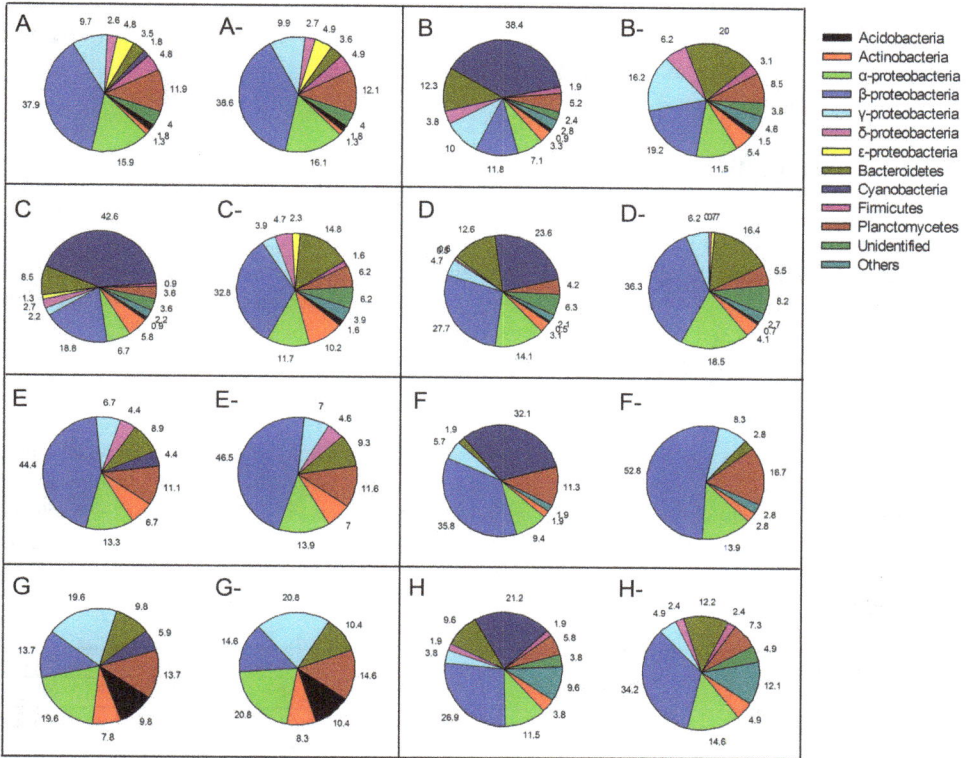

Fig. 5. Composition of associated bacterial communities. A~H, bacterial communities with *Cyanobacteria*; A- ~ H-, bacterial communities without *Cyanobacteria*. Others include *Gemmatimonadetes, Verrucomicrobia, OP10, Nitrospira* and *Chloroflexi*. A~H, represent different OA samples (detailed information is presented in Table 2).

4.3 Ecology of main phyla of attached bacterial communities

Studies have demonstrated that the *β-proteobacteria, Bacteroidetes* (formerly known as the *Cytophaga-Flavobacteria-Bacteroides*), *γ-proteobacteria* and *α-proteobacteria* are the most abundant bacterial cluster habitat on aggregates in most aquatic environments (see above). Despite the recognition that bacteria occupy a prominent role in aggregation processes and greatly impact biogeochemical processes, the bacterial taxa participating in these activities remain largely undescribed.

The *Bacteroidetes* comprise a large proportion of particle-associated bacteria (Table 3). Previous findings have highlighted the importance of *Bacteroidetes* during phytoplankton blooms (Eiler and Bertilsson 2004, Riemann and Winding 2001). In Lake Taihu, members of the *Bacteroidetes* formed a substantial component of the bacterial communities associated with algae-based OA. The high abundance of *Bacteroidetes* in the detritus-associated communities is related to their physiological characteristics. First, surface-dependent gliding motility is an important and widespread characteristic of these bacteria. Second, *Bacteroidetes* are chemoorganotrophic and efficiently degrade a variety of high molecular weight compounds such as protein, cellulose, pectin and chitin, and other polysaccharides (Kirchman 2002, DeLong et al. 1993). This finding might be related to their strong dependency on organic matter load or phytoplankton blooms

5. Ecological significance of OA-attached bacteria

Organic aggregates are involved in many ways in the nutrients cycling in aquatic ecosystems, such as decomposition of OA, cycling flux of particulate organic matter (POM), releasing of dissolved organic matter (DOM) and inorganic nutrients to the surrounding water. Bacteria clearly play an important role in most of the activities. POM solubilization, substrate hydrolysis and uptake, bacterial production, respiration and DOM release into the surrounding water are the major microbial processes.

5.1 Bank of nutrients

OA are composed of component particles consisting of living, senescent and dead algae, mainly diatoms in the ocean and *Cyanobacteria* in eutrophic lakes, but also of coccolithophorids, cysts of thecate dinoflagellates, phytodetritus, diatom frustules, bacteria, protozoans, zooplankton molts and carcasses, abandoned larvacean houses, pteropod webs, fecal pellets, macrophyte detritus, clay and silt minerals, calcite and other particles scavenged from surrounding water (Simon et al. 2002).

Sampling site	Site 10	Site 3	Site 8	Site 14	Mean
OA abundance /mg ·L^{-1}	48.49±41.17	47.41±41.01	58.65±41.92	39.32±44.23	48.47
OA-POM /mg ·g^{-1}	265.99±192.36	208.64±119.44	162.18±75.73	248.83±161.36	221.41
OA-C/mg ·g^{-1}	133.00±96.18	104.32±59.72	81.09±37.86	99.52±58.65	104.48
OA-N /mg ·g^{-1}	26.75±29.29	19.88±10.56	14.26±8.74	18.51±12.79	19.85
OA-P /mg ·g^{-1}	3.69±2.18	2.3±0.93	1.73±0.62	2.48±1.23	2.55
C_{OA-N}/TN (%)	17.17	17.47	15.40	16.10	16.54
C_{OA-P}/TP (%)	34.83	28.80	36.10	73.48	43.30
OA-C:N:P	36:7:1	45:9:1	47:8:1	40:8:1	41:8:1

Table 4. Chemical composition of OA in different sampling sites (see Fig. 1) in Lake Taihu.

OA are rich of organic matter, for organic carbon constitutes 10 to 40% of the total aggregates dry weight in marine and up to 66% in limnetic aggregates. In marine systems, Grossart and Simon (1993) found that soluble reactive phosphorus (SRP) concentrations in the matrix water of aggregates were > 1,000 times higher than in the surrounding water. Measurements of ammonia, nitrite, nitrate, and phosphate indicate significant differences in nutrient concentrations associated with OA from those of the surrounding water (Blackburn et al. 1998, Shanks and Trent 1979). But in shallow seas, estuaries and shallow lakes, macrophyte debris, resuspended minerals and inorganic particles also are important component of OA, so the relative percentage of organic matter are actually low in this kind of systems. According to our study, POM constitutes 10.45% of OA dry weight in Lake Taihu and the highest value (75%) occur in summer during *Cyanobacteria* blooms. OA are also rich in other nutrients, such as nitrogen (N) and phosphorus (P), as OA-N and OA-P represented up to 16.54% and 43.30% of total N and P in water column in Lake Taihu (Table

4). In Lake Kinneret, Israel, Grossart et al. (1998) revealed that aggregates with associated microorganisms are not only sites of vertical fluxes, centers of rapid and efficient recycling of POM, and a source of DOM, but also a potentially important food source for higher trophic levels. These findings indicate that organic aggregates are important sites for nutrient regeneration in aquatic systems and should be included in the conceptual framework of aquatic ecosystem studies.

5.2 Bacterial production and respiration

OA harbor a rich community of heterotrophic microbes, predominantly bacteria, which contribute only a small part of the total carbon of OA, but they are most important for OA decomposition (Bearon 2007, Grossart and Ploug 2000, Grossart and Simon 1998). Kirchman and Mitchell (1982) examined the abundance and heterotrophic uptake of bacteria attached to particulate matter suspended in five coastal ponds and two marshes, and found that although the number of particle-bound bacteria was <10%, these bacteria incorporated >40% of [^{14}C]glucose and [^{14}C]glutamate in the systems. In laboratory conditions, Grossart et al. (2007) found that natural bacterial assemblages attached to model aggregates (agar spheres) had threefold higher cell-specific bacterial protein production (BPP) than their free-living counterpart. Eiler et al. (2006) demonstrated that phycosphere-associated bacteria contributed between 8.5 and 82%of total bacterial secondary production.

In some aquatic systems, the growth rate of associated bacteria were found only slightly higher or in the same range as rates of free-living ones, and the turnover time of OA particulate organic carbon (POC) estimated by OA-associated bacteria production is so long that months to years would be required. Hence, other processes such as enzymatic hydrolysis and respiration are considered to be major microbial processes in OA decomposition (Becquevort et al. 1998). For example, the protease activity of model aggregates-attached bacteria was 10–20 times higher than that of free-living bacteria (Grossart et al. 2007). The increased protease activity allow attached bacteria to quickly exploit aggregate resources, which may accelerate remineralization of marine snow and reduce the downward carbon fluxes (Grossart et al. 2007). In fact, high solubilization rates and net release of nutrients to the surrounding water, which exceeded greatly the carbon demand of OA-associated bacteria, were found to be more reasonable to explain the rapid decomposition of POC in OA (Grossart and Simon 1998).

5.3 Nutrient recycling in shallow eutrophic lakes

The aggregate microniche is recognized as an area of nutrient enrichment containing higher densities of both active and dead phototrophic and heterotrophic planktonic cells than in the surrounding water (Turley 2000). Because of distinct differences of hydrodynamic forcing among pelagic marine, deep lake, turbid riverine, and shallow lake systems, the origin, size, composition and destination of the aggregates and the major groups of microorganisms colonized on them, may exhibit pronounced differences. In Lake Taihu, OA were mainly come from the detritus of *Cyanobacteria* bloom and resuspended sediment, which dominate the whole lake nutrient cycling of Lake Taihu. In summer and in the phytoplankton dominated areas of Lake Taihu, the OA were contributed mainly by the detritus of phytoplankton, while in the period of wind-induced water turbulence, the OA were mainly contributed by resuspended inorganic sediment.

There are significantly seasonal differences of the characteristics of OA in Lake Taihu, for the organic matter content of OA was significantly high during the summer of *Cyanobacteria* bloom than that in winter.

To investigate the relationship between wind induced turbulent and OA nutrient recycling in shallow eutrophic lake, in situ experiments were conducted in Meiliang Bay, located in the north part of Lake Taihu. The physical and chemical characteristics were continually monitored during a wind course (Fig. 6) one time per day, which continued for 10 days, and other water physical and chemical parameters were concomitantly monitored. OA abundance was significantly higher during wind period than calm stage and the maximum of OA abundance, which occurred the same day when wind speed was highest, was 29 times higher than that in the calm stage (Fig. 7). Although OA-C, OA-N and OA-P concentration were decreased during wind period the total OA C, N and P in water column were increased for the increased OA abundance (Fig. 8). Additionally, total suspended soil (TSS), total nitrogen (TN) and total phosphorus (TP) concentration during wind period were significantly higher than that of clam stage, but total dissolved phosphorus (TDP), total dissolved nitrogen (TDN) and soluble reactive phosphorus (SRP) were insignificantly different. During the wind period, OA alkaline phosphatase activity (OA-APA) and OA enzymatically hydrolysable phosphorus (OA-EHP) were both increased significantly, which accelerated organic phosphorous mineralization and SRP release (Fig. 9). The results indicate that SRP release induced by wind in shallow lakes may come from suspended matter, especially OA release rather than directly comes from sediment.

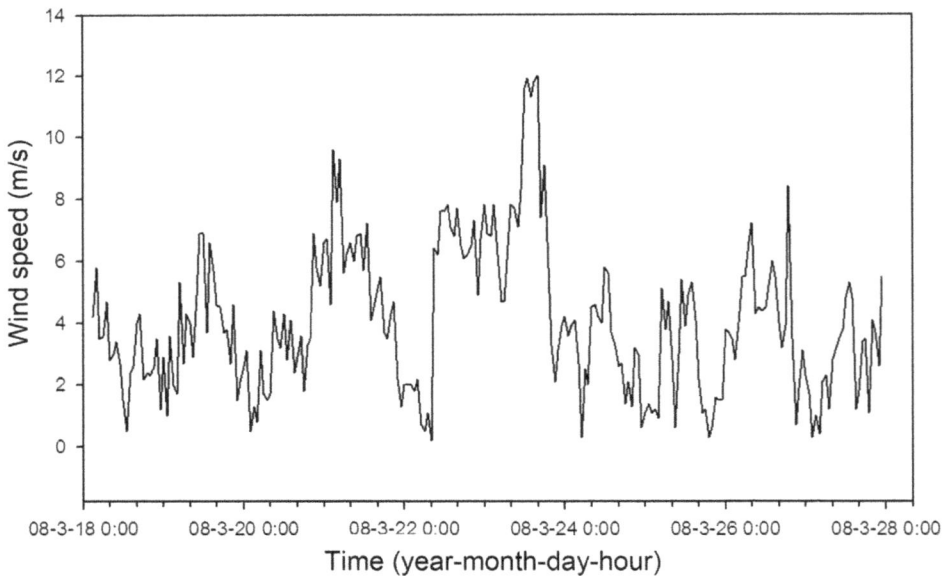

Fig. 6. Ten min averaged wind speed in the observation point during the study period.

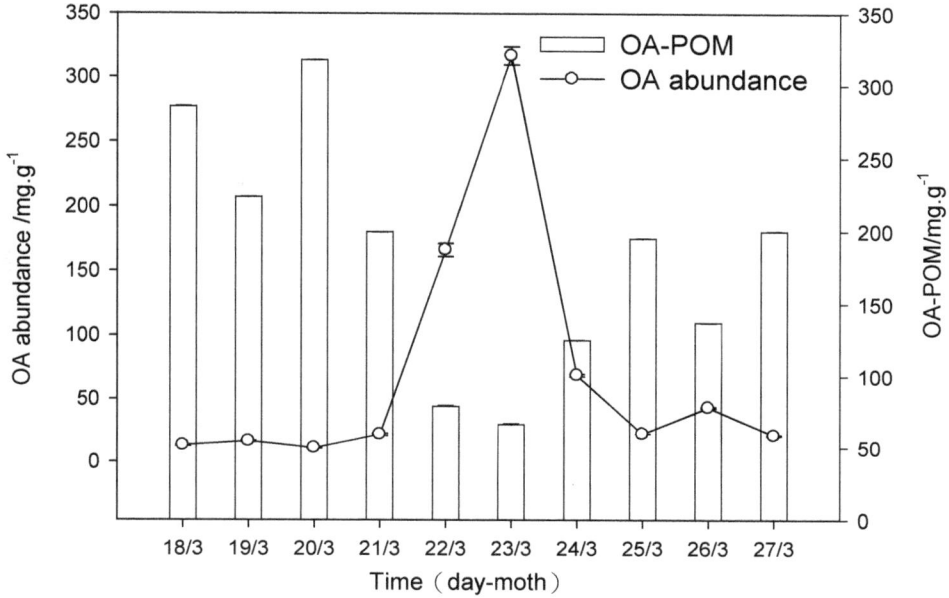

Fig. 7. Daily variability of OA abundance and OA-POM concentration during the 10-days observation in Meiliang Bay of Lake Taihu.

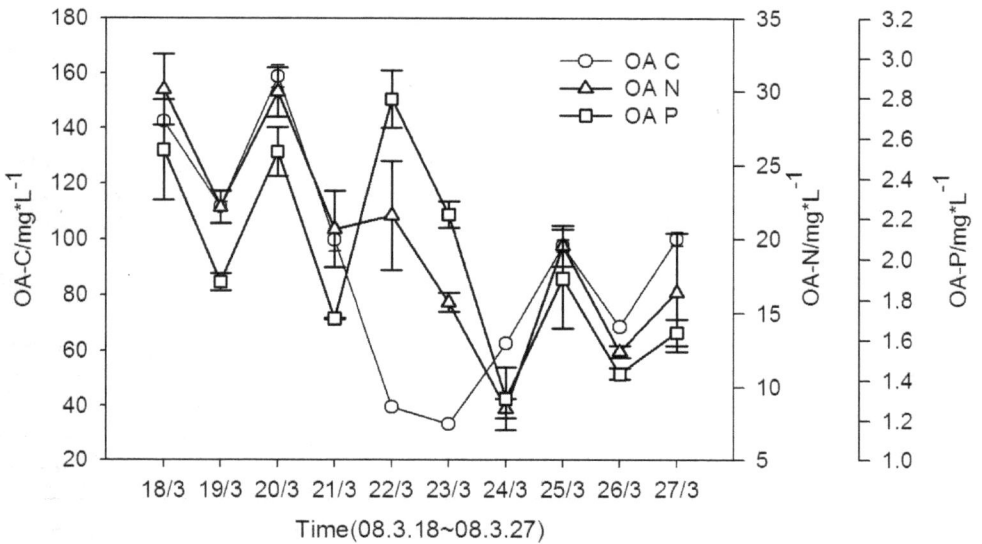

Fig. 8. Carbon, nitrogen, and phosphorous content of OA during wind period.

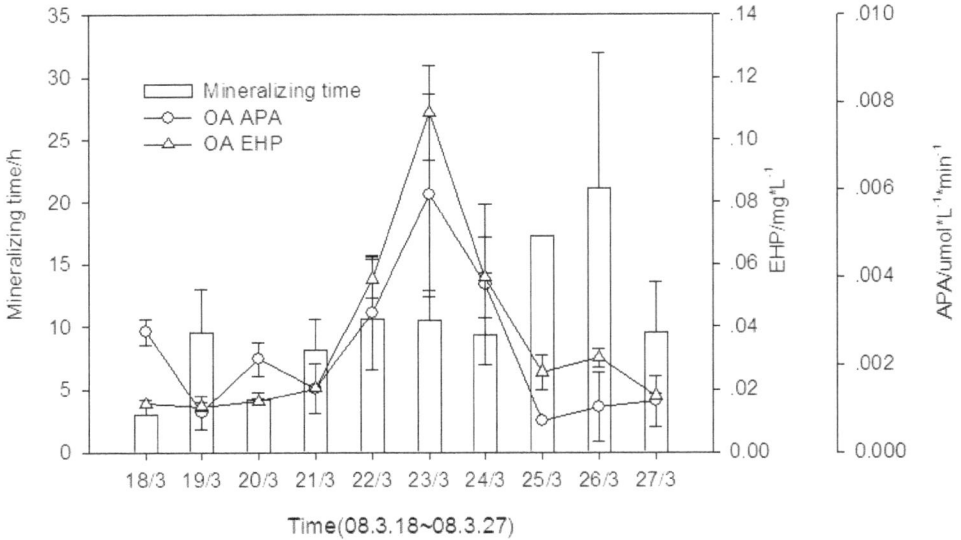

Fig. 9. Changes in OA-APA, OA-EHP concentration and OA-EHP mineralizing time.

5.4 Cyanobacterial blooms

During phytoplankton blooms in eutrophic lakes, primary productivity and its processing by the food web create a heterogeneous environment of particulate, colloidal, and dissolved organic matter in a continuum of size classes and concentrations. Blooms also have effect on the concentration and nutrient cycling of OA in water column (Turner 2002). Phosphorus is vital biogenic element in freshwater ecosystems such as lakes, reservoirs and rivers. Excessive P can eutrophicate freshwater bodies and further bring about harmful algal blooms, such as *Cyanobacteria* blooms in Lake Taihu (Paerl et al. 2011, Xu et al. 2010). The P-biogeochemical cycle plays crucial roles in freshwater ecosystems, and P release during the decomposition of cyanobacterial-bloom-formed OA is one of the most important processes involved.

To investigate P release during phytoplankton bloom, we carried out a stimulant experiment which focused on OA formation, decomposition and dissolved P release during cyanobacterial bloom in Lake Taihu. Considering the important role of turbulence, sediment and light on the decomposition, 6 treatments was set (Table 5). All treatments were conducted at a constant temperature $30 \pm 0.5°C$. The experiments lasted for 15 days, and the OA abundance, OA POM, Chl-*a*, SRP and TDP were measured on day 1, 2, 4, 6, 9, 12, 15.

Treatments	Turbulence	Sediment	Light
Treatment 1 （T1）	With	Without	Light
Treatment 2 （T2）	Without	Without	Light
Treatment 3 （T3）	With	With	Light
Treatment 4 （T4）	Without	With	Light
Treatment 5 （T5）	With	With	Dark
Treatment 6 （T6）	Without	With	Dark

Table 5. Different treatments of cyanobacterial decomposition experiments.

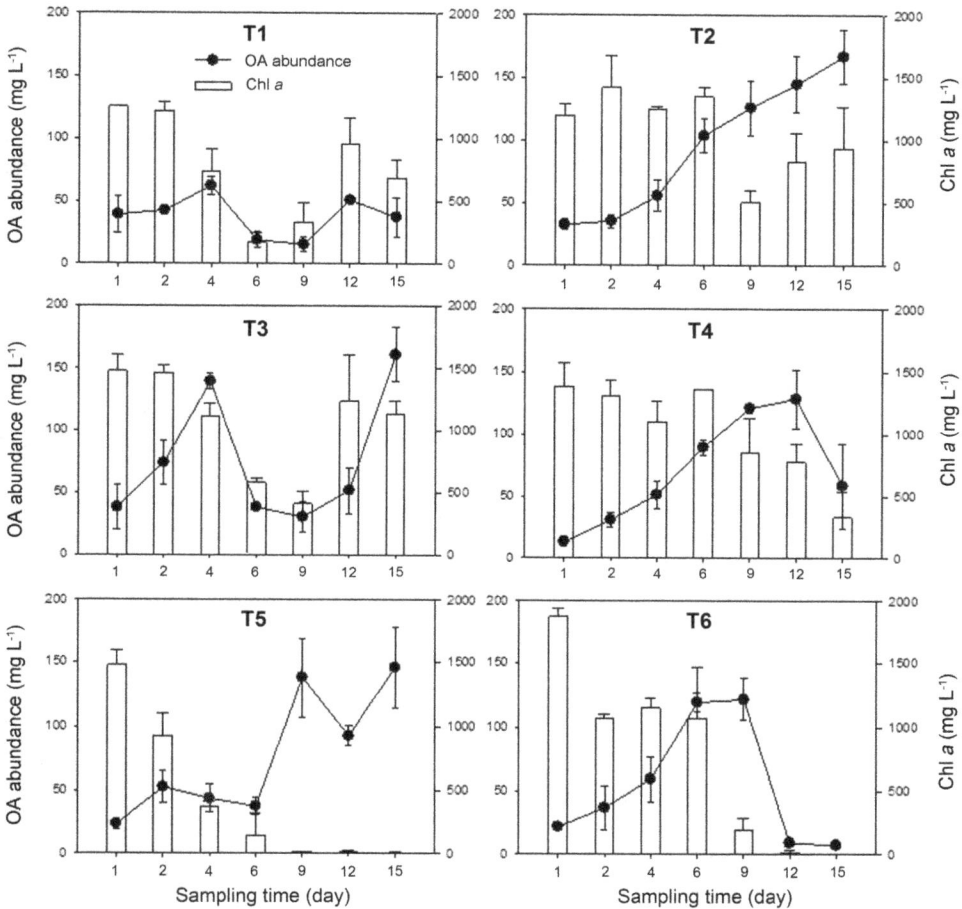

Fig. 10. Changes of OA abundance and the concentrations of Chl-*a* during experiment period.

According to our study, the OA abundances were high during the study period, averaging 68.26 mg L^{-1} (Fig. 10) and organic matter constitutes 67.8% of OA dry weight, which is much higher than that of the average of Lake Taihu. The OA enzymatic hydrolysable phosphorus (OA-EHP) occupies 77% of OA phosphorus (OA-P). This part of P can be hydrolyzed quickly by phosphatase to SRP, and released to surrounding water, which can be used directly by phytoplankton.

6. Conclusions, open questions and prospects

Given the facts that lakes have been described as early indicators of both regional and global environmental change (Williamson et al. 2008), and that lakes and other inland waters play a more critical role in the global carbon budget than previously recognized (Prairie et al. 2007), the role of microbes in these processes is of renewed interest (Newton et al. 2011). Based on the comparisons of abundance, diversity, community dynamic and potential function of aggregates-attached bacteria in marine, estuaries and rivers, deep lakes and shallow eutrophic lakes, we know that aggregates-attached bacteria are more abundant and show more phylogenetic diversity in more productive systems, such as rivers and shallow lakes. Already, the application of molecular techniques provides us with unprecedented access to the diversity and composition of OA-attached bacteria in time and space. However, the detailed ecological functions of these bacterial taxa participating in biogeochemical processes remain largely undescribed, especially in shallow eutrophic lakes. In our view, the major open issues, concerning the microbial ecology of attached bacteria in shallow eutrophic lakes, are the following: 1) Investigation of interaction between algae and their attached bacteria; 2) Identification and quantitative assessment of the role of attached bacteria in cyanobacterial decomposition and the regeneration of *Cyanobacteria* blooms; and 3) Examination of the importance of exchanging between aggregates-attached and free-living bacterial communities. Further work, in particular on specific details with respect to the interaction among attached bacteria, free-living bacteria and algal blooms on both small and large temporal and spatial scales, will certainly improve our understanding of the microbial ecology of organic aggregates-attached bacteria in aquatic ecosystems, with respect to processes and mechanisms.

7. Abbreviation

ANOSIM - Analysis of similarity; BPP - bacterial protein production; CCA - canonical correspondence analysis; CSP - Coomassie-stained particles; DAPI - 4' 6-diamidino-2-phenylindole dihydrochloride; DGGE - denaturing gradient gel electrophoresis; DNA - desoxynucleic acid; EHP – enzymatically hydrolysable phosphorus; DOM - dissolved organic matter; FISH - fluorescent in situ hybridization; DOM – dissolved organic matter; OA - organic aggregates; OABC - organic-aggregate-associated bacterial communities; OTUs - operational taxonomic units; PCR - polymerase chain reaction; POC – particulate organic carbon; POM –particulate organic matter; rRNA - ribosomal ribonucleic acid; SEM - scanning electron microscopy; SRB - sulfate-reducing bacteria; SRP - soluble reactive phosphorus; TEP - transparent exopolymer particles; TP - total phosphorus; T-RFLP - terminal restriction fragment length polymorphism; TSS - total suspended solids

8. Acknowledgment

This work was supported by the National Basic Research Program of China (grant 2008CB418103) and National Water Pollution Control and Management of Science and Technology Major Projects (grant 2009ZX07101-013).

9. References

Alber, M. & Valiela I. (1994) Production of microbial organic aggregates from macrophyte-derived dissolved organic material. *Limnology and Oceanography*, 39, 37-50.

Alldredge, A., Cole J. J. & Caron D. A. (1986) Production of heterotrophic bacteria inhabiting macroscopic organic aggregates (marine snow) from surface waters. *Limnology and Oceanography*, 31, 68-78.

Alldredge, A. L. & Silver M. W. (1988) Characteristics, dynamics and significance of marine snow. *Progress in Oceanography*, 20, 41-82.

Allgaier, M. & Grossart H.-P. (2006) Seasonal dynamics and phylogenetic diversity of free-living and particle-associated bacterial communities in four lakes in northeastern Germany. *Aquatic Microbial Ecology*, 45, 115-128.

Azam, F. & Richard A. L. (2001) Sea snow microcosms. *Nature*, 414, 495-498.

Böckelmann, U. 2001. Description and characterization of bacteria attached to lotic organic aggregates (river snow) in the Elbe River of Germany and the South Saskatchewan River of Canada. Berlin: der Technischen Universität Berlin.

Böckelmann, U., Manz W., Neu T. R. & Szewzyk U. (2000) Characterization of the microbial community of lotic organic aggregates (`river snow') in the Elbe River of Germany by cultivation and molecular methods. *FEMS Microbiology Ecology* 33, 157-170.

Böckelmann, U., Manz W., Neu T. R. & Szewzyk U. (2002) Investigation of lotic microbial aggregates by a combined technique of fluorescent in situ hybridization and lectin-binding-analysis. *Journal of Microbiological Methods*, 49, 75-87.

Bearon, R. N. (2007) A model for bacterial colonization of sinking aggregates. *Bulletin of Mathematical Biology*, 69, 417-431.

Becquevort, S., Rousseau V. & Lancelot C. (1998) Major and comparable roles for free-living and attached bacteria in the degradation of Phaeocystis-derived organic matter in Belgian coastal waters of the North Sea. *Aquatic Microbial Ecology*, 14, 39-48.

Berger, B., Hoch B., Kavka G. & Herndl G. J. (1996) Bacterial colonization of suspended solids in the River Danube. *Aquatic Microbial Ecology*, 10, 37-44.

Bidle, K. D. & Azam F. (2001) Bacterial control of silicon regeneration from diatom detritus: Significance of bacterial ectohydrolases and species identity. *Limnology and Oceanography*, 46, 1606-1623.

Bidle, K. D. & Fletcher M. (1995) Comparison of Free-Living and Particle-Associated Bacterial Communities in the Chesapeake Bay by Stable Low-Molecular-Weight Rna Analysis. *Applied and Environmental Microbiology*, 61, 944-952.

Blackburn, N., Fenchel T. & Mitchell J. (1998) Microscale nutrient patches in planktonic habitats shown by chemotactic bacteria. *Science*, 282, 2254-2256.

Brachvogel, T., Schweitzer B. & Simon M. (2001) Dynamics and bacterial colonization of microaggregates in a large mesotrophic lake. *Aquatic Microbial Ecology*, 26, 23-35.

Caron, D. A., Davis P. G., Madin L. P. & Sieburth J. M. (1982) Heterotrophic bacteria and bacterivorous protozoa in oceanic macroaggregates. *Science*, 218, 795-797.

Crump, B. C., Armbrust E. V. & Baross J. A. (1999) Phylogenetic analysis of particle-attached and free-living bacterial communities in the Columbia river, its estuary, and the adjacent coastal ocean. *Applied and Environmental Microbiology* 65, 3192-3204.

Crump, B. C., Baross J. A. & Simenstad C. A. (1998) Dominance of particle-attached bacteria in the Columbia River estuary, USA. *Aquatic Microbial Ecology*, 14, 7-18.

DeLong, E. F., Franks D. G. & Alldredge A. L. (1993) Phylogenetic diversity of aggregate-attached vs. free-living marine bacterial assemblages. *Limnology and Oceanography*, 38, 924-934.

Eiler, A. & Bertilsson S. (2004) Composition of freshwater bacterial communities associated with cyanobacterial blooms in four Swedish lakes. *Environmental Microbiology* 6, 1228-1243.

Eiler, A., Olsson J. A. & Bertilsson S. (2006) Diurnal variations in the auto- and heterotrophic activity of cyanobacterial phycospheres (*Gloeotrichia echinulata*) and the identity of attached bacteria. *Freshwater Biology*, 51, 298-311.

Ghiglione, J. F., Conan P. & Pujo-Pay M. (2009) Diversity of total and active free-living vs. particle-attached bacteria in the euphotic zone of the NW Mediterranean Sea. *FEMS Microbiology Letters*, 299, 9-21.

Ghiglione, J. F., Mevel G., Pujo-Pay M., Mousseau L., Lebaron P. & Goutx M. (2007) Diel and seasonal variations in abundance, activity, and community structure of particle-attached and free-living bacteria in NW Mediterranean Sea. *Microbial Ecology*, 54, 217-231.

Grossart, H.-P. & Ploug H. (2000) Bacterial production and growth efficiencies: Direct measurements on riverine aggregates. *Limnology and Oceanography*, 45, 436-445.

Grossart, H. P., Berman T., Simon M. & Pohlmann K. (1998) Occurrence and microbial dynamics of macroscopic organic aggregates (lake snow) in Lake Kinneret, Israel, in fall. *Aquatic Microbial Ecology*, 14, 59-67.

Grossart, H. P., Hietanen S. & Ploug H. (2003a) Microbial dynamics on diatom aggregates in Øresund, Denmark. *Marine Ecology Progress Series*, 249, 69-78.

Grossart, H. P., Kiørboe T., Tang K. & Ploug H. (2003b) Bacterial colonization of particles: Growth and interactions. *Applied and Environmental Microbiology*, 69, 3500-3509.

Grossart, H. P., Kiørboe T., Tang K. W., Allgaier M., Yam E. M. & Ploug H. (2006) Interactions between marine snow and heterotrophic bacteria: aggregate formation and microbial dynamics. *Aquatic Microbial Ecology*, 42, 19-26.

Grossart, H. P. & Simon M. (1993) Limnetic marcoscopic organic aggregates (lake snow): Occurrence, characteristics, and microbial dynamics in Lake Constance. *Limnology and Oceanography*, 38, 532-546.

Grossart, H. P. & Simon M. (1998) Bacterial colonization and microbial decomposition of limnetic organic aggregates (lake snow). *Aquatic Microbial Ecology*, 15, 127-140.

Grossart, H. P., Tang K. W., Kiørboe T. & Ploug H. (2007) Comparison of cell-specific activity between free-living and attached bacteria using isolates and natural assemblages. *FEMS Microbiology Letters* 266, 194-200.

Hempel, M., Blume M., Blindow I. & Gross E. M. (2008) Epiphytic bacterial community composition on two common submerged macrophytes in brackish water and freshwater. *BMC Microbiology*, 8, 10.

Jürgens, K. & Matz C. (2002) Predation as a shaping force for the phenotypic and genotypic composition of planktonic bacteria. *Antonie van Leeuwenhoek*, 81, 413-434.

Jürgens, K. & Sala M. M. (2000) Predation-mediated shifts in size distribution of microbial biomass and activity during detritus decomposition. *Oikos*, 91, 29-40.

Kellogg, C. & Deming J. (2009) Comparison of free-living, suspended particle, and aggregate-associated bacterial and archaeal communities in the Laptev Sea. *Aquatic Microbial Ecology*, 57, 1-18.

Kiørboe, T., Grossart H. P., Ploug H. & Tang K. (2002) Mechanisms and rates of bacterial colonization of sinking aggregates. *Applied and Environmental Microbiology*, 68, 3996-4006.

Kiørboe, T., Grossart H. P., Ploug H., Tang K. & Auer B. (2004) Particle-associated flagellates: swimming patterns, colonization rates, and grazing on attached bacteria. *Aquatic Microbial Ecology*, 35, 141-152.

Kiørboe, T. & Jackson G. A. (2001) Marine snow, organic solute plumes, and optimal chemosensory behavior of bacteria. *Limnology and Oceanography*, 46, 1309-1318.

Kiørboe, T., Tang K., Grossart H. P. & Ploug H. (2003) Dynamics of microbial communities on marine snow aggregates: Colonization, growth, detachment, and grazing mortality of attached bacteria. *Applied and Environmental Microbiology*, 69, 3036-3047.

Kirchman, D. & Mitchell R. (1982) Contribution of particle-bound bacteria to total microheterotrophic activity in 5 ponds and 2 marshes. *Applied and Environmental Microbiology*, 43, 200-209.

Kirchman, D. L. (2002) The ecology of *Cytophaga-Flavobacteria* in aquatic environments. *FEMS Microbiology Ecology* 39, 91-100.

Lemarchand, C., Jardillier L., Carrias J. F., Richardot M., Debroas D., Sime-Ngando T. & Amblard C. (2006) Community composition and activity of prokaryotes associated to detrital particles in two contrasting lake ecosystems. *FEMS Microbiology Ecology*, 57, 442-451.

Luef, B., Aspetsberger F., Hein T., Huber F. & Peduzzi P. (2007) Impact of hydrology on free-living and particle-associated microorganisms in a river floodplain system (Danube, Austria). *Freshwater Biology*, 52, 1043-1057.

Methé, B. A., Hiorns W. D. & Zehr J. P. (1998) Contrasts between marine and freshwater bacterial community composition: Analyses of communities in Lake George and six other Adirondack lakes. *Limnology and Oceanography*, 43, 368-374.

Mevel, G., Vernet M., Goutx M. & Ghiglione J. F. (2008) Seasonal to hour variation scales in abundance and production of total and particle-attached bacteria in the open NW Mediterranean Sea (0-1000 m). *Biogeosciences*, 5, 1573-1586.

Newton, R. J., Jones S. E., Eiler A., McMahon K. D. & Bertilsson S. (2011) A guide to the natural history of freshwater lake bacteria. *Microbiology and Molecular Biology Reviews*, 75, 14-49.

Paerl, H. W. (1974) Bacterial uptake of dissolved organic matter in relation to detrital aggregation in marine and freshwater systems. *Limnology and Oceanography*, 19, 966-972.

Paerl, H. W. & Prufert L. E. (1987) Oxygen-poor microzones as potential sites of microbial N_2 fixation in nitrogen-depleted aerobic marine waters. *Applied and Environmental Microbiology*, 53, 1078-1087.

Paerl, H. W., Xu H., McCarthy M. J., Zhu G. W., Qin B. Q., Li Y. P. & Gardner W. S. (2011) Controlling harmful cyanobacterial blooms in a hyper-eutrophic lake (Lake Taihu, China): The need for a dual nutrient (N & P) management strategy. *Water Research*, 45, 1973-1983.

Peduzzi, P. & Luef B. (2008) Viruses, bacteria and suspended particles in a backwater and main channel site of the Danube (Austria). *Aquatic Sciences,* 70, 186-194.

Ploug, H. & Grossart H. P. (2000) Bacterial growth and grazing on diatom aggregates: Respiratory carbon turnover as a function of aggregate size and sinking velocity. *Limnology and Oceanography,* 45, 1467-1475.

Ploug, H., Grossart H. P., Azam F. & Jorgensen B. B. (1999) Photosynthesis, respiration, and carbon turnover in sinking marine snow from surface waters of Southern California Bight: implications for the carbon cycle in the ocean. *Marine Ecology-Progress Series,* 179, 1-11.

Prairie, Y. T., Cole J. J., Caraco N. F., McDowell W. H., Tranvik L. J., Striegl R. G., Duarte C. M., Kortelainen P., Downing J. A., Middelburg J. J. & Melack J. (2007) Plumbing the global carbon cycle: Integrating inland waters into the terrestrial carbon budget. *Ecosystems,* 10, 171-184.

Qin, B. Q. 2008. *Lake Taihu, China: dynamics and environmental change.* Springer Netherlands.

Rath, J., Wu K. Y., Herndl G. J. & DeLong E. F. (1998) High phylogenetic diversity in a marine-snow-associated bacterial assemblage. *Aquatic Microbial Ecology,* 14, 261-269.

Riemann, L. & Winding A. (2001) Community dynamics of free-living and particle-associated bacterial assemblages during a freshwater phytoplankton bloom. *Microbial Ecology,* 42, 274-285.

Rink, B., Martens T., Fischer D., Lemke A., Grossart H. P., Simon M. & Brinkhoff T. (2008) Short-term dynamics of bacterial communities in a tidally affected coastal ecosystem. *FEMS Microbiology Ecology* 66, 306-319.

Schweitzer, B., Huber I., Amann R., Ludwig W. & Simon M. (2001) α- and β-*Proteobacteria* control the consumption and release of amino acids on lake snow aggregates. *Applied and Environmental Microbiology,* 67, 632-645.

Selje, N. & Simon M. (2003) Composition and dynamics of particle-associated and free-living bacterial communities in the Weser estuary, Germany. *Aquatic Microbial Ecology,* 30, 221-237.

Shanks, A. L. & Trent J. D. (1979) Marine snow - microscale nutrient patches. *Limnology and Oceanography,* 24, 850-854.

Silver, M. W., Coale S. L., Pilskaln C. H. & Steinberg D. R. (1998) Giant aggregates: Importance as microbial centers and agents of material flux in the mesopelagic zone. *Limnology and Oceanography,* 43, 498-507.

Silver, M. W., Shanks A. L. & Trent J. D. (1978) Marine snow: a microplankton habitat and source of smallscale patchiness in pelagic populations. *Science,* 201, 371-373.

Šimek, K., Hornak K., Jezbera J., Nedoma J., Vrba J., Straskrabova V., Macek M., Dolan J. R. & Hahn M. W. (2006) Maximum growth rates and possible life strategies of different bacterioplankton groups in relation to phosphorus availability in a freshwater reservoir. *Environmental Microbiology* 8, 1613-1624.

Simon, M., Alldredge A. L. & Azam F. (1990) Bacterial Carbon Dynamics on Marine Snow. *Marine Ecology-Progress Series,* 65, 205-211.

Simon, M., Glockner F. O. & Amann R. (1999) Different community structure and temperature optima of heterotrophic picoplankton in various regions of the Southern Ocean. *Aquatic Microbial Ecology,* 18, 275-284.

Simon, M., Grossart H. P., Schweitzer B. & Ploug H. (2002) Microbial ecology of organic aggregates in aquatic ecosystems. *Aquatic Microbial Ecology*, 28, 175-211.

Smith, D. C., Simon M., Alldredge A. L. & Azam F. (1992) Intense hydrolytic enzyme activity on marine aggregates and implications for rapid particle dissolution. *Nature*, 359, 139-141.

Tang, K. W., Grossart H. P., Yam E. M., Jackson G. A., Ducklowl H. W. & Kiørboe T. (2006) Mesocosm study of particle dynamics and control of particle-associated bacteria by flagellate grazing. *Marine Ecology-Progress Series*, 325, 15-27.

Tang, X. 2009. Spatio-temporal dynamics of organic aggregates and its associated bacterial communities in a large, shallow, eutrophic lake (Lake Taihu, China). In *Division of Lake Biology and Ecology*. Nanjing: Nanjing Institute of Geography and Limnology, Chinese Academy of Sciences.

Tang, X., Gao G., Chao J., Wang X., Zhu G. & Qin B. (2010) Dynamics of organic-aggregate-associated bacterial communities and related environmental factors in Lake Taihu, a large eutrophic shallow lake in China. *Limnology and Oceanography*, 55, 469-480.

Tang, X., Gao G., Qin B., Zhu L., Chao J., Wang J. & Yang G. (2009) Characterization of bacterial communities associated with organic aggregates in a large, shallow, eutrophic freshwater lake (Lake Taihu, China). *Microbial Ecology*, 58, 307-322.

Turley, C. (2000) Bacteria in the cold deep-sea benthic boundary layer and sediment–water interface of the NE Atlantic. *FEMS Microbiology Ecology*, 33, 89-99.

Turley, C. & Mackie P. (1994) Biogeochemical significance of attached and free-living bacteria and the flux of particles in the NE Atlantic Ocean. *Marine Ecology-Progress Series*, 115, 191-191.

Turner, J. T. (2002) Zooplankton fecal pellets, marine snow and sinking phytoplankton blooms. *Aquatic Microbial Ecology*, 27, 57-102.

Wörner, U., Zimmerman-Timm H. & Kausch H. (2000) Succession of protists on estuarine aggregates. *Microbial Ecology*, 40, 209-222.

Weiss, P., Schweitzer B., Amann R. & Simon M. (1996) Identification in situ and dynamics of bacteria on limnetic organic aggregates (lake snow). *Applied and Environmental Microbiology*, 62, 1998-2005.

Williamson, C. E., Dodds W., Kratz T. K. & Palmer M. A. (2008) Lakes and streams as sentinels of environmental change in terrestrial and atmospheric processes. *Frontiers in Ecology and the Environment*, 6, 247-254.

Wu, Q. L., Zwart G., Wu J., Agterveld M., Liu S. & Hahn M. W. (2007a) Submersed macrophytes play a key role in structuring bacterioplankton community composition in the large, shallow, subtropical Taihu Lake, China. *Environmental Microbiology* 9, 2765-2774.

Wu, Q. L. L., Chen Y. W., Xu K. D., Liu Z. W. & Hahn M. W. (2007b) Intra-habitat heterogeneity of microbial food web structure under the regime of eutrophication and sediment resuspension in the large subtropical shallow Lake Taihu, China. *Hydrobiologia*, 581, 241-254.

Xu, H., Paerl H. W., Qin B., Zhu G. & Gao G. (2010) Nitrogen and phosphorus inputs control phytoplankton growth in eutrophic Lake Taihu, China. *Limnology and Oceanography*, 55, 420-432.

Zimmermann-Timm, H. (2002) Characteristics, dynamics and importance of aggregates in rivers - an invited review. *International Review of Hydrobiology*, 87, 197-240.

Zimmermann-Timm, H., Holst H. & M'¹ller S. (1998) Seasonal dynamics of aggregates and their typical biocoenosis in the Elbe Estuary. *Estuaries and Coasts,* 21, 613-621.

Zimmermann, H. (1997) The microbial community on aggregates in the Elbe Estuary, Germany. *Aquatic Microbial Ecology,* 13, 37-46.

Zwisler, W., Selje N. & Simon M. (2003) Seasonal patterns of the bacterioplankton community composition in a large mesotrophic lake. *Aquatic Microbial Ecology,* 31, 211-225.

10

Oxygenated Hydrocarbons in Coastal Waters

Warren J. de Bruyn, Catherine D. Clark and Lauren Pagel

School of Earth and Environmental Sciences,
Schmid College of Science and Technology,
Chapman University, Orange, California,
USA

1. Introduction

Low molecular weight (LMW) carbonyl compounds (also termed oxygenated hydrocarbons) are produced photochemically from chromophoric dissolved organic matter (CDOM) in natural waters (Kieber and Mopper, 1987; Kieber et al, 1990; de Bruyn et al., 2011). CDOM refers to the optically-active (i.e. light absorbing and hence colored) portion of dissolved organic matter (DOM). DOM is the largest organic carbon reservoir in the ocean $(700 \times 10^{15}$ g C), of comparable size to the atmospheric carbon dioxide (CO_2) pool (Hedges, 2002). Studying the distribution and dynamics of marine DOM is of interest to oceanographers because oceanic DOM is a significant component of the global carbon cycle, in addition to affecting the bioavailability of chemical species through absorption, contributing to the spectral properties of seawater, acting as a water source tracer and a microbial food source (Hedges, 2002). DOM is primarily derived from decaying terrestrial plant matter in fresh and coastal waters, with production in oceanic waters from marine organisms via viral or bacterial lysis, grazing and microbial degradation (Perdue, 1998; McKnight and Aiken, 1998; Hessen and Tranvik, 1998). DOM is a complex macromolecular humic-type material, which has been extensively studied by marine scientists over the last 4 decades using analytical techniques including chromatographic separation, size fractionation, radioisotopes, elemental analyses, mass spectroscopy and nuclear resonance mass spectrometry (Perdue, 1998; Benner, 2002; Sharp, 2002; Whitehead, 2008). Optical techniques like UV-VIS absorption and fluorescence spectroscopy are used to study CDOM, including the identification of components via 3 dimensional excitation-emission matrix spectroscopy (EEMs) and parallel factor analysis (PARAFAC) computational methods (see for example Yamashita et al., 2008).

CDOM absorbs sunlight to produce photo-excited states which undergo a series of primary and secondary reactions in natural waters that produce reactive species like peroxides and singlet oxygen, in addition to low molecular weight carbon-containing compounds (Miller, 1994). These photochemical processes play a significant role in the global carbon cycle through remineralization of dissolved organic carbon (DOC) to carbon dioxide (Bano et al., 1998). As a major energy source for microorganisms, LMW carbonyls also influence CO_2 levels indirectly (Clark et al., 2004). LMW carbonyls in seawater also have the potential to influence atmospheric chemistry via air-sea exchange between surface waters and the atmosphere (Zhou and Mopper, 1990; Jacob et al., 2002). Oxygenated hydrocarbons are ubiquitous in the atmosphere (Singh et al., 2004) where they react rapidly with OH radicals (Singh et al., 2001) and produce other reactive HO_x radicals, ozone, carbon monoxide, peroxyacetyl nitrate and formaldehyde (see for eg. de Gouw et al., 2005; Dufour et al., 2007; Millet et al., 2010). As an OH sink and an atmospheric HO_x and ozone source, oxygenated hydrocarbons have a direct impact on the oxidative capacity of the atmosphere. Over the last decade there have been a number of attempts to inventory sources and analyze atmospheric budgets of these species (Singh et al., 1995; Singh et al., 2001; Jacob et al., 2002; Millet et al., 2008; Millet et al., 2010; Naik et al., 2010). While these budget calculations have improved over time, large uncertainties remain, with the role of the oceans often the largest uncertainty.

In spite of their potential significance, the database of LMW carbonyl and aldehyde measurements in seawater is extremely small. In Table 1 we show the current database for formaldehyde (CH_3O), acetaldehyde (CH_3COH) and acetone (CH_3COCH_3), the oxygenated hydrocarbons measured in this study. Previous measurements have been conducted in tropical and northern regions of the Atlantic and Pacific Oceans. The most in-depth studies have been those by Kieber and Mopper in the Atlantic Ocean, primarily in and around the southeastern coastal region of the United States (west coast of Florida, Mopper and Stahovec 1986; Biscayne Bay off Florida and Caribbean Sea, Kieber and Mopper 1990; Biscayne Bay, Sargasso Sea and Bahamas, Zhou and Mopper 1997). In general, measured ambient acetone concentrations in seawater range from 3.00 to 40 x 10^{-9} mol. L^{-3} (nM), measured acetaldehyde concentrations range from 1.38 nM to 30 nM and measured formaldehyde concentrations range from 4 to 98 nM. In general concentrations are higher at coastal sites where CDOM levels are higher. Zhou and Mopper (1997) also measured oxygenated hydrocarbon levels in the surface microlayer in the Atlantic Ocean of 54.8 and 15.7 nM for acetone and acetaldehyde respectively significantly higher than bulk levels. More recent studies have found similar levels in the North Atlantic and Pacific Oceans (Table 1).

With a photochemical production mechanism, one would expect to see well defined diel cycles in concentrations of these species if 1) the photochemical source is dominant , 2) the sinks are relatively constant over the course of a day, and 3) the photochemical precursor does not vary significantly over the diel cycle. Takeda et al. (2006) reported higher concentrations of acetaldehyde and formaldehyde in Hiroshima Bay at noon relative to midnight (by a factor of 3), consistent with a photochemically driven process. Stahovec and Mopper (1986) also report data suggesting a photochemical production mechanism, specifically diurnal fluctuations by a factor of 10 from 3 to 30 nM for acetaldehyde, and weaker fluctuations from 15 to 40 nM for formaldehyde. Similarly, Zhou and Mopper (1997) observed well defined diel cycles in acetaldehyde and formaldehyde, with midday maxima in Hatchet Bay, Bahamas. The formaldehyde and acetaldehyde cycles ranged from 15 to 30 nM and 2 to 12 nM respectively. On the other hand, Marindino et al. (2005) did not observe well defined diurnal cycles in acetone in the tropical Atlantic.

We report here ambient concentrations and diel cycling of three LMW oxygen-containing compounds (specifically acetone, acetaldehyde, and formaldehyde) measured in coastal waters of the Pacific Ocean on the southwest coast of the USA. These studies represent the first measurements of ambient concentrations of LMW compounds in this global region (near-shore Pacific waters of the western USA), where coastal waters are predominantly hydrologically linked to brackish tidally flushed salt marsh systems rather than riverine inputs (Clark et al., 2008a) as in the bulk of the prior studies on the east coast of the USA (Table 1).

Region	[CH₃COCH₃]	[CH₃COH]	[CH₃O]	Method	Source
West Tropical Atlantic	4-6	2-30	15-50	HPLC DNPH	Mopper & Stahovec, 1986
West Tropical Atlantic		4-15	10-30	HPLC DNPH	Kieber & Mopper, 1990
West Tropical Atlantic	3.00	1.38	3.88	HPLC DNPH	Zhou & Mopper, 1997
Tropical Atlantic	17.6			PTR-MS	Williams et al. 2004
Tropical Pacific	14.5			APCIMS	Marindino et al., 2005
North Pacific		7-9	31-98	HPLC DNPH	Takeda et al., 2006
North Atlantic	5.5-6.9			PFBHA SPME GC/MS	Hudson et. al. 2007
West NorthPacific	10-40	<2-10		MEM PTR-MS	Kameyama et al., 2010

Table 1. Previous measurements of acetone (CH₃COCH₃), acetaldehyde (CH₃COH) and formaldehyde (CH₃O) in seawater. All concentrations are in nM. Technique abbreviations: Atmospheric Pressure Chemical Ionization Mass Spectrometry (APCIMS); pre-column 2,4-dinitrophenylhydrazine derivitization high performance liquid chromatography (HPLC DNPH) ; Proton Transfer Mass Spectrometry (PTR-MS); membrane proton transfer mass spectrometry (MEM PTR-MS), PFBHA derivitization solid phase micro-extraction gas chromatography mass spectrometry (PFBHA SPME GC/MS).

2. Methods

2.1 Sampling sites and procedures

Most samples were collected at Huntington State Beach (HSB), Orange County, Southern California (33.3833 N 117.5846 W), generally within a 2 h period around solar noon in January and February 2009. Other beaches sampled were Newport Beach (NB 33.3631 N 117.5542 W) and Seal Beach (SB 33.7414 N 118.104 W). This covers a 30 mile stretch of coastline in Orange County. Two Upper Newport Back Bay (UNBB 33.642243 N 117.8867235 W) estuarine source water samples were also included for comparison. The UNBB empties into the ocean at Newport Beach, down-coast from HSB (Jeong et al., 2005). Beach water samples were collected from ankle-deep surf-zone waters on an incoming wave. All samples were vacuum filtered with minimal pressure differential at the lab through glass fiber filters (GF/F; nominal pore size 0.7 μm; Whatman International Ltd) and ambient oxygenated hydrocarbon concentrations measured within 20 min to 2 h of collection. Samples were kept in sealed air-tight amber bottles during transit to the laboratory where they were derivatised and analysed by HPLC

A 28 hour diel study was conducted at Crystal Cove State Beach, Orange County, Southern California (33.574 N 117.840 W) from 8 to 9 July in 2009. Samples were collected every 2 hours and filtered as described above, but the water samples were derivatised in situ at the field site laboratory (see method described below) and extracted onto C18 Sep-Pak cartridges, stored at 4°C and analyzed at the conclusion of the diel study (within 48 h of collection).

2.2 HPLC analysis

Formaldehyde, acetaldehyde and acetone concentrations were quantified with a pre-column 2,4-dinitrophenylhydrazine (DNPH) derivitization HPLC method (Agilent 1100; Novapak C-18 (4μm) column; UV detection at 370nm; as described by Kieber and Mopper (1990) and Zhou and Mopper (1997)). 20 mg of re-crystallized DNPH (Sigma Aldrich) was dissolved in 15 mL of a solution of concentrated hydrochloric acid (~12M; Pharmco; ACS Reagent grade), deionised water (DI) and acetonitrile (ACN) mixed in a volume ratio of 2:5:1. Any carbonyl contamination in the resultant DNPH solution was removed by 2 successive extractions with carbon tetrachloride (Sigma Aldrich Chromasolv for HPLC; 99.9%) just prior to use.

To derivatize natural water samples, 200 μL x 10^{-6} of DNPH solution was added to a 20 mL water sample in a 22 mL Teflon vial and the reaction allowed to proceed for 60 min before extraction and pre-concentration on C18 Sep-Pak cartridges (Supelco). Prior to use, cartridges were cleaned with ACN and distilled water. The derivatized sample was passed through the conditioned extraction cartridge at a flow rate of 10-15 mL min^{-1}. Excess reagent was washed off the cartridge with 25 mL of a 17% ACN (v/v) aqueous solution followed by 5 mL of DI. Carbonyl hydrazones were eluted from the cartridge with 1 mL ACN into Teflon vials.

Prior to HPLC analysis, extracts were reduced to dryness with a stream of carbonyl free nitrogen gas at room temperature and re-dissolved in 2 mL of a 10% ACN (v/v) solution in DI for a 10-fold enrichment. 2 mL of the enriched sample was injected directly onto the Novapak C-18 column (Waters). Carbonyl hydrazones were eluted using a two-solvent gradient: solvent A was 10% ACN (v/v) solution in DI adjusted to a pH of 2.6 with 10 N sulfuric acid (Sigma Aldrich); solvent B was 100% ACN. The gradient was: isocratic at 35% B for 2 min; 35% to 53% B in 4 min; isocratic at 40% B for 8 min; 40% to 80% B in 10 min; and isocratic at 100% B for 15 min. Column flow rate was 1.5 mL min^{-1}; column temperature was controlled at 25°C.

Derivitization steps were carried out in a fume hood in a solvent-free laboratory to prevent potential contamination. A reagent blank was obtained as per Kieber and Mopper (1990) by injecting un-derivitized reagent directly i.e. the solution was treated exactly the same but the derivitization time was zero. No reagent blanks were observed for acetaldehyde and acetone. Formaldehyde had a reagent blank of 9 ± 4 nM. Based on triplicate measurements of each sample, the average precision for formaldehyde, acetaldehyde and acetone was 15, 9, and 10% respectively. Detection limits were estimated to be approximately 2, 0.5 and 0.5 nM for formaldehyde, acetaldehyde and acetone respectively.

2.3 Absorbance measurements

The absorbance of natural water samples is frequently used as a proxy for the level of CDOM (for example Green and Blough, 1994; Seritti et al., 1998; Gallegos et al., 2004), with higher absorbance corresponding to waters with higher concentrations of CDOM and organic carbon (Moran et al., 2000; Stedmon and Markager, 2003; Kowalcuk et al., 2010). Absorbance spectra were measured with a diode-array UV-visible spectrometer (Agilent Technologies 8453) from 200-700 nm in a quartz sample cell (path length = 10 cm) with a deionized water blank. Absorbance was transformed to absorption coefficient (a, in m^{-1}) by multiplying the measured absorbance at 300 nm by 2.303 and dividing by the path length in m (Hu et al. 2002). The absorption coefficient at 300 nm is commonly reported for CDOM studies for inter-comparisons (Miller 1998).

3. Results and discussion

3.1 Ambient concentrations

Seawater formaldehyde levels ranged from 7.5 to 88 nM with an average of 27 ± 25 nM. Acetaldehyde and acetone levels ranged from 2.7 to 19.9 nM and 2.7 to 12.5 nM respectively, with average levels in seawater of 9 ± 4 nM and 8 ± 2 nM respectively. Measured seawater levels for all compounds were not significantly different from the source water levels in the estuary. Overall, the ambient levels measured here in the Pacific waters of the Southwestern USA (Table 2) are consistent with the limited database of measurements in the literature for Atlantic and Pacific waters (Table 1).

Location	Date	Time	[CH$_3$O]	[CH$_3$COH]	[CH$_3$COCH$_3$]
Wetlands					
UNBB	01/16	11:00	42 ± 3	11.5 ± 0.9	12.5 ± 0.3
	02/15	11:00	10.3 ± 0.3	10.4 ± 0.2	7.5 ± 0.6
Beaches					
SB	02/10	11:00	57 ± 1	11.7 ± 0.7	6.7 ± 0.3
NB	02/11	11:00	15 ± 1	7.2 ± 0.5	12.5 ± 0.5
HSB	01/20	11:00	23 ± 13	2.7 ± 0.9	2.7 ± 1.8
	01/21	11:00	88 ± 11	9.0 ± 1.0	8.4 ± 0.6
	01/26	17:00	10 ± 1	7.2 ± 0.5	7.3 ± 0.3
	01/27	07:00	13 ± 5	9.4 ± 1.7	9.4 ± 0.6
	01/30	13:00	14 ± 1	8.1 ± 0.7	7.9 ± 0.1
	02/01	15:00	6.5 ± 0.3	7.1 ± 0.1	9.4 ± 0.1
	02/02	19:00	25 ± 1	7.5 ± 0.3	7.5 ± 0.5
	02/04	09:00	32 ± 1	19.9 ± 0.8	6.9 ± 0.3
	02/08	11:00	7.5 ± 0.8	7.4 ± 0.95	7.3 ± 0.9
Average			27 ± 25	9 ± 4	8 ± 2

Table 2. Ambient concentrations (in nM) of formaldehyde, acetaldehyde and acetone in water samples from Upper Newport Back Bay (UNBB), Seal Beach (SB), Newport Beach (NB) and Huntington State Beach (HSB) in Orange County, Southern California, USA, 2009. Concentrations are the mean of 3 measurements ± 1σ. Average values do not include the wetland samples.

In general, oxygenated hydrocarbon levels reported in the literature decrease from the coast into the open ocean as CDOM levels decrease based on decreasing absorption coefficients. Levels are also higher in the surface microlayer compared to underlying waters. The acetone levels of 8 ± 2 nM measured in this study are lower than the 30 nM reported in one coastal study in a coastal zone with significant fresh water inputs and high CDOM levels (Zhou and Mopper, 1997), and on the lower end of the range of previous ocean studies in environments with low CDOM levels. For example, Zhou and Mopper (1997) reported bulk open ocean concentrations of 3 nM , whereas Williams et al. (2004) reported open ocean acetone levels of 17 nM for the tropical Atlantic and Marindino et al. (2005) reported levels of 3-60 and 2-18 nM for the tropical and northern Pacific oceans respectively. The database of formaldehyde and acetaldehyde measurements in seawater for comparison purposes is more limited. Zhou and Mopper (1997) reported open ocean formaldehyde and acetaldehyde levels of 4 and 3 nM respectively with higher coastal levels of 15-42 nM for formaldehyde and 1-12 nM for acetaldehyde in the Atlantic. Their coastal values are consistent with the average values of 27 ± 25 nM for formaldehyde and 9 ± 4 nM for acetaldehyde we obtained in this study for coastal Pacific waters.

Although the ambient levels we measured in our study are within the range of coastal levels in the literature, it is important to note a key difference. Namely, coastal waters in this semi-arid region are dominated by inputs from tidal flushing of salt marshes for much of the year in the absence of limited and seasonal rain events (Clark et al., 2008) i.e. this is a coastal environment with low CDOM levels. Previous coastal studies focused on humic-rich coastal environments with significant freshwater riverine inputs and consequently high concentrations of CDOM. For example, Kieber et al. (1990) measured LMW concentrations for natural waters ranging from 10 to 20 m^{-1} in absorption coefficient (used as a proxy for the amount of CDOM). These absorption values are 10 to 50 times greater than the values of 0.27 to 2.4 m^{-1} we measured for these coastal waters on the Southwest coast of the USA. Given the much lower values for absorption coefficients and hence CDOM levels in our study, the similarity in ambient concentration levels suggests that there must be significant differences in production efficiencies and/or loss processes from salt-marsh derived CDOM in this study vs. the riverine CDOM in prior studies. Our previous laboratory-based study (de Bruyn et al., 2011) showed that the apparent quantum yields (i.e. efficiencies) of photochemical production for the 3 LMW carbonyls discussed here increased by an order of magnitude in going from wetland to near-shore coastal waters. These changes correlated linearly with spectral slopes, one optical measure of the aging of CDOM (Tardowski and Donaghay, 2002; Tzortziou et al., 2007; Helms et al. 2008).

3.2 Diel study

Results from the 28 hour diel study at Crystal Cove State Beach are shown in Figure 1. Relatively small concentration ranges were observed for acetone and acetaldehyde, from 6.6 to 8.5 nM for acetone and 2.0 to 10.6 nM for acetaldehyde with average levels for acetone and acetaldehyde of 5.5 ± 2 nM and 7.5 ± 0.5 nM respectively.

Both the range and average concentrations are consistent with the levels measured at the other beach sites earlier in the year (Table 1). However, formaldehyde levels exhibited a wider range of concentrations from 27.2 to 98.6 nM, with an average of 47.2 ± 25 nM which was higher than the average concentration measured at the other beach sites. Diel cycles consistent with a photochemically driven process have been previously observed for oxygenated hydrocarbons in bulk open seawaters (Zhou and Mopper, 1997) and Florida coastal waters (Mopper and Stahovec, 1986).

We observed some evidence for photochemical production of acetone and acetaldehyde, which both showed an increase in levels prior to noon followed by a decrease. For acetone, the midday maximum is the dominant feature; however there are also maxima during the night. For acetaldehyde, the midday maximum is not the dominant feature as there are 2 larger maxima during the evening and night. This contrasts with Mopper and Stahovec (1986) who observed strong diel cycling in acetaldehyde from 2 to 3 nM at night to 20 to 30 nM in early afternoon. For formaldehyde, there is no evidence of a photochemically driven process. Rather the diel cycle is dominated by a maximum of 100 nM (this abbreviation is used throughout rest of paragraph and is defined earlier) in the early morning. Mopper and Stahovec (1986) reported weak diurnal fluctuations in formaldehyde ranging from 15 to 50 nM.

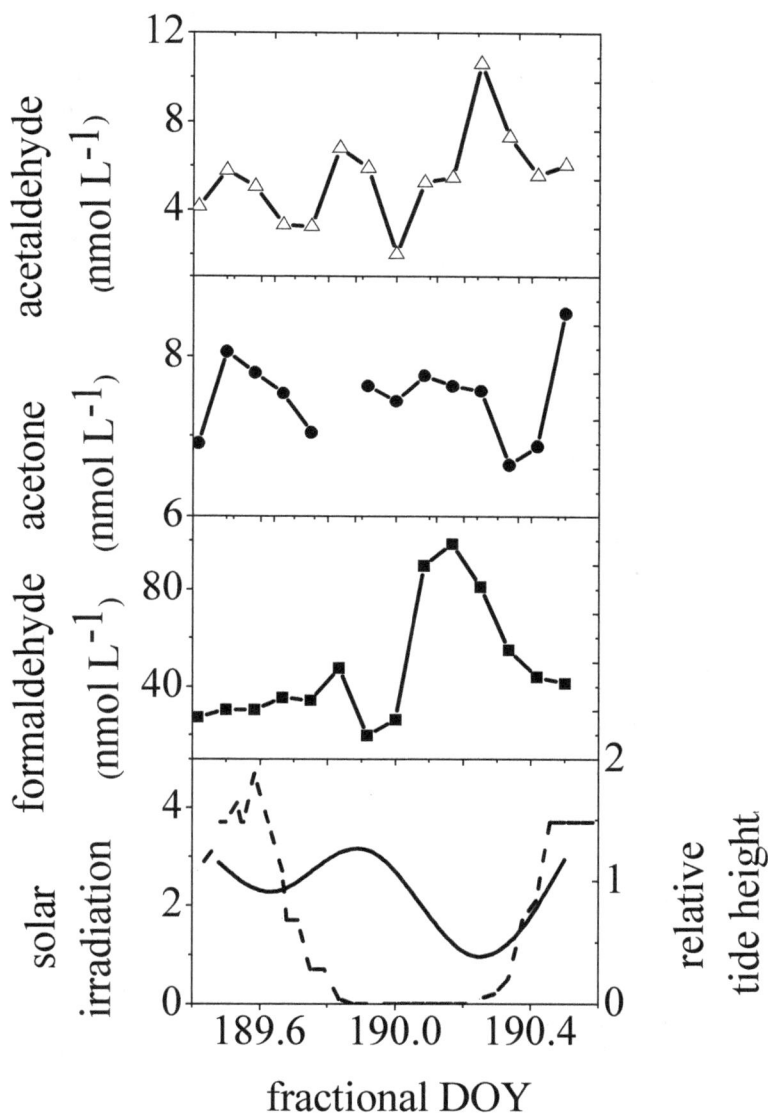

Fig. 1. A diel plot of formaldehyde (■), acetaldehyde (Δ) and acetone (●) concentrations in nmol L^{-1} (nM) as a function of local time (from 10 am 8th July to 2 pm 9th July 2009). Time on the x-axis is shown as fractional day of the year (DOY). The lower panel shows the tidal height (solid line; in m) and solar irradiation levels (dashed line; in W m^{-2}) over the same time period.

The lack of well defined photochemically-driven diel cycles in our study could be due to a combination of multiple factors: 1) additional variability in the photochemical source due to variability in the CDOM levels , 2) variability in the oxygenated hydrocarbon sinks or 3) additional non-photochemical sources.

We will discuss first factor 1, differences in CDOM levels, as a potential cause of the observed weak diel cycles. Absorption coefficient values varied by an order of magnitude over the 28 h period from 0.75 to 7.0 .m^{-1}, suggesting that CDOM levels varied by about the same factor during the study. Absorption coefficient variability in the surf-zone showed rapid oscillations on the time-scale of hours during this diel study (see data figures in Clark et al., 2010). This is consistent with an earlier diel study we conducted at a beach up-coast from this study site (Clark et al., 2009), where these dynamic oscillations were attributed to the passage of different parcels of water through long-shore and rip currents (Grant et al., 2005). Hydrogen peroxide (H_2O_2) was measured independently at this site over the same time period (Clark et al., 2010). This is also a photochemical product of sunlight irradiated CDOM in surface waters (Hoigne et al., 1988). By contrast to the LMW carbonyl compounds, H_2O_2 showed a well defined diurnal cycle driven by sunlight in multiple studies at this site (Clark et al., 2010). The rapid cycling in the absorption coefficients (and hence presumably CDOM levels) did not affect the regularity of the diurnal signal observed, suggesting sunlight levels were the dominant factor. The diel correlation observed for hydrogen peroxide (another CDOM photochemical product) suggests that the lack of a well defined diel cycle in oxygenated hydrocarbons observed here is not a result of the variability in CDOM levels i.e. factor 1 is not an issue.

The well-defined peroxide cycle obtained over the same time period also suggests that if we assume that peroxide and oxygenated hydrocarbon sinks are similar (both are believed to be primarily biological or particle driven (Clark et al., 2008b, and references therein)) then the limited diurnal cycle measured here is probably not due to sink variability i.e. factor 2 is not an issue. This leaves factor 3, or additional non-photochemical sources, as an explanation for the lack of well-defined diel cycles observed for the LMW carbonyls.

3.3 Additional sources

Additional non-CDOM sources for production of hydrogen peroxide in coastal waters have been suggested by Clark et al. (2009, 2010). These include photochemical production from metal species in the water and beach sediments, and non-photochemical sources from decaying plant matter in the intertidal zone. The night-time maxima observed here for the LMW carbonyls occurred at or approaching the lowest point of an ebbing tide, consistent with a non-photochemical source of oxygenated hydrocarbons in the intertidal zone. We hypothesize that this is due to senescent plant wrack releasing oxygenated hydrocarbons. Specifically, this is a marine reserve site with significant coverage by giant kelp beds in the near-shore region, which have been re-established by an aggressive reforestation program over the last decade by local environmental organizations (Orange County Coastkeeper; http://www.cacoastkeeper.org/news/kelp!-it-needs-somebody; accessed 6 June 2011). Giant kelp is a fast-growing type of brown algae (Jackson, 1977 and 1987); prior studies have shown a correlation between aldehyde levels and algal blooms and the production of aldehydes from phytoplankton (Pohnert et al., 2002; Sinha et al., 2007), suggesting biological production from kelp may be possible.

To test this hypothesis, experiments were carried out to see if kelp can produce oxygenated hydrocarbons. Fresh kelp collected from the intertidal zone at Crystal Cove State Beach was weighed and 5g immersed in 500 ml of de-ionized water for 3 hours. 20 ml samples were removed every 30 minutes, filtered, derivitized and analysed for acetone, acetaldehyde and formaldehyde. Results are shown in Figure 2. In general, the acetone and acetaldehyde concentrations increased by 8 nM from about 4 nM to 12 nM over 3 hours. Formaldehyde concentrations showed a much larger increase of about 300 nM up to a maximum of 450 nM over the same time period. The acetone and acetaldehyde levels reached after 60 minutes are comparable to the increases observed when the samples were irradiated for one hour with a low power mercury-xenon lamp system (Oriel, 150 W) with a 300 to 400 nm band pass filter (De Bruyn et al., 2011). In contrast, formaldehyde concentration changes were significantly higher than the levels of 240 nM obtained from 3 hours of irradiation.

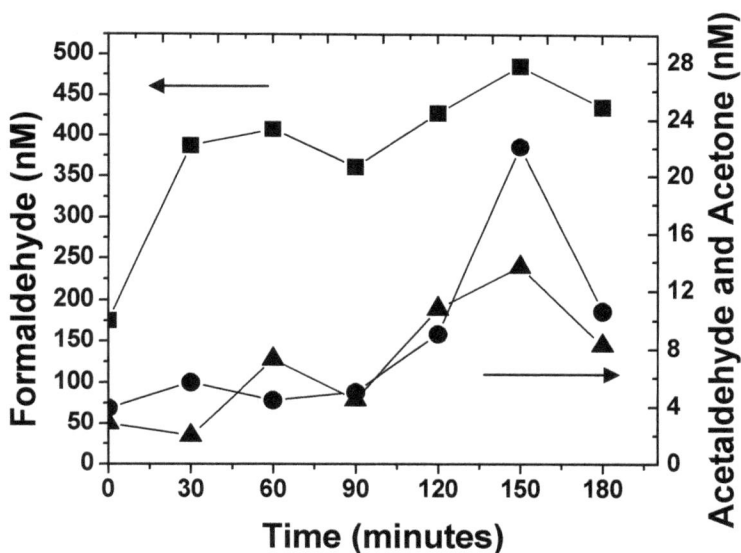

Fig. 2. Aqueous-phase concentrations of formaldehyde (■), acetaldehyde (●), and acetone (▲) in nM as a function of immersion time (minutes) for a senescent kelp sample. Note that the formaldehyde concentration axis on the left hand side covers a concentration range an order of magnitude higher (0 to 525 nM) than that for acetone and acetaldehyde on the right hand side (0 to 30 nM).

While this bench-top experiment is not the real world and does not take into consideration real oceanographic kelp mass to seawater volume ratios, it does suggest that formaldehyde leaching from kelp could be a significant source of oxygenated hydrocarbons in these waters. For formaldehyde, this source could potentially be more significant than the photochemical source.

4. Conclusions and future work

Formaldehyde, acetaldehyde and acetone concentrations were measured at a number of coastal water sites in southern California with low CDOM levels. Seawater formaldehyde levels ranged from 7.5 to 88×10^{-9} mol L^{-1} (nM) with an average of 27 ± 25 nM. Acetaldehyde and acetone levels ranged from 2.7 to 19.9 nM and 2.7 to 12.5 nM respectively, with average levels in seawater of 9 ± 4 nM and 8 ± 2 nM respectively. These ranges are consistent with the limited dataset of existing measurements in the literature, and are consistent with levels observed during a 28 hour diel study.

Increased concentrations were observed near solar noon, consistent with photochemical production for acetone and acetaldehyde but well defined diel cycles were not observed, most likely due to other non-photochemical sources of oxygenated hydrocarbons. Preliminary leaching experiments suggest that these compounds, particularly formaldehyde, can be produced in the dark from decaying kelp on the order of or greater than the levels that can be produced photochemically and were measured in these coastal waters. Since kelp beds are distributed globally through temperate and polar coastal regions, these could form significant LMW producers on a regional to global scale. Potentially, other near-shore biological sources like seagrass beds could also contribute to LMW production.

Future studies should focus on production from living kelp beds in situ in the near-shore region and senescent kelp in the intertidal zone, as well as contributions from other plant species. It would also be useful to extend these studies to the previously unstudied Pacific Northeast region, and also to other regions with low rainfall globally where salt marshes dominate inputs of organic matter to coastal waters, as opposed to the previous studies in riverine-dominated waters.

5. Acknowledgements

The authors thank the National Science Foundation for funding (OCE grant #072528433), students Benjamin Brahm, Paige Aiona and Charlotte Hirsch for assistance with sampling and Harry Helling of the Crystal Cove Alliance and the California State Parks for access to their beach sites and research facility.

6. References

Bano N.; Moran, M.A. & Hodson, R.E. (1998). Photochemical formation of labile organic matter from two components of dissolved organic carbon in a freshwater wetland. *Aq. Microb. Ecol.*, Vol. 16, pp 95-102.

Benner, R.H. (1998). Cycling of dissolved organic matter in the ocean. In *Ecological Studies: Aquatic Humic Substances*, vol. 133, Hessen, D.O. & Tranvik, L.J., Chapter 12, pp.317-331, Springer-Verlag, ISBN 978-3-642-08362-4, Heidelberg, Germany.

Benner R.H. (2002). Chemical composition and reactivity. In *Biogeochemistry of marine dissolved organic matter*, Hansell, D.A. & Carlson, C.A., Chapter 3; pp 59-85, Academic Press, ISBN 0-12-323841-2, San Diego, USA.

Clark C.D.; Hiscock, W.T., Millero, F.J., Hitchcock, G., Brand, L., Del Vecchio, R, Blough, N., Miller, W., Ziolkowski, L., Chen, R.F. & Zika, R. G. (2004). CDOM Distribution and Carbon Dioxide Production on the Southwest Florida Shelf. *Mar. Chem.* Vol. 89, pp 145.

Clark C.D.; Litz, L. P. & Grant, S.B.(2008a). Saltmarshes as a source of chromophoric dissolved organic matter to Southern California coastal waters. *Limnol. Oceanogr.*, Vol. 53, pp 1923.

Clark C. D.; De Bruyn, W.J. , Jakubowski , S. D. & Grant S.B. (2008b).Hydrogen peroxide production in marine bathing waters: implications for fecal indicator bacteria mortality. *Marine Pollution Bulletin*, Vol. 56, pp 397.

Clark C.D.; De Bruyn, W. & Jones, J. (2009). Photochemical production of hydrogen peroxide in size fractionated Southern California coastal waters. *Chemosphere*, Vol. 76, pp 141.

Clark C.D.; De Bruyn, W., Hirsch, C. M. & Jakubowski, S. (2010). Hydrogen peroxide measurements in recreational marine bathing waters in Southern California, USA, *Wat. Res.*, Vol. 44, pp 2203-2210.

Clark C.D.; De Bruyn, W. J., Hirsch, C.M. & Aiona, P. (2010). Diel cycles of hydrogen Peroxide in marine bathing waters in Southern California, USA: *in situ* surf-zone measurements. *Marine Pollution Bulletin*, Vol , ppp.

De Bruyn W.J.; Clark, C. D., Pagel, L, & Takahara, C,. (2011). The photoproduction of formaldehyde, acetaldehyde and acetone from dissolved organic matter in natural waters. Submitted to *Photochemistry Photobiology*.

De Gouw J.A.; Middlebrook, A.M., Warneke, C., Goldan, P.D., Kuster, W.C., Roberts, J.M, Fehsenfeld, F. C., Worsnop, D. R., Canagaratna, M.R., Pszenny, A.A.P., Keene, W.C., Marchewka, M., Bertman, S. B. & Bates, T.S.(2005). Budget of organic carbon in a polluted atmosphere: results from the New England Air Quality Study in 2002. *J. Geophys. Res.*, Vol. 110: D16305, doi: 10.1029/2004JD005623

Dufour G.; Szopa, S., Hauglustaine, D.A., Boone, C.D., Rinsland, C.P. & Bernath, P.F. (2007). The influence of biogenic emissions on upper-tropospheric methanol as revealed from space. *Atmos. Chem. Phys.*, Vol. 7, pp 6119-6129.

Gallegos C.L.; Jordan, T.E., Hines, A.H. & Weller, D.E. (2005). Temporal variability of optical properties in a shallow, eutrophic estuary: seasonal and interannual variability. *Estuar. Coast. Shelf Sci.*, Vol. 64, pp 156-170. doi: 10.1016/j.ecss.2005. 01.013.

Grant S.B. ; Kim, J. H., Jones, B.H., Jenkins, S.A., Wasyl, J. & Cudaback, C. (2005). Surf zone entrainment, along-shore transport, and human health implications of pollution from tidal outlets. *J. Geophys. Res.*, Vol. 110, pp C10025-C10045.

Green S.A. & Blough, N.V. (1994). Optical absorbance and fluorescence properties of chromophoric dissolved organic matter in natural waters. *Limnol. Oceanogr.*, Vol. 39, pp 1903-1916.

Hedges J. (2002). Why dissolved organics matter? In *Biogeochemistry of marine dissolved organic matter*, Hansell, D.A. & Carlson, C.A., Chapter 1, pp 1-27, Academic Press, ISBN 0-12-323841-2, San Diego, USA.

Helms J.R.; Stubbins, A., Ritchie, J.D., Minor, E.C., Kieber, D.J. & Mopper, K. (2008). Absorption spectral slopes and slope ratios as indicators of molecular weight, source and photobleaching. *Limnol. Oceanogr.*, Vol. 53, pp 955-969.

Hessen D.O. & Tranvik, L.,J. (1998). Aquatic humic matter: from molecular structure to ecosystem stability. In In *Ecological Studies: Aquatic Humic Substances*, vol. 133, Hessen, D.O. & Tranvik, L.J., Springer-Verlag, ISBN 978-3-642-08362-4, Heidelberg, Germany.

Hoigne J.; Fau, B.C., Haag, W.R., Scully, F.E. & Zepp, R.G. (1988). Aquatic humic substances as sources and sinks of photochemically produced transient reactants. Ch 23. In Aquatic humic substances, vol. 219, pp 363-381.

Hu C.; Muller-Karger, F.E. & Zepp, R.G. (2002). Absorbance, a(300) and apparent quantum yield: a comment on common ambiguity in the use of these optical concepts. *Limnol. Oceanogr*, Vol. 47, pp 1261-1267.

Hudson E.D.; Okuda, K. & Ariya P.A. (2007). Determination of acetone in seawater using derivitization solid-phase microextraction. *Anal. Bioanal. Chem.*, Vol. 388, pp 1275-1282.

Jackson G.A. (1977). Nutrients and production of giant kelp, *Macrocytis pyrifera*, off Southern California. *Limnol. Oceanogr.*, Vol. 22, pp 979-995.

Jackson G.A. (1987). Modelling the growth and harvest yield of the giant kelp *Macrocytis pyrifera*. *Mar. Biol.*, Vol. 4, pp 611-624 dooi: 10.1007/BF00393105

Jacob D.J.; Field, B. D., Jin, E. M., Bey, I., Li, Q. , Logan, J. A. & Yantosca R. M. (2002). Atmospheric budget of acetone, 2002. *J. Geophys. Res.*, Vol. 107: D10, 4100, doi 10.1029/2001JD000694.

Jeong Y.; Grant, S.B., Ritter, S., Pednekar, A., Candelaria, L. & Winant, C. (2005). Identifying pollutant sources in tidally mixed systems: case study of fecal indicator bacteria from marinas in Newport Bay, Southern California. *Environ. Sci. Technol.*, Vol.39, pp 9083-9093 doi:10.1021/es0482684

Kameyama S.; Tanomoto, T., Inomata, S., Tsunogai, U. Ooki, A., Takeda, S., Obata, H., Tsuda, A. & Uematsu, U. (2010). *Mar. Chem.*, Vol. 122, pp 59-73 doi:10.1016/j.marchem.2010.08.003

Kieber D.J. & Mopper, K. (1987). Photochemical formation of glyoxylic and pyruvic acids in seawater. *Mar. Chem.*, Vol. 21, pp 135-149.

Kieber R.J.; Zhou, X. & Mopper, K. (1990). Formation of carbonyl compounds from UV-induced photodegradation of humic substances in natural waters: fate of riverine carbon in the sea. *Limnol. Oceanogr.*, Vol. 35, pp 1503-1515.

Kieber R.J. & Mopper, K. (1990). Determination of picomolar concentrations of carbonyl compounds in natural waters, including seawater, by liquid chromatography. *Environ. Sci. Technol.*, Vol. 24, pp 1477-1481

Kowalczuk P.; Cooper, W.J., Durako, M.J., Kahn, A.E., Gonsior, M. & Young, H. (2010). Characterization of dissolved organic matter fluorescence in South Atlantic Bight with use of PARAFAC model: relationships between fluorescence and its components, absorption coefficients and organic carbon concentrations. *Mar. Chem.* Vol. 118, pp 22-36.

Marandino C.A.; De Bruyn, W. J., Miller, S.D., Prather, M. J, Saltzman, E. S. (2005). Oceanic uptake and the atmospheric carbon budget. *Geophys. Res. Lett*, Vol. 32, L15806.

McKnight D.M. & Aiken, G.R. (1998). Sources and age of aquatic humus. In *Ecological Studies: Aquatic Humic Substances*, vol. 133, Hessen, D.O. & Tranvik, L.J., Chapter 1, pp.9-39, Springer-Verlag, ISBN 978-3-642-08362-4, Heidelberg, Germany.

Miller W.L. (1994). In *Aquatic and Surface Photochemistry*, CRC Press, Inc.; Chap. 7; pp.111-127.

Miller W.L. (1998). Effects of UV radiation on aquatic humus: photochemical principles and experimental considerations. In *Ecological Studies: Aquatic Humic Substances,* vol. 133, Hessen, D.O. & Tranvik, L.J., Chapter 6, pp 125-141, Springer-Verlag, ISBN 978-3-642-08362-4, Heidelberg, Germany.

Millet D.B.; Jacob, D.J. , Custer, T.G., de Gouw, J.A., Goldstein, A.H., Karl T., Singh, H.B., Sive, B.C. , Talbot, R.W., Warneke, C. & Williams, J. (2008). New constraints on terrestrial and oceanic sources of atmospheric methanol. *Atmos. Chem. Phys.*, Vol. 8, pp 6887-6905.

Millet D.B., Guenther, A., Siegel, D.A., Nelson, N. B., Singh, H.B., de Gouw, J.A., Warneke, C. , Williams, J. , Erdekens, G. , Sinha, V., Karl, T., Flocke, F., Apel, E., Riemer, D.D., Palmer, P.I. & Barkley, M. (2010). Global atmospheric budget of acetaldehyde: 3-D model analysis and constraints from in situ and satellite observations. *Atmos. Chem. Phys.*, Vol. 10, pp 3405-3425.

Mopper K. & Stahovec. (1986). Sources and Sinks of low molecular weight organic carbonyl compounds in seawater, *Mar. Chem.*, Vol. 19, pp 305-321.

Mopper K. & Kieber, D.J. (2002). Photochemistry and the cycling of carbon, sulfur, nitrogen and phosphorus. In *Biogeochemistry of marine dissolved organic matter*, Hansell, D.A. & Carlson, C.A., Chapter 9; pp 456-503, Academic Press, ISBN 0-12-323841-2, San Diego, USA.

Moran M.A.; Sheldon Jr., W.M. & Zepp, R.G. (2000). Carbon loss and optical property changes during long-term photochemical and biological degradation of estuarine dissolved organic matter. *Limnol. Oceanogr.*, Vol. 45, pp 1254-1264.

Naik V.; Fiore, A.M., Horowitz, L.W., Singh, H.B., Wiedinmyer, C., Guenther, A., de Gouw, J.A., Millet, D.B., Goldan, P.D., Kuster, W.C., & Goldstein, A. (2010). Observational constraints on the global atmospheric budget of ethanol. *Atmos. Chem. Phys. Discuss.*, Vol. 10, pp 925-945.

Perdue E.M. (1998). Chemical composition, structure and metal binding properties. In *Ecological Studies: Aquatic Humic Substances*, vol. 133, Hessen, D.O. & Tranvik, L.J., Chapter 2, pp. 9-37, Springer-Verlag, ISBN 978-3-642-08362-4, Heidelberg, Germany.

Pohnert G.; Lumineau, O., Cueff, A., Adolp, S., Cordevant, C., Lange, M. & Poluet, S. (2002). Are volatile unsaturated aldehydes from diatoms the main line of chemical defense against copepods? *Mar. Ecol. Prog. Series*, Vol. 245, pp 33-45.

Seritti A; Russo, D., Nannicini, L. & Del Vecchio, R. (1998). DOC, absorbance and fluorescence properties of estuarine and coastal waters of the Northern Tyrrhenian Sea. *Chem. Spec. Bioavail.*, Vol. 10, pp 95-106.

Sharp J. (2002). Analytical methods for total DOM pools. *In* Hansell, D.A. & Carlson, C.A. Biogeochemistry of marine dissolved organic matter, Academic Press, pp 35 - 54.

Singh H.B; Kanakidou, M., Crutzen, P.J., Jacob, D.J. (1995). High concentrations and photochemical fate of oxygenated hydrocarbons in the global troposphere. *Nature*, Vol. 378, pp 50.

Singh H.; Chen, Y., Staudt, A., Jacob, D., Blake, D., Heikes, B. & Snow J. (2001). Evidence from the Pacific troposphere for large global sources of oxygenated organic compounds. *Nature*, Vol. 410, pp 1078-81.

Singh H.B., Slas, L.J., Chatfield, R.B., Czech, E., Fried, A., Walega, J., Evans, M.J., Field, B.D., Jacob, D.J., Blake, D., Heikes, B. , Talbot, R. , Sachse, G., Crawford, J.H., Avery, M.A., Sandholm, S. & Fuelberg, H. (2004). Analysis of atmospheric distribution sources and sinks of oxygenated volatile organic chemicals based on measurements over the Pacific during TRACE-P. *J. Geophys. Res.*, Vol. 109, D15S07, doi: 10.1029/ 2003JD003883

Sinha V.; Williams, J., Meyerhofer, M., Riebesell, U., Paulino, A.I. & Larsen, A. (2007). Air-sea fluxes of methanol, acetone, acetaldehyde, isoprene, and DMS from a Norwegian fjord following a phtyoplankton bloom in a mesocosm experiment. *Atmos. Chem. Phys.*, Vol. 7, pp 739-755.

Stedmon C.A.; Markager, S. & Kaas, H. (2001). The optics of chromophoric dissolved organic matter (CDOM) in the Greenland Sea: an algorithm for differentiation between marine and terrestrially derived organic matter. *Limnol. Oceanogr.* Vol. 46, pp 2087-2093.

Stedmon C.A. & Markager S. (2003). Behavior of the optical properties of colored dissolved organic matter under conservative mixing. *Estuar. Coastal Shelf Sci.*, Vol. 57, pp 973-979 doi: 10.1016/S0272-7714(03)00003-9

Takeda K; Katoh, S., Nakatani, N. & Sakugawa, H. (2006). Rapid and highly sensitive determination of low-molecular weight carbonyl compounds in drinking water and natural water by preconcentration HPLC with 2,4-dinitrophenylhydrazine. *Anal. Sci.*, Vol. 22, pp 1509-1514.

Tardowski M.S. & Donaghay, P.L. (2002). Photobleaching of aquatic dissolved materials: absorption removal, spectral alteration and their relationship. *J. Geophys. Res.*, Vol. 107, 10.1029/1999JC000281

Tzortziou M; Osburn, C. L. & Neale, P. J. (2007). Photobleaching of dissolved organic material from a tidal marsh-estuarine system of the Chesapeake Bay. *Photochem. Photobiol.* Vol. 83, pp 782-792, doi 10.1562/2006-09-28-RA-1048.

Whitehead K. (2008). Marine organic geochemistry. In *Chemical oceanography and the marine carbon cycle*, Emerson , S. & Hedges, J., chapter 8, pp 261-294, Cambridge University Press,ISBN 978-0-521-83313-4, Cambridge, United Kingdom.

Williams J; Holzinger, R. , Gros, V., Xu, X., Atlas, E. & Wallace, D.W.R. (2004). Measurements of organic species in air and seawater from the tropical Atlantic. *Geophys. Res. Lett.*, Vol. 31, L23S06 doi: 10.1029/2004GL020012

Yamashita Y; Jaffe, R., Maie, N. & Tanoue, E. (2008). Assessing the dynamics of dissolved organic matter (DOM) in coastal environments by excitation emission matrix fluorescence and parallel factor analysis (EEM-PARAFAC). *Limnol. Oceanogr.*, Vol. 53, pp 1900-1908.

Zhou X. & Mopper, K. (1990). Apparent partition coefficients of 15 carbonyl compounds between air and seawater and between air and freshwater: implications for air-sea exchange. *Environ. Sci. Technol.*, Vol. 24, pp 1864-1869.

Zhou X. & Mopper K. (1997). Photochemical production of low-molecular-weight carbonyl compounds in seawater and surface microlayer and their air-sea exchange. *Mar. Chem.*, Vol. 56, pp 201-213.

Part 4

Applied Ecology

11

Ecology and Zoogeography of Parasites

Ewa Sobecka
West Pomeranian University of Technology, Szczecin
Poland

1. Introduction

Parasitism is a ubiquitous phenomenon that is probably as old as heterotrophic organisms themselves. It is one of the major types of symbiotic relationship between organisms of different species.

The development of the biological sciences brought about a broader definition of parasitism. By the end of the nineteenth century, it was known that parasites thrive at the expense of their hosts and use the host as their dwellings. The definition of parasitism emphasized the damage parasites inflicted on hosts, the metabolic dependence of parasites on hosts, and ecological interactions between populations of these two species of living organisms. It is also known that parasites can lower the fitness of hosts by exploiting nutritional resources, habitats, and dispersal. Parasites can also modify host behavior and lead to their castration (Levri, 1998).

Parasites can have varied impacts on hosts. The attachment organs of the parasites can cause mechanical damage to host tissues. Acanthocephalan spikes and cestode hooks and suckers can cause intestinal pathology, while monogenean hooks and clamps can damage the structures of fish gill filaments. Large internal parasites or those occurring in great numbers can block digestive tracts. Host tissues can sustain mechanical damage as parasites migrate or pass from one developmental stage to the next. While migrating into the host body cavity, roundworm larvae from the family Anisakidae break through stomach walls or pyloric caecae of the fish, which are the intermediate host. Tapeworm larvae of the genus *Ligula*, which settle in the body cavity, exert pressure on the fish gonads causing castration. They feed either on as yet undigested food or the components of food that has been broken down by enzymes; thus depleting hosts of protein, lipids, carbohydrates, vitamins, and enzymes. The parasites themselves produce enzymes that damage host cells and inhibit physiological processes. Parasitic metabolic products can be toxic to hosts; some parasites produce substances that impede blood clotting (leeches), destroy epithelial cells (cercariae Digenea), which creates the conditions necessary for secondary infection.

In this chapter, parasitism will be examined from the perspective of the close relationship between two organisms: the parasite and the host. The parasite lives at the expense of the host, but it is also dependent on its host in many other aspects. Parasitism is distinguishable from predation by its more extensive reproductive potential, its smaller size and thus limited visibility, and in its representing a lower level of evolutionary advancement.

If the relationship described above is only temporary, then it is referred to it as temporary parasitism. In many cases, this is limited to one-time contact with the host that is nonetheless indispensable for further parasite development. Temporary parasitism is very similar to predation, especially when a single parasite species interact with numerous host species. In this instance, these parasites can be vectors of the infectious stages of other parasites.

Parasitic relationships that lack metabolic dependence are referred to as nesting or social parasitism, which indicate that organisms using other organisms as nesting sites or for rearing offspring.

2. Ecology of parasitism

Ecology has been defined as the study of interactions between organisms and their environments and among the organisms inhabiting these environments. This is a complex field of study, which is why it should be investigated on various levels. In the ecology of parasites the niche, which is the entirety of the parasite-host relationship and fragmentation, is of fundamental importance (Combes, 1995).

A niche is the entire set of biotic and abiotic factors (variables) that provides a population with living conditions. The number of variables that combine to shape a niche is large, and includes the environmental physicochemical parameters (salinity, temperature, light), host populations (abundance, fluctuations), variability of hosts occurring successively, the occurrence of parasitoids, and the character of the host resources exploited (Rohde, 1994). This last variable occurs repeatedly in the subsequent divisions of the habitat, e.g., on fish gills where parasites that feed on epithelial cells and mucous can occur simultaneously with parasites that draw blood from the blood vessels. Overlapping niches are the main parameter in competition among species.

Despite the large number of potential hosts, the parasite-host system can only exist when conditions are appropriate. The host must inhabit the same ecosystem as the parasite, have contact with the parasite, and offer it appropriate conditions for both life and survival. All of these conditions together are the selection mechanism that operates at the level of the parasite and host genome. Parasites can infect single or numerous hosts during the different developmental stages of their biological cycles; however, each stage has to overcome host metabolic and immunological barriers (Combes, 1995). The parasites that have, through evolution, developed the greatest ability to adapt to environments and to avoid defense mechanisms are the most successful at infecting and exploiting hosts. Success also depends on the creation of survival stages that can endure for certain periods of time in disadvantageous environmental conditions; these include protozoan and myxosporean cysts and helminth eggs. Developmental stages can also be dependent on the host to a certain degree. The parasitic stages of digeneans are sporocysts, rediae, adult stages, free-living miracidium, cercariae, and metacercariae (Buczek, 2003).

The physiological processes of parasites are dependent on the character of the site they inhabit, which is also its source of food. The mutual relationship between parasites and hosts is referred to as specificity, and it can be either narrow (e.g., Monogenea), or wide (e.g., most Nematoda). The older parasites are in phylogenetic development, the more specific they become. The functioning of the parasite-host system is possible when there is convergence between the activities of the particular developmental stages and the availability of food. This is linked also to the secretion by the parasites of digestive enzymes and metabolic regulation depending on their energy demands, which can be markedly decreased under disadvantageous conditions. An important factor to survival is the possibility of exploiting food reserves accumulated by previous developmental stages.

Both sides participate in the creation of the parasite-host system. Initial contact usually ends in parasite death which results from host defense reactions. Instances of host death are also noted when they are introduced into a new ecosystem. Thus, the younger the parasite-host system is, the more pathogenic are the parasites and the less tolerant of them are the hosts. The aim of the formation and evolution of the parasite-host system is, among others, the reproductive success of both sides (Combes, 1995).

Evolution is a continual process comprising progressive changes in species characters of subsequent generations through the elimination and the selection of individuals in a population. Customarily, this process is viewed on a scale of whole organisms or populations, but all the changes observed in a given organism are molecular. Variability underlies evolution , and it sometimes only manifests as genetic variability. Genetic changes within one species cannot be transferred to another since there is no possibility for lateral gene transfer among most organisms that reproduce sexually. Changes in phenotype are a consequence of gene mutation; the evolution of each partner is conditioned by the pressure of choice from the other partner. One must not forget, however, that choice alone is insufficient as a factor causing co-evolution. Parasite and host selection run in different directions. The best adapted are the parasites, which when they find better hosts and exploit them the best attain great reproductive success. Hosts, in turn, avoid contact with parasites by adopting new behaviors, and if contact is made, they fight it. The costs of these activities are higher for the hosts. Mounting a defense requires energy which is designated for other life processes of the organism. The parasite-host relationship is based on an evolutionary race heading towards "meeting/avoiding" and "survival/killing" (Niewiadomska et al., 2001).

The parasite-host relationship is a dynamic one for both partners. This is linked to morphological, physiological, and structural adaptations that render easier the competition for resources that are more beneficial under given environmental conditions and to biological and ecological factors that ease contact among organisms in the environment. Parasites treating host organisms as environments to inhabit results in disruptions in host physiology, susceptibility to the impacts of toxins, and damage or the stimulation of pathological development of some tissues, and the depletion of nutritional substances. The degree to which these changes occur depends on many factors on the sided of both the parasites and the hosts. Where and how parasites infect hosts, parasite sizes, how they feed, and how they migrate are all significant. Hosts defend themselves against parasites with non-specific (phagocytes, inflammation) and specific (cellular or humoral reactions) immune responses (Niewiadomska et al., 2001). One must bear in mind that parasites do not associate with just one host for their lifespans. In most cases the reproductive stage of the developmental cycle requires changes of hosts and environments. In each instance, a parasite-host system is established, although this relationship is not usually as long as that between parasites and definitive hosts. When there is a transition from one generation to the next during developmental cycles, new generations have to establish their own systems with hosts. In consequence, this has led to various types of parasitism that require different adaptation strategies and various paths for that generation to become parasitic.

The concept of fragmentation is defined as how parasites are distributed throughout the biosphere. All parasite species, just as other species do, form different populations. A population is a group of organisms of the same species living simultaneously in the same environment and exerting a mutual influence on each other with equal likelihood of interbreeding. Such groups, or populations, are isolated geographically so that gene exchange among populations does not occur. In the case of parasites, it is difficult to draw the precise lines that demarcate population distributions, because it does happen that the same parasite species infects different host populations within one geographical region (e.g., in the same water basin) and they do not exchange gene pools. Parasites with a complex developmental cycle can be transported into another region by its intermediary hosts, and only those parasites with simple developmental cycles cannot move farther than their hosts. A set of populations of different species that occupy, at least partially, the same habitat and can impact each other is referred to as a community (Blondel, 1986). In ecological parasitology this term is, in a certain sense, the equivalent of the term "biocenosis".

The proper ecosystem functioning of populations and biocenoses has a defined structure. The structure of a population comprises it abundance, the distribution of organisms in the biotope, and the numerical ratio of juveniles to adults in the biocenoses. The structure of the biocenosis to the abundance, composition, and proportion of species that belong to it, the character of their links to the varied ecological factors, and trophic dependencies combine to create a whole that remains in a state of dynamic equilibrium in the natural environment.

'Infrapopulation' and 'metapopulation' are both terms linked to parasite populations. Infrapopulation refers to all the parasites of one species infecting a given individual host at a given time (Bush et al., 2002). An example of an infrapopulation would be all of the nematode *Hysterothylacium aduncum* (all developmental stages) infecting one cod. A metapopulation (also referred to as a xenopopulation) is the term used to refer to a group of a given parasite species within a host population (Riggs & Esch, 1987; Combes, 1995). These are all infrapopulations that occur within the whole host population. A good example is the nematode *Anisakis simplex* among seals.

The group of all individuals (all developmental stages) living among host populations in an ecosystem form suprapopulations. This concept includes all infrapopulations that occur within an ecosystem (Esch et al., 1975).

The solutions discussed above refer to single-species groups that reflect the concept of population with regard to free-living organisms. Within the category of biocenosis and based on the categories designated for populations, the concept of multi-species groups is proposed. Infracommunity refers to a group of infrapopulations of various parasite species infecting one host species; component community is defined as a group of metapopulations of various parasite species occurring within one host species population. Groups of suprapopulations of various parasite species infecting different host species within a given ecosystem form compound communities (Holmes and Price 1986). In practice, the term infracommunity is often limited to groups of parasites occurring in one organ (Złotorzycka, 1998). The terminology presented here describes groups of parasites at the exclusion of dispersion stages that occur in the external environment. They play an important role in gene flow and the regulation of population numbers since parasites are exposed to high mortality in the external environment. They are food components of free living organisms.

One of the primary elements of population structure and multi-species communities of parasites is their fragmentation. Infrapopulations and infracommunities occur in single hosts, and are often limited to a specific organ.

The locations where parasites live, regardless of whether these are hosts of the external environment, are their habitats. Macrohabitats (biotopes or ecosystems) are the living environments of hosts. Microhabitat is the name used to refer to organs or sections of organs infected by parasites. The conditions of the various habitats determine whether or not particular parasites can settle there. Every organ is characterized by many morphological and physicochemical parameters, and differences are noted among them even within a single organ. Despite this, the habitats of parasites can nearly always be determined with great precision. The existence of a habitat is the result of the heterogeneity of each living organism. Different genes are expressed in different organs, which is why the environments offered to parasites differ in different parts of hosts (Combes, 1995). Helminths are most frequently particular about the places they choose to settle; these include designated regions of the intestines or the gills. There are, however, many species of parasite that are not 'picky' regarding habitat. These are most frequently cellular parasites. Some habitats are not always immediately available to parasites or their developmental stages, which is why migration is necessary, and makes it imperative to overcome defensive barriers that are also presented by the host organism. The location of parasite habitats impacts the migration process within host organisms, but also on the processes of releasing parasite eggs or larvae into the external environment. If external parasites make direct contact with the external environment, and mesoparasites do so thanks to organs which they inhabit and from which their eggs or larvae are released through natural channels, then internal parasites inside cells, tissues, or organs have no such possibilities. However, this process requires damaging the habitat, and vectors are useful for this, and it is indeed thanks to this process that some leech species transmit flagellates and haemogregarina.

Numerous studies have indicated that parasites do not occur in all the individuals of a species inhabiting a given ecosystem, and that parasites will occur in different numbers among species in which they can develop. The term prevalence is employed to describe this concept; it is the ratio of the number of infected host individuals to the number of individuals examined, and it is expressed in percentages. The mean number of parasites per infected host is the mean intensity of infection, while the number of parasites per individual in a population is the mean parasite density. The distribution of the parasite population among the host population is determined based on these parameters.

Each infracommunity is characterized by a determined parasite species structure that inhabits host organisms. However, studies indicate that some species occur frequently, while others do so less frequently. However, parasite infracommunities are characterized by the occurrence of dominant species that are a constant element of its composition, and they also occur most abundantly. A permanent component are also influents, which do not occur as abundantly as the dominants, and accessory species that appear only sporadically. The range of the frequency of occurrence of parasites is also described by specific terminology that is also used too describe free-living organisms (Bush & Holmes, 1986). Core species are those that occur frequently and abundantly. This term is the equivalent of a dominant species. Intermediate species do not occur as frequently as do core species, while satellite species are those that occur the least frequently and at low abundances.

The qualitative evaluation of the structural quality of communities is based on the terms of richness and diversity. The fist of these is expressed in the number and abundance of the species comprising the community, while the second describes the quantitative relationships between the species in an community.

The richness and diversity of parasite communities within a single host or in a host population undergo cyclical fluctuations that are dependent on the structure of the host population. This refers primarily to intermediate and satellite species which can appear only during determined seasons of the year or host life stages (Niewiadomska et al., 2001). These changes are linked primarily to the host lifestyle and dietary requirements.

The food requirements of the parasites necessitates introducing yet more terminology. Groups of species exploiting sources in similar ways are referred to as guilds. The way in which resources are exploited and their source determine the type of guild (Bush et al., 2002). Groups of parasites if different species that occur in a single host are known as an infraguild. A component gild describes all of the infraguilds of a local host population (Sousa 1990). Such relationships occur when the host intestine is infected by an community of parasites – one feeds on blood, another on products of host digestion, and others that feed on the epithelium.

The range of parasite specificity for hosts is described with the terms 'specialist ', 'generalist', and 'intercepted specialists'. The first group includes parasites that are linked to a small number of related host species, and often by just one host species (Price, 1980). Generalists are parasites that are not specialized, and have the ability to colonize many unrelated hosts. The third group comprises parasites that have a narrow specificity, which means they can only overtake and develop in a non-specific host only under specific conditions (Holmes, 1990).

The character of the development cycle is also linked by two different terms. The entire life cycle of autogenic parasites occurs within a single ecosystem. In the aquatic environment, such parasites have simple developmental cycles and the aquatic organism is also the definitive host. The second type of parasite has a complex developmental cycle and both invertebrates and vertebrates serve as intermediate or definitive hosts. Allogenic parasites are those in which only part of the developmental cycle runs its course in the studied ecosystem; the aquatic organism is the intermediate host and the definitive host is a terrestrial vertebrate, which can become infected in one ecosystem and then transfer the parasite to another (Esch et al., 1988).

3. Transmission of marine parasites

Parasites invade hosts in various ways: ectoparasites occur on the exterior of hosts , while endoparasites occur internally. Sometimes it is difficult to assign a parasite to one of these two categories, which is why there is a third category – mesoparasites. Different types of parasitism are presumed to have evolved differently, and were affected by feeding modes, suitable conditions, or pure chance. Studies of parasitism among different taxonomic groups of living organisms indicate there is high diversity in its advance, type, and evolutionary origin. Relationships are also noted among the structures of organisms in a given taxonomic group and the type of parasitism that evolved. All parasitic nematodes and the majority of flatworms are endoparasites, while the majority of arthropods are ectoparasites

Not all host organisms are colonized by parasites to the same extent. The host species that is predominantly and most abundantly infected within a given biocenosis is referred to as the principal host. Hosts that are infected sporadically, and often without the parasite completing development or suffering mortality, is called an accidental host. Organism in which parasites achieve sexual maturity and often also reproduce is known as a definitive host. Organism in which parasites pass through the earlier stages of development is called an intermediate host (Matthews, 1998). Similarly, parasite life cycles also can be classified as different types. Parasites with direct life cycles have a single host, while those with an indirect life cycle have several hosts.

Any marine organism is a potential host since becoming a host depends on the occurrence of suitable conditions. These include site, season, environmental conditions, the presence of suitable developmental stages, the possibility of contact between the two organisms, and the acceptance of such a relationship (Rohde & Rohde, 2005).

One of the characteristic traits of parasites is often the necessity of changing habitats repeatedly. In the case of most parasites, the same genome must be able to build several generations of differing organisms that are able to find and exploit several different habitats. If the transmission of parasites occurs between different host individuals this is called lateral transmission, while if it is from the parents (mammals) to their immediate offspring, then it is referred to as vertical transmission (Shoop, 1991).

The transmission of parasites from one host to another is known as direct, but it is rarely actual physical contact, but more often the close proximity in the water of two organisms is sufficient. This mode of transmission is used most often by external parasites with direct life cycles. An example are the monogeneans of the species *Isancistrum subulatae* which infects squid by passing from one partner to the other as their tentacles draw closer to copulate (Llewellyn, 1984). A particular type of transmission is vertical, which means that the mother passes parasites to her offspring. The intestinal nematodes *Uncinaria* spp. of the California sea lion, *Zalophus californianus*, and the northern fur seal, *Callorhinus ursinus*, infect young seals as they nurse and take in larval parasites along with milk (Lyons et al., 2000).

The infectious parasite stages are usually juveniles and rarely adults. This mode of transmission is facilitated by the high densities of host populations in aquaculture facilities.

Parasites are transmitted through active, free-living stages which have evolved special mechanisms that enable them to find hosts, attach and/or penetrate, and find a suitable habitat. These mechanisms include the following: phototaxis, chemotaxis, adequate hatching time (emergence), or simply being in the right place at the right time. A good example are monogeneans, the pelagic larvae of which attach to the gills or fins of fish, or isopods that penetrate the gill or mouth cavities of the (Ravichandran et al., 2009).

Passive infection occurs when parasites enter the host organism through the body openings, by direct contact, or through trans-uterine or trans-ovarian transmission.

Oral transmission is typical for parasites with complex life cycles, and individual developmental stages are linked through the food chain. The links in this chain are the intermediate hosts that harbor parasite developmental stages, and the definitive host where parasites achieve sexual maturity. Parasites are transmitted when current hosts are consumed by the next host that is suitable the development of more advanced stages.

In the marine environment, the majority of free-living developmental stages are active and alive for a span of 24 - 48 hours. During this period they must find a host and overcome numerous obstacles posed by the external environment. Infectious parasite stages have energy reserves to actively pursue these goals. They are neither morphologically nor physiologically capable of acquiring food, as this would distract them from finding hosts.

The first intermediate host is usually organisms with short life spans, such as copepods, oligochaetes, or molluscs. The likelihood of infecting such host by a minute, delicate juvenile parasite stage is very low, which renders the risk of parasite life cycle disruption immensely high. Longer-lived, predatory decapods, chaetognaths, and small fish are the subsequent links in the food chain. They are intermediate, definitive, or paratenic hosts successively. The paratenic host (otherwise known as the transport host) is not intended to support any further morphological or physiological development of the parasite. It simply serves as a convenient link between the intermediate and definitive hosts. Parasite numbers gradually increases in the paratenic host so their role is not only to transmit but also to increase the level of infection (through parasite build-up in the body of the paratenic hosts) (Marcogliese, 1995, 2002).

Passive transmission through skin, or mucous membranes, is aided by vectors, which are also known as transmitting organisms. Most frequently, these organisms feed on blood, and parasites undergo a prescribed developmental phase in them. Aquatic organism parasites from the genus *Trypanosoma* with some species of leech playing the role of vector are examples of this life strategy.

As mentioned previously, parasites can successfully exploit many different host species throughout development. Component biological cycles use hosts successively; the simplest, holoxenic, has just one type of host, while the most complex, heteroxenic, can include up to four hosts. The number of host species in the developmental history of parasites is usually stable. However, there are exceptions to this rule that do not occur because of evolutionary changes. Shortened cycles can happen for several different reasons. One possibility is that parasites omit one of the intermediate host or definitive host if sexual maturation occurs in the intermediate host; however, the two cycles exist simultaneously with a smaller and constant number of hosts. Other possibilities include achieving sexual maturation earlier (larval stages producing eggs), or when stages that usually occur in different types of host occur successively in one host without themselves undergoing transformation. Shortened cycles are also noted when host and/or developmental stages are eliminated. These processes underscore the plasticity of biological cycles and adaptations that aim to avoid the most sensitive stage which is changing hosts.

Extending cycles is accomplished by including paratenic hosts, in which larval stages occur, but not necessarily to the end of the developmental cycle. Paratenic hosts frequently become reservoirs for invasive parasite stages, which allows parasites to survive when conditions are inappropriate. They also ensure possibilities for infecting hosts through different pathways; younger individuals eat smaller intermediate hosts, while older ones eat larger paratenic hosts. The transmission of some groups of parasites always runs the same course with intermediary and paratenic hosts even if, theoretically, the latter is not necessary.

Many marine species have a wide range of hosts. Low parasite host-specificity of allows them to spread faster and more widely in environments. The presence of parasites in environments depends on the presence of hosts. Parasites, therefore, can be regarded as indicators of food chain structure, or as indicators of changes occurring in the quality of aquatic environments.

4. Zoogeography of marine parasites

Biogeography as a scientific discipline began over two centuries ago when Alexander von Humboldt discovered latitudinal gradients in species diversity during an expedition to South America in 1799. He also suggested that temperature might provide an explanation. In the subsequent years, von Humboldt & Bonpland (1805) and his students were the first to name and evaluate the role climate plays in the distribution and structure of vegetation in the world. For many years, zoogeographic study remained a secondary pursuit, until W. L. Sclater (1858) proposed dividing the world into faunal regions based on the distribution of birds. The divisions he proposed are very similar to those that are currently applied.

Zoogeography is, in fact, a branch of biogeography, as it studies the geographic distribution of animals in the past, now, and predicts likely scenarios of this phenomenon for the future. It also includes the study of factors that explain current distributions of animals by drawing on knowledge from other scientific disciplines These include climatology, ecology, physiography, evolution history, oceanography, limnology, paleontology, systematics, and evolution, and consequently these are able to explain the evolution of relationships between specific species. Zoogeography is based on the following observations:

- every species and higher taxa of animal has a non-random distribution in time and space;
- different geographical regions have sets of typical animals that co-exist, e.g., the fauna of the Baltic Sea is different from that of the Black Sea;
- these differences (and similarities) cannot be explained only by the distances between regions. They are readily observable patterns that refer to species composition and higher animal taxa that developed at the same time and place.

One must remember that the fauna differed distinctly in all geographic regions from that occurs now. Studies have indicated that animals resembling those occurring today or their ancestors could have previously lived in locations that are far removed from their current ranges of occurrence.

Zoogeographic studies of parasites began in 1891 with the work of von Ihering, who noted the significance of the role of parasites in the zoogeography of hosts. Although he was not a parasitologist, von Ihering noted a strict relationship between commensalistic turbellarians and freshwater crustaceans inhabiting both sides of the Andes Mountains and in New Zealand. This provided evidence of the previous biogeographic link between these two lands.

The zoogeography of marine parasites was also studied by Manter (1940, 55, 63). He focused on the relationships and geographic distribution of Digenea along the Atlantic and Pacific coasts of Central America, the biogeographic aspects of marine fish digeneans of America, and the same group of parasites of freshwater fish in South America. Most of these biogeographic arguments lacked foundations stemming from the continental drift theory. Nevertheless, Manter prompted a new approach to studies of marine parasites. In the late 1970s, the earlier term of biogeography was linked with the already accepted continental drift theory and cladistics to create a new research program with repeatable methods that focused on interpreting biogeographic and evolutionary models (Brooks and McLennan, 1993).

Over time, two main trends emerged in zoogeographic models: historical and ecological.

The historical trend recognizes that each geographic region has different species aggregations, and that some systematic groups of organisms exhibit tendencies to aggregate in the same geographic regions where the simultaneous impacts of climate, location, and evolutionary processes are responsible for the general directions of the development of fauna over long periods. Emphasis in this kind of research is placed on the dynamics and statics of the main geographical and geological events affecting vast areas over significant periods spanning as much as millions of years. This approach is based on different groups of animals that evolved simultaneously.

The ecological trend attempts to elucidate the patterns of current distributions primarily through the range of ecological requirements of animals with a particular focus on environmental parameters, physiological tolerance, and the adaptive abilities of the organisms. Special emphasis is placed on the statistic and dynamics of current events or those which occurred not long ago, but little attention is paid to the time or space of the phenomenon studied.

Initially, zoogeographic studies of marine organism parasites focused mainly on the waters of the northern hemisphere. The first wide-ranging studies of freshwater parasite fauna in the USSR were conducted by Šul'man (1961) and Polyanski (1961). In the 1980s ecological and zoogeographic studies were initiated by Rohde (1981, 1982, 1984), who drew attention to the fact that the parasite distribution in the seas and oceans is uneven. Rohde noted that the Indo-Pacific has more parasite species than does the Atlantic Ocean, and that the species composition in these two areas also differs. He also described the role of latitudinal gradients in parasite species diversity emphasizing that not only is there an increase in the absolute numbers of teleost monogeneans towards the equator, there is likely also a relative increase in the number of parasite species because of the greater number of host species. He also compared the prevalence and intensity of Monogenea infection at various geographic latitudes and concluded that frequencies of occurrence are higher in the tropical zone, while intensity of infection differed significantly and did not exhibit any trends. Rohde explained that in the tropical zone more species can co-exist since their niches are smaller.

Within the context of zoogeography, discussions about temperature center around its significance to the distribution of parasites and to the phenomenon of fluctuations of parasite infection in warm and cold seas, and differences in the species composition between parasite fauna in deep and shallow waters, and the occurrence of relic and endemic species in isolated basins. According to the continental drift theory, it has been confirmed that the Pacific is older than the Atlantic by about 150,000 years, as is probably explained by the occurrence of more species in the Pacific as there has been more time for them to originate and accumulate. The role of zoogeography is crucial in studies of the significance of host migration to parasites, and the possibility of using parasites as biological indicators of host movement and the existence of separate breeding stocks of the same host species. The origin and co-evolution of parasites, their host species, and host specificity are also studied. The zoogeography of parasites is also useful in studies of the origin of host genera (Williams & Jones, 1994).

Different groups of factors and processes can impact the geographic distribution of parasites and their hosts depending on the scale considered. On a global scale, the following can impact distribution:

-   the division and movements of land and the appearance of natural geographic barriers that divide and separate biotas (vicariance). For marine organisms, this can be the appearance of new isthmuses or changes in water currents;
-   the ability of organisms spread either passively or actively – with parasites, many free-living stages spread actively, while only eggs and internal parasites spread passively;
-   the time-frames within which vicariance and spreading occur; in the majority of these types of cases, it was the consequences of both processes.

The success of colonization is determined by many biotic and abiotic factors that influence the reproduction and survival of organisms. For many free-living organisms, competition, predation, and parasitism are important biotic factors impacting their distribution. For parasites this is probably the host biology. On a smaller scale, these factors become more specific. In the marine ecosystem the most important factors are temperature, salinity, depth, and currents. The use of parasites as biological tags in population studies of marine fish and mammals has proved to be a successful tool for discriminating stocks for all species to which it has been applied. Parasitological data are valuable when conventional tagging are not feasible or it costly and can give misleading results. The distribution patterns of marine parasites are determined mainly by temperature–salinity profiles and by their association with specific masses of water. Analyses of distribution patterns of some parasite species in relation to gradients in environmental (oceanographic) conditions showed that latitudinal gradients in parasite distribution are probably directly related to water temperature. Indeed, temperature, which is a good predictor of latitudinal gradients of richness and diversity of species, shows a latitudinal pattern in south-western Atlantic coasts, decreasing southwards, due to the influence of subtropical and subantarctic marine currents flowing along the edge of the continental slope. This pattern also determines the distribution of zooplankton, with a characteristic specific composition in different water masses. The gradient in the distribution of parasites determines differential compositions of their communities at different latitudes, which makes possible the identification of different stocks of their hosts (Timi, 2007). The eggs and cysts (propagules) of many parasites can be dispersed by the host as it moves and by moving water. Ocean current may be an effective mechanism of dispersion over larger areas and increase the parasites spatial distribution and their chance of coming in contact with an appropriate host (Bush et al., 2002).

5. Host range and distribution

5.1 Protista and Myxozoa

The Protista are characterized by a vast diversity of form and function. To date, about 100,000 species representing 40 families of free-living and parasitizing organisms have been identified, the most important of which are as follows.

The Microsporidia are strictly intracellular eukaryotic parasites (2 - 12 µm size range) characterized by simple ellipsoid unicellular spores containing sporoplasma and single coiled polar tubes, through which, following extrusion, the infectious sporoplasma enters the host cell. Microsporidia infect vertebrates and invertebrates, and the parasite enters the host body through ingestion. Microsporidia can reproduce sexually or asexually, and the life cycles of different species vary.

To date about 100 species belonging to 16 genera parasitizing fish have been identified throughout the world (Lom, 2002). Depending on the location the parasite settles, it causes cell and tissue damage that delays development, growth, or reduces fecundity. It can also cause the mortality of the host.

The majority of Mastigophora are free-living aquatic organisms, but representatives of the families Dinoflagellata and Kinetoplastida are marine organism parasites. They are mostly uninucleated organisms that are equipped at some stage in their life cycles with one or more flagella that permit them to move. Reproduction is either asexual (usually by longitudinal splitting) or sexual.

To date about 4,000 species of Dinoflagellata, including 140 species of external parasites of crustaceans and fish, have been identified. Most of these have chloroplasts, some contain dyes that accumulate in the tissues of fish or filter-feeding shellfish. These dyes are neurotoxins for the mammals that consume the fish or shellfish. The greatest impact of parasitic dinoflagellates is noted in warm-water marine fish, especially in intensive mariculture (Paperna, 1984).

Among the approximately 500 Kinetoplastida species that have been described to date, many parasitize the blood of vertebrates. Their development proceeds with either one or two hosts – fish that are infected by leech vectors. These parasites are distributed widely throughout the world.

Most of the terrestrial and aquatic species of Ciliophora are free-living organisms. Over 150 species, most of which are ectoparasites, occur in fish. Their systematic classification is based mainly on many morphological characters, the course of the developmental cycle, habitats, and life style. These organisms have vegetative macronuclei and reproductive micronuclei, cilia at certain stages of life, and cell membranes. Asexual reproduction occurs through horizontal splitting . Some species also reproduce through conjugation. Marine and brackish-water parasites include sessile ciliates which can live singly or stalked and bearing a colony (Lom & Dykova, 1992). The juvenile stages are free-swimming thanks to cilia that are deployed in a wreath-like structure.

Peritrichous ciliates are often noted on the skin and gills of freshwater and marine fish from the northernmost to the southernmost latitudes of the temperate and tropical zones (Fig. 1.). Their adhesive discs have crowns of hard denticles (Fig. 2.). The shape of these discs as well as the number and structure of the hard denticles are the most useful character for determining species. Transmission is direct among these parasites with low specificity.

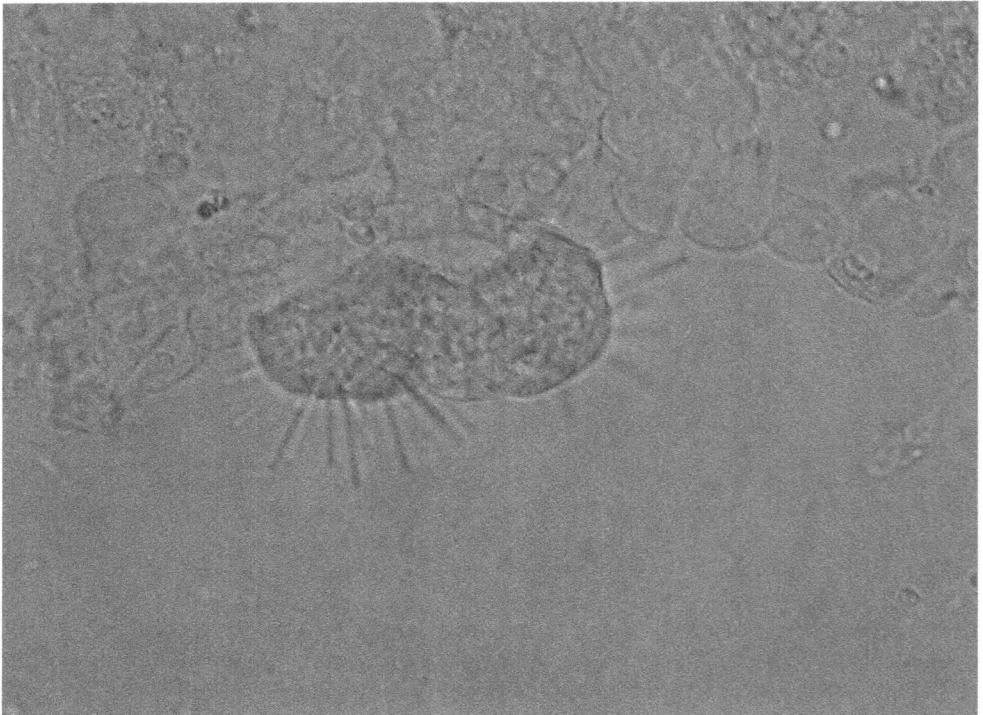

Fig. 1. Peritrichous ciliates on the gills of Chinese mitten crab (microscopic slide, magnification 400x, photograph by E. Sobecka)

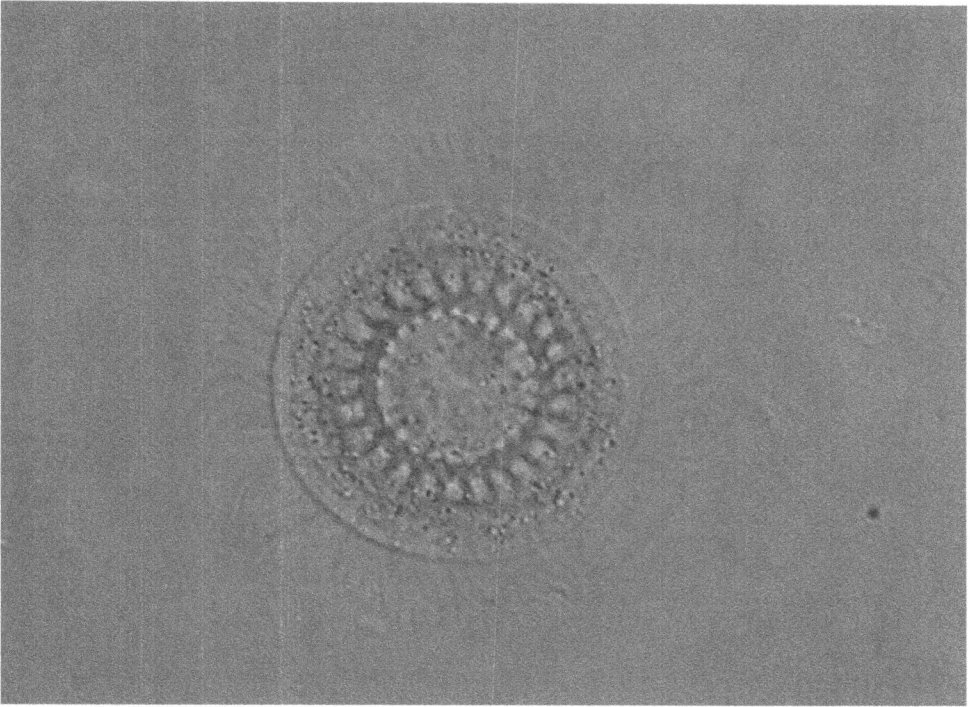

Fig. 2. Adhesive disc of *Trichodina* sp. (microscopic slide, magnification 400x, photograph by E. Sobecka)

Ciliates of the orders Prorodontida and Dysteriida are dangerous parasites of fish. The first is a holotrichous ciliate, which occurs in the seas of the tropical, subtropical, and temperate zones. It infects the epithelium of the skin, gills and eyes causing excessive mucus production, epithelial hyperplasia, damage to the continuity of the external covering with the possibility of repeat infection, and impaired gas exchange. Ciliates from the order Dysteriida are cosmopolitan parasites that usually occur in small numbers in the tropical zone; however, when conditions are especially good, it is a cause of high fish mortality (Gallet de Saint Aurin et al., 1990).

Myxozoa are commonly noted fish and invertebrate parasites. To date, about 2,800 species belonging to 61 genera, of which the vast majority are fish parasites, annelid worms, or bryozoans. The first description of a parasite of this group dates to the early nineteenth century (Leydig, 1851). The discovery that *Buddenbrockia* sp., a nematode-like parasite of freshwater bryozoans is a myxozoan, and that the myxozoans originate from bilaterians allowed for a better understanding of the known endoparasites of fish and worms in the class Myxosporea (Okamura et al., 2002). Phylogenetic and genetic studies conducted at this time led researchers to the conclusion that Myxozoa are highly specialized Metazoa that are the cause of several diseases occurring in aquaculture and among wild stocks that are of economic consequence (Feist, 2008). The phylum Myxozoa has two classes – Myxosporea and Malacosporea. The Myxozoa classification depends on spore morphology, but these are increasingly linked with the results of molecular studies (Fiala, 2006; Lom & Dykova, 2006).

Many species have complex life cycles; annelids are definitive hosts whereas vertebrates are intermediate hosts for Myxosporea (Fiala, 2006). They are characterized by multicellular spores with polar capsules that contain extrudible polar filaments that attach to host cells (Fig. 3). These are parasites inhabit tissues or cavities (histozoic), or many organs (coelozoic). Myxozoans occur in fresh, brackish, and marine water throughout the world. They have been noted to infect teleost fish, elasmobranchs, amphibians, reptiles, and, in single cases, octopuses (Arthur & Lom 1985, Heupel and Bennet 1996, Helke & Poynton, 2005; Yokoyama & Masuda 2001). The actinospore stage is noted in polychaetes and sipunculids (Køie et al., 2004). Thanks to their specificity in marine fish and the fact that infections can persist for a lifetime, myxosporeans are used in studies of stock discrimination.

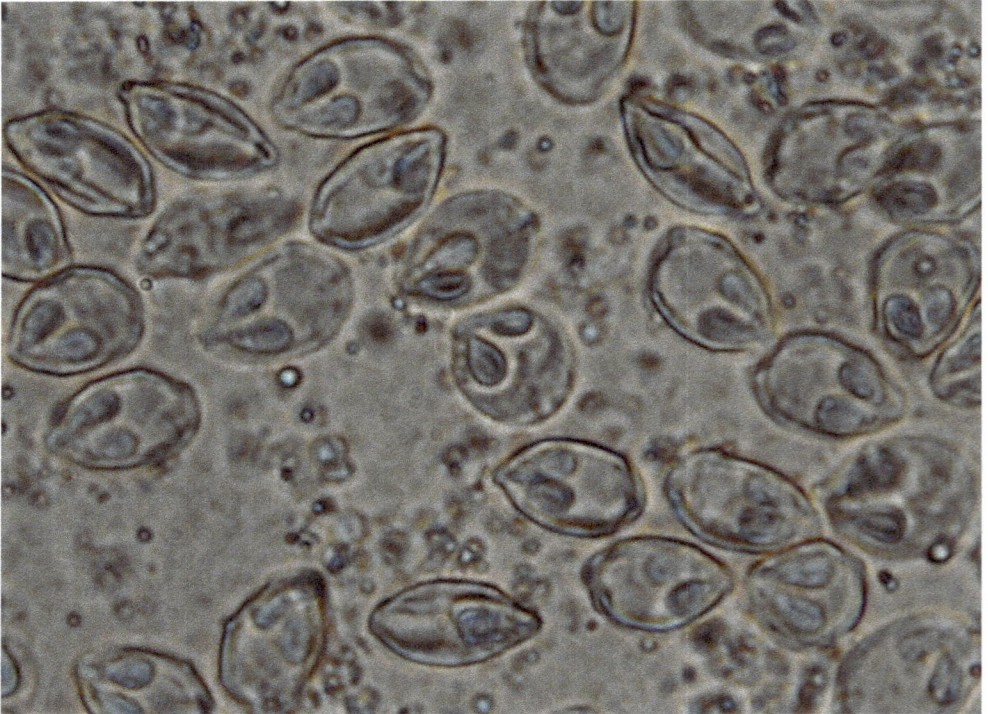

Fig. 3. *Myxobolus* sp. spores on the gills of fish (microscopic slide, magnification 1000x, photograph by E. Sobecka)

5.2 Monogenea

The range of marine animals parasitized by monogeneans includes chimaeras, sharks, rays, Cyclostomata, lungfish, and teleosts, but they infect amphibians and reptiles less frequently (Cone, 1995; Kearn, 1963; Poynton et al., 1997). They settle on the gills, skin, fins, and sometimes in the mouth and nasal cavities and in the urogenital system. Most species occur on one host species, so their range of occurrence coincides perfectly with that of the host species. These are small hermaphroditic organisms (0.1 - 20 mm length range) with a dorsal-ventral flattening of the body that is usually elongated. It is easy to distinguish them from other flat parasitic worms because of the posterior attachment organ – the opisthaptor, which is its main attachment organ (Fig. 4). The highly varied build and armaments is one of the main taxonomic characters that distinguishes two subclasss – Monopisthocotylea (opisthaptor with hooks) and Polyopisthocotylea (opisthaptor with suckers). The anterior end of monogeneans also have an attachment organ (prohaptor) which can have suckers, warts, pits, growths and attachment glands. The prohaptor is used as the parasite moves. In addition to the differences mentioned earlier in the build of the opisthaptor, the systematic division into two subclasss is based on micro-habitats inhabited, diets, and parasite mobility (Boeger & Kritsky, 2001). Approximately 1,000 species of marine Monopisthocotylea are known. They can inhabit different micro-habitats in fish, feed on mucus and epithelial cells, and very actively move around the body surface of the host (Whittington, 2004).

Polyopisthocotylea, of which to date 1,000 species have been described, usually settle on the gills of fish, feed on blood, and generally spend the majority of their adult lives in one place.

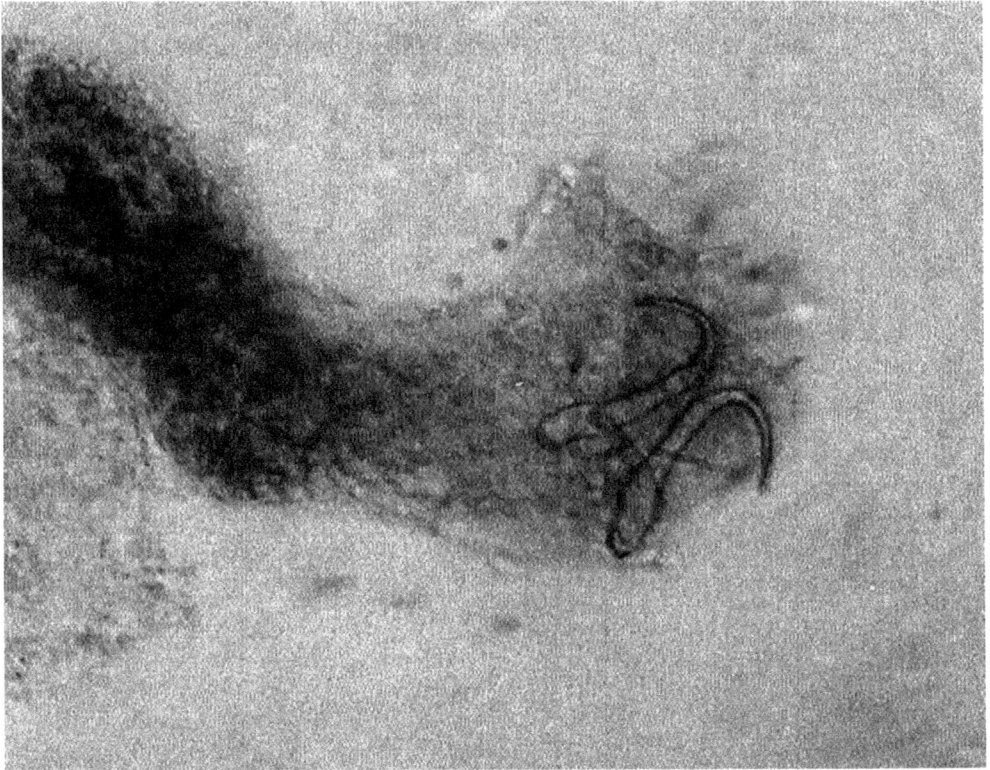

Fig. 4. Posterior attachment organ of monogenean belonging to the subclass Monopisthocotylea (microscopic slide, magnification 400x, photograph by E. Sobecka)

The monogeneans have a single-host reproductive cycle without a intermediate host. This oviparous species releases eggs into the water. When they hatch, the slow-moving, ciliated larvae have to find a host relatively quickly. With viviparous monogeneans, the uterus of the mother releases a fully-formed parasite of the next generation which carries an embryo of the subsequent generation.

Marine monogeneans from both subclasss occur in both brackish and oceanic waters from the littoral zone to the open waters of all the climate zones.

5.3 Digenea

Digeneans, or flukes, are a very abundant group of flatworms (Platyhelminthes) and one of the most abundant groups of parasites. Along with a small group of species from the subclass Aspidogastrea, they comprise the class Trematoda. Approximately 5,000 species from 70 families are fish parasites, the most frequently noted of which are trematodes belonging to Hemiuridae, Fellodistomidae, Bucephalidae, Derogenidae, and Lecithasteridae (Fig. 5.). Sea turtles and iguanas, semi-marine crocodilians, and sea snakes are all infected with trematodes (Yamaguti, 1971; Lafferty, 1993). While heavy parasite infection is not normally noted in sharks, rays, skates, and chimaeras, cetaceans, seals, sirenians, and other marine mammals suffer serious infections from these parasites. It is difficult not to include sea birds in this group of hosts since digeneans from several families mature in them.

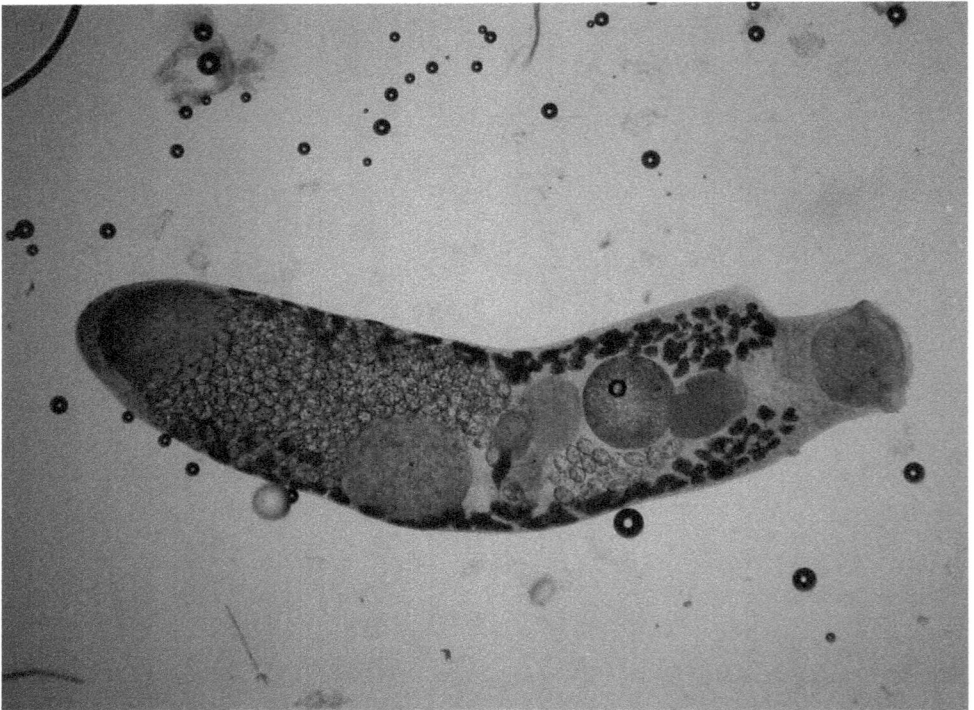

Fig. 5. Adult fluke from digestive tract of fish (microscopic slide, magnification 100x, photograph by E. Sobecka)

These are generally parasites of the digestive tract, but in fish they also settle in the bladder, swim bladder, abdominal cavity, gall bladder, ovaries, and blood vessels. These organisms have highly differentiated builds and sizes. They measure from 250 μm to 12 cm in length, and the body is oval like an elongated leaf and usually with dorsal-ventral flattening. Most species have an oral sucker at the anterior end of the body which surrounds the oral opening. The ventral sucker is only for attachment, and its presence and location are both important taxonomic characters. The body surface area can be smooth or it can be covered with scales of different shape of size.

Most Digeneans are hermaphrodites, with only a few families (e.g., Schistosomatidae) that are gonochoristic. They have a complex developmental cycle; after one generation reproduces sexually, the next few generations reproduce asexually (or parthenogenetically, as some researchers prefer). The asexual generations are created with one fertilized egg which develops with ciliated miracidium equipped with eye spots and a group of reproductive cells.

The first intermediate host of the flukes, into which the miracidium must actively penetrate, are snails, mussels, or polychaete annelids. Once there, they transform into mother sporocysts that resemble bags without a hint of either reproductive or digestive systems. One should bear in mind that this is not a larval stage, but is sometimes referred to as the first intramolluscan generation (Rohde & Rohde, 2005). The germinative cells develop into the second asexual intramolluscan generation, known as the daughter sporocyst, or the mother rediae. Rediae settle in the liver-pancreas of the host where they produce a generation of rediae, and then finally cercariae.

The development of sex generations can have two larvae: cercaria and metacercaria. The first is equipped with a tail and swims easily searching for a host (intermediate – invertebrate or vertebrate) into which it usually actively penetrates and settles into the organ that is typical of its species and where it transforms into a metacercaria. This is a small fluke with an immature reproductive system. If the subsequent (definitive) host consumes the infected intermediate host, the metacercaria will develop into a mature digenean in the digestive tract, and less frequently in other organs.

The typical digenean developmental cycle require three hosts, of which molluscs (first intermediate host) are the hosts of the asexual generation, invertebrates of vertebrates (the second intermediate host), in which metacercariae live, and the vertebrates (definitive host), in which hermaphroditic adults mature. The fluke life cycle can be shortened if one of its hosts is unavailable (but not that in which the asexual generation lives). Along with asexual reproduction in the first intermediate host, the reproductive success of digeneans increases, and they are considered to be a perfect evolutionarily-adapted group of parasites.

The different digenean families exhibit different specificity regarding definitive hosts, and rarely are they limited to just one group (e.g., manatee and dugongs). Most of them do not exhibit such tendencies, and the choice of host is usually determined by the composition of its food. The specificity range with regard to the first intermediate host is narrower and does not extend beyond a few species of molluscs. Digeneans occur in the seas of all the climate zones.

5.4 Cestoda

Cestodes, members of the Platyhelminthes, include about 5,000 species from two groups: Cestodaria, which comprises two orders – Gyrocotylidea and Amphylinidea, and

Eucestoda, which comprises 11 orders. This second group will be the focus of this section. Three of these 11 orders are exclusive marine organism parasites. Three others are primarily marine organisms. Eucestoda are parasites with bodies composed of three sections : the attachment organ or scolex, the shape of which and the presence or absence of armaments are characteristic for the representatives of different orders, the short neck where growth occurs, and the ribbon-like body (strobila) and segregated into a linear series of compartments (proglottids) comprising one or more sets of reproductive organs (Caira & Reyda, 2005). In this way, tapeworms meet their reproductive role as there are as many individuals as there are compartments in its strobila. The number and position of the various elements of the reproductive system are subsequent taxonomic characters in the systematics of these parasites. The length of tapeworms ranges from several millimeters to nearly 20 meters. Tapeworms lack a digestive system, and the surface area of the body is capable of absorption. Adult parasites inhabit digestive tracts, and rarely the neighboring organs.

The development of marine eucestodes is not fully understood, and the full developmental cycle has been described only for tapeworms from five orders. The life cycle is complex and begins when an embryonated egg containing a larva (hexacant) is expelled by the definitive host into the water (Chervy, 2002). Here it is swallowed by its first intermediate host, which can include any of the following: Polychaeta; Gastropoda; Euphausiacea; Amphipoda; Copepoda; Cirripedia; Branchiopoda. Some Trypanorhyncha and Pseudophyllidea have free-living larvae (coracidium). The egg or coracidium that is swallowed by the intermediate host develops into the second larva (cercoid). Further development happens in the subsequent host, which can be any of the following: Cnidaria; Nemertea; Ctenophora; Molusca; Arthropoda; Echinodermata; Chaetognatha; Agnatha; Chondrichthyes; Osteichthyes; Chelonia; Cetacea; Pinnipedia. This group also includes paratenic hosts, and for some groups of tapeworms, this is also the definitive host, in which the parasite develops into adult form. For others is it simple another intermediate host, in which the third larva develops (metacercoid, Fig. 6.). These hosts include Chondrichthyes (Chimaeriformes, Batoidea, Galea), Aves, Mustelidae, Pinnipedia, and Cetacea (Caira & Reyda, 2005). Thus, the only developmental form that inhabits the external environment is the hexacant, while the others inhabit subsequent hosts and shape at each stage of development parasite-host relationships.

5.5 Nematoda

Nematodes (roundworms) are one of the numerous types in the animal kingdom with 256 families and in excess of 40,000 species (Anderson, 2000). Many species are free-living, but there are also plant and animal parasites. Nematodes have adapted to every possible inhabitable environment from the tropics to the polar regions. They are even noted in desert sand and natural hot springs. The phylum Nematoda comprises the classes Adenophorea and Secernentea (Anderson, 2000), and parasitic nematodes belong to both. Marine nematodes are grouped into seven orders. Parasite nematodes are generally larger, deposit more eggs, and live longer than do free-living nematodes. They are characterized by the constant number of cells that comprise the adult organism. They measure from 1 mm to 1 m. The nematode body is cylindrical and elongated, but it can be filiform and symmetrical bilaterally. The digestive tract has three sections, and its structure as well as that of the mouth opening are important taxonomic features. Nematodes (roundworms) are mostly dioecious, and females and males are distinctly different from each other. Eggs are deposited at various stages of development (Fig. 7.). Post-embryonic development include four larval stages, and the juvenile nematode will then undergo four molts before it becomes an adult. During each of these, the larvae shed their old covering in favor of a newly made one. Each moult is preceded by a change in structure regarding the development of digestive and reproductive tracts. Among parasitic nematodes, there are species that require just one host for their developmental cycles (monoxenic life cycle). The hosts for monoxenic marine nematodes include annelids, molluscs, echinoderms, and crustaceans. The development with a heteroxenic life cycle is possible with the participation of intermediate hosts and paratenesis (the use of paratenic or transport hosts). These are crustaceans, jawless fish, Chondrichthyes, and other fish. Birds and marine mammals are definitive hosts for this group of nematodes. Larval nematodes tend to be more widely distributed than are adults, probably because they are less specific in their choice of host.

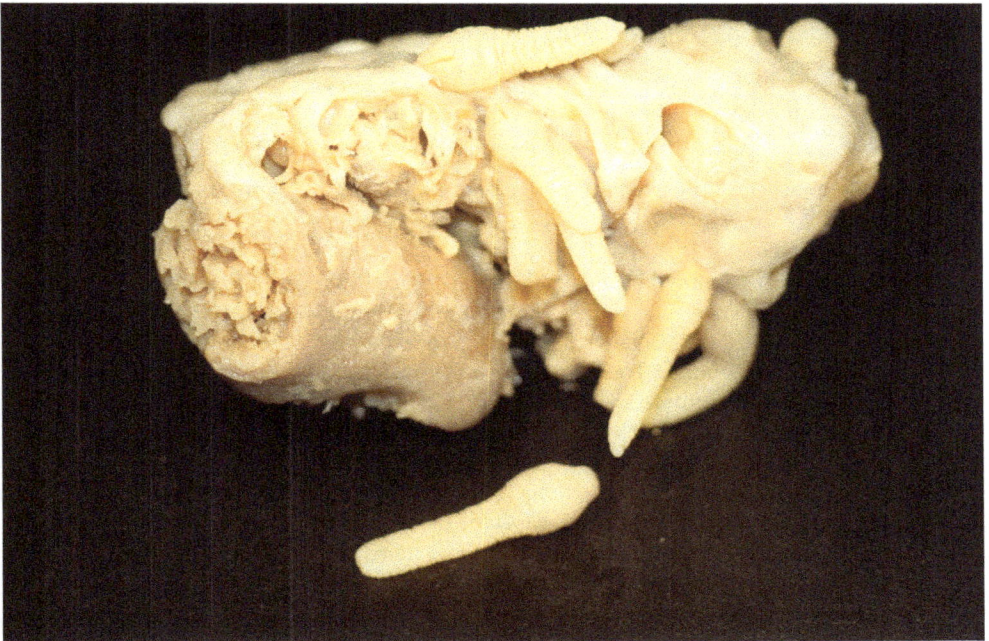

Fig. 6. Trypanorhynch *Hepatoxylon trichiuri* larvae from the liver of snoek, *Thyrsites atun* (photograph by S. Keszka; specimens from the collection of the Division of Fish Diseases, West Pomeranian University of Technology in Szczecin, Poland)

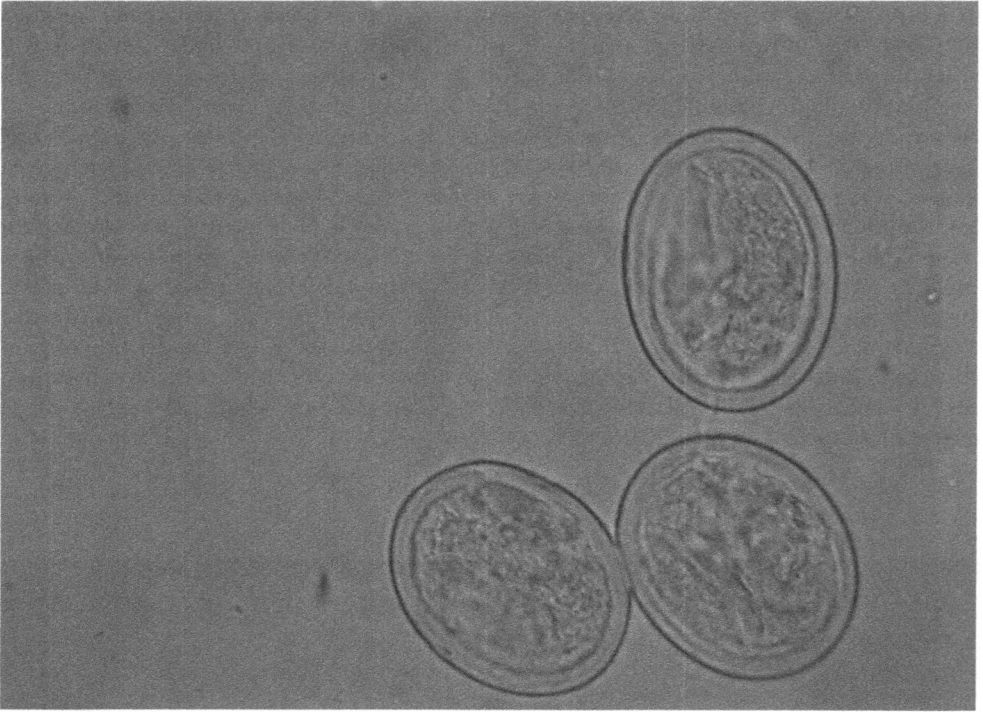

Fig. 7. Eggs of nematodes (microscopic slide, magnification 200x, photograph by E. Łuczak)

5.6 Acanthocephala

Acanthocephalans, also known as thorny or spiny-headed worms, are exclusively parasitic organisms. Over 1,000 species have been noted, and of these nearly half are parasites of the digestive tracts of fish (Taraschewski, 2005). The remaining are noted in amphibians, reptiles, birds, and mammal. The body of the acanthocephalans is cylindrical, slightly flattened, and divided into three parts: the front (praesoma) with its proboscis: a protrusible, armed attachment organ, the neck and the longest posterior section (metasoma) where the internal organs are. The proboscis is spindle shaped, cylindrical, or spherical and has rows of hooks, the number, shape, and arrangement of which are very important taxonomic features. Some systematists insist that because of the likeness of the attachment organs, the lack of a digestive system, and similar excretion systems and developmental cycles mean that all acanthocephalans and tapeworms can be combined. However, body shapes, a similar muscular build, and a pseudocel similarity to the nematodes. Phylogenetic studies have shown that, along with Rotatoria, they do form one taxon (Herlyn et al. 2003). Acanthocephalans from the order Archiacanthocephala are parasites of terrestrial animals while representatives of Eoacanthocephala and Palaeacanthocephala inhabit aquatic vertebrates.

Mature parasites measure from several mm to about 60 cm. These organisms are dioecious, and females are larger than males. The developmental cycle is complex; its eggs contain fully-formed larvae (acanthors), which, one in the aquatic environment, are swallowed by crustaceans, the intermediate hosts. In the body cavity of the crustacean, the acanthor transforms into the second larval stage (acanthella), which, in turn, makes a cyst and transforms itself into a into a cystacanth, the larval third stage that invades the definitive host (Fig. 8.). The larvae that are swallowed along with the intermediate host attaches itself to the mucus membranes of the intestines and it matures. The paratenic host actually plays a role in the development of some marine species of acanthocephalans. If the definitive host is a bird or a marine mammal, the amphipod serves as an intermediate host and fish as the paratenic host. The wide specificity with regard to both intermediate host as well as to the definitive host makes acanthocephalans cosmopolitan parasites.

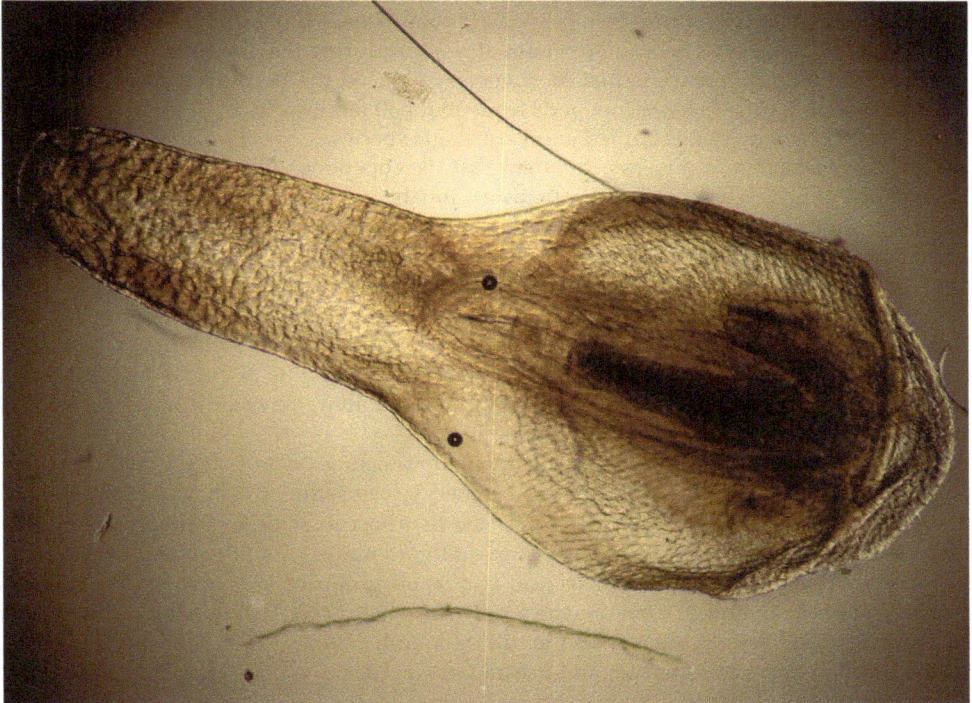

Fig. 8. Cystacanth, the larval third stage of *Corynosoma* sp. from the body cavity of fish (photograph by E. Łuczak)

5.7 Crustacea
Crustacea comprises a group of animals that is omnipresent and the most morphologically differentiated. Most of the species are marine. Among the approximately 50,000 species identified to date, many are parasitic and infect a wide spectrum of host from cnidarians to vertebrates (Bush et al., 2002). The classification of parasitic crustaceans based on Martin and Davis (2001) includes four subclasss, of which the most numerous is Copepoda (11,500 species, 29 orders). These are mostly endo- or ectoparasites of fish and invertebrates, and, to a lesser extent, of marine mammals.

The complexity of adaptations of copepods to the parasitic way of life is positively correlated with the length of time of their evolutionary association with the host. Those copepods that, in size and shape, are similar to free-living 'relatives', have slightly altered builds, retained the ability to swim, and their association with the host is only constant at the adult stage. Well-adapted species modified their build, biology and developmental cycle to such a degree that they almost entirely eliminated free-living developmental stages and increased the numbers of eggs released (Piasecki & Avenant-Oldewage, 2008). Parasite membership in the subclass Copepoda can sometimes only be confirmed thanks to the presence of two egg sacs in the female and nauplii. The males are usually smaller and less altered, with adult stages located near or attached to the females. Among the parasitizing copepods, some species are very small (about 1 mm) and very large (up to 60 cm in length). The developmental cycle of free-living copepods comprises two phases – naupliar (6 stages) and copepodit (5 stages). These same phases also occur during the development of parasitic copepods; the only difference is that most species reduced the number of naupliar stages. The parasites of marine fish are copepods from about 30 families. Most of them have a very narrow host specificity that is limited to even a single species.

The Isopoda parasite group is less abundant that the copepods with just about 450 species, the majority of which are marine ectoparasites inhabiting tropical and subtropical waters (Möller & Anders 1986) (Fig. 9). Juvenile and adult isopod stages from the families Cymothoidae and Aegidae parasitize fish, and also elasmobranchs and crustaceans in small numbers. Gnathiidae males and females are free living as adults. Cymothoidae are forms of protandrous hermaphrodites, which indicates that the specimens develop and function first as males and can then become females. The presence of mature females hampers the further development of males (Grabda, 1991). They measure from 1 to 6 cm in length Gravid females release eggs into a brood pouch or 'marsupium' formed from their ventral oostegites, where all of larval development occurs.

Gnathiidae is a small group of parasitic isopods that includes about 50 species. They only parasitize during the larval stage. Freshly hatched larvae settle on fish, moult three times before maturing into adults, and then leave the fish host to settle in pairs in the muddy sediments of an aquatic basin where they stop eating, reproduce, and then finally die.

Branchiura, commonly known as fish lice, is a small group of crustaceans that includes about 175 species of parasite from one the family (Boxshal, 2005). They inhabit the surface of the fish body, and less frequently in the gill or mouth cavities. Very few species inhabit marine coastal waters or estuaries, and it is not observed in the open waters of the oceans. The dorsal-ventral flattening of their bodies is pronounced, which allows them to better cling to the host. They are nearly transparent or milky white. The head of the parasite is fused to the first trunk segment, and the large, deep-set movable eyes are visible. On each free trunk segment there are a pair of legs, thanks to which the parasite can switch hosts. The females deposit their eggs in a mucus cocoon that they attach to rocks or vegetation. Swimming larvae hatch from these eggs, find a host, and being to feed on the host's blood.

The discussion of parasitic crustaceans should include Ascothoracida, Tantulocarida, and Cirripedia. The first of these, Ascothoracida, are ecto-, endo-, mezoparasite echinoderms and cnidarians. Tantulocarida is a group of small (up to 2 mm in length) marine crustaceans that are parasites of many other crustacean species. Rhizocephalan Cirripedia are also crustacean parasites. They are considered to be the most modified of all crustaceans, and are an example of excellent adaptation to the parasitic life strategy (Bush et al., 2002).

Fig. 9. Adult isopod belonging to the family Cymothoidae from the gill chamber of fish (photograph by E. Łuczak)

6. References

Anderson, R.C.(2000). *Nematode parasites of vertebrates. Their development and transmission.* 2nd edn. CABI Publishing, ISBN 9780851994215, Wallingford.

Arthur, J. R., Lom, J. (1985). *Sphaerospora araii* n. sp. (Myxosporea: Sphaerosporidae) from the kidney of a longnose skate (*Raja rhina* Jordan and Gilbert) from the Pacific Ocean off Canada. *Canadian Journal of Zoology,* Vol. 63, pp. 2902-2906. ISSN 0008-4301

Blondel, J. (1986). Biogéografie évolutive. Masson, ISBN 2-225-80801-5, Paris.

Boeger, W. A. & Kritsky, D. C. (2001). Phylogenetic relationships of the Monogenoidea, In: *Interrelationships of the Platyhelminthes,* Littlewood D. T. J. & Bray R. A., pp. 92-102, Taylor & Francis, ISBN 0-7484-0903-3, London, New-York

Boxshal, G.A. (2005). Crustacean parasites, In: *Marine Parasitology,* Rohde K., pp. 123 – 149, Oxon, U.K., CSIRO Publ. and CAB International, ISBN 0643090258,Melbourne and Wallingford

Brooks, D. R., & McLennan, D. A. (1993). *Parascript: Parasites and the language of evolution.* Smithsonian Institution Press, Washington, D.C., URL:http://phthiraptera.info/content/parascript-parasites-and-language-evolution

Buczek, A. (2003). *Choroby pasożytnicze. Epidemiologia, diagnostyka, objawy.* LIBER, ISBN 83-89373-16-5, Lublin

Bush, A.O. & Holmes, J.C. (1986). Intestinal helminthes of lesser scaup ducks: patterns of association. *Canadian Journal of Zoology,* Vol. 64, pp.132–141, ISSN 0008-4301

Bush, A.O., Fernandez, J.C., Esch, G.W. & Seed, J.R.(2002). *Parasitism. The diversity and ecology of animal parasites,* Cambridge University Press, ISBN 0521664470, Cambridge

Caira, J.N. & Reyda, F.B. (2005). Eucestoda (True tapeworms), In: *Marine Parasitology,* Rohde K., pp. 92 – 104, Oxon, U.K., CSIRO Publ. and CAB International, ISBN 1 84593 053 3, Melbourne and Wallingford

Chervy, L. (2002). The terminology of larval cestodes or metacestodes. *Systematic Parasitology, Vol.* 52, No. 1, pp. 1-33, ISSN: 0165-5752

Combes, C. (1995). Interaction durable. Ecologie et évolution du parasitisme. Masson, Editeur, ISBN 2-10-005753-7, Paris

Cone, D. K. (1995). Monogenea (Phylum Platyhelminthes), In: *Fish diseases and disorders, Protozoan and metazoan infections,* Vol. 1, In: Woo P. T. K., pp. 289–327, CAB International, ISBN 9780851988238, Cambridge

Esch, G.W, Kennedy, C.R., Bush, A.O. & Aho J.M. (1988). Patterns in helminths communities in freshwater fish in Great Britain: alternative strategies for colonisation. *Parasitology,* Vol. 96, No. 3, pp. 519–532, ISSN 0031-1820

Esch, G.W., Gibson, J.W.& Bourque, J.E. (1975). An analysis of the relationship between stress and parasitism. *American Midland Naturalists,* Vol. 93,No. 2, 339–353, ISSN: 00030031

Feist, S. (2008). Metazoan Diseases, In: *Fish Diseases.* Vol 2. J.C. Eiras, H. Segner, T. Wahl and B.G. Kapoor, pp. 613–682, Science Publishers, ISBN 978-1-57808-528-6, Enfield, Jersey, Plymouth

Fiala, I. (2006). The phylogeny of Myxosporea (Myxozoa) based on small subunit ribosomal RNA gene analysis. *International Journal for Parasitology,* Vol. 36, No. 14, pp.1521-1534, ISNN 0020-7519

Gallet Desaint Aurin D., Raymond J., Vianas V. (1989). Marine finfish pathology: Specific problems and research in the French West Indies, In: Advances in Tropical Aquaculture, Tahiti (French Polynesia), 20 Feb - 4 Mar 1989, pp. 143–160, access on 21.04.2011, available from: http://archimer.ifremer.fr/doc/00000/1478/

Grabda, J. (1991). *Marine fish parasitology, an outline.* PWN—Polish Scientific Publishers, Warszawa and VCH—Verlagsgesellschaft mbH, ISBN 3-527-26898-7, Weinheim

Helke, K. L., & Poynton, S. L. (2005). *Myxidium mackiei* (Myxosporea) in Indo-gangetic flap-shelled turtles *Lissemys punctata andersonii*: Parasite-host interaction and ultrastructure. *Diseases of Aquatic Organisms* Vol. 63, No 2-3, pp.215–230, ISSN 0177-5103

Herlyn, H., Piskurek, O., Schmitz, J., Ehlers, U. & Zischlera, H. (2003). The syndermatan phylogeny and the evolution of acanthocephalan endoparasitism as inferred from 18S rDNA sequences. *Molecular Phylogenetics and Evolution* , Vol. 26, No. 1, 155-164, ISSN 1055-7903

Heupel, M.R. & Bennett, M.B. (1996). A myxosporean parasite (Myxosporea: Multivalvulida) in skeletal muscle of epaulette sharks, *Hemiscyllium ocellatum* (Bonnaterre, 1788) from the Great Barrier Reef. *Journal of Fish Diseases*, Vol.19, No. 2, pp. 189-191, ISSN 0140-7775

Holmes, J.C. (1990). Helminth communities in marine fish, In: *Parasite communities patterns and processes*. Esch, G., Busch, A. & Aho, J., pp. 101-130, Chapman and Hall, ISBN 0412335409 , London, New York

Holmes, J.C.& Price, P.W. (1986). Communities of parasites, In: *Community Ecology: pattterns and processes*. Kikkawa, I J. & Anderson, D.J. pp. 187–213, Blackwell Scientific Publications, ISBN 0867932724, Oxford

Von Humboldt, A. & Bonpland, A. (1805). *Essai sur la géographie des plantes: accompagne d'un tableau physique des regions equinoxiales*. Levrault, Schoell et Compagne, Librairies, Paris

Von Ihering, H. (1891). On the Ancient Relations between New Zealand and South America. *Transactions and Proceedings of the New Zealand Institute*, Vol. 24, pp. 431-445

Kearn, G. C. (1963). Feeding in some monogenean skin parasites: *Entobdella soleae* on *Solea solea* and *Acanthocotyle* sp. on *Raia clavata*. *Journal of the Marine Biological Association of the United Kingdom*, Vol. 43, No. 3, pp.749–767, ISSN 0025-3154

Køie, M., Whipps, C.M. & Kent, M.L. (2004). *Ellipsomyxa gobii* (Myxozoa: Ceratomyxidae) in the common goby *Pomatoschistus microps* (Teleostei: Gobiidae) uses *Nereis* spp. (Annelida: Polychaeta) as invertebrate hosts. *Folia Parasitologica*, Vol. 51, No. 1, pp.14 –18, ISSN 0015-5683

Lafferty, K.D. (1993). The marine snail, *Cerithidea californica*, matures at smaller sizes where parasitism is high. *Oikos*, Vol. 68, No. 1, 3 –11, ISSN 0030-1299

Leydig, F. 1851. Ueber Psorospermien und Gregarinen. Müller's *Archiv für Anatomie und Physiologie* , pp. 221-233, Leipzig

Levri, E.P. (1998). Parasite-induced change in host behavior of a freshwater snail: parasitic manipulation or byproduct of infection? *Behavioral Ecology*, Vol.10. No. 3, pp. 234-241, ISSN 1045-2249

Llewellyn, J. (1984). The biology of *Isancistrum subulatae* n. sp., a monogenean parasitic on the squid, *Alloteuthis subulata*, at Plymouth. *Journal of the Marine Biological Association of United Kingdom*, Vol. 64, No. 2, pp. 285 – 302, ISSN 0025-3154

Lom, J. 2002. A catalogue of described genera and species of microsporidians parasitic in fish. *Systematic Parasitology*, Vol. 53, No. 2, pp. 81-99, ISSN 0165-5752

Lom, J. & Dykova, I. (1992). *Protozoan parasites of fish*. Elsevier, ISBN 0444894349, Amsterdam

Lom J. & Dykova, I. (2006). Myxozoan genera: definition and notes on taxonomy, life-cycle terminology and pathogenic species. *Folia Parasitologica*, Vol. 53,No. 1, pp. 1–36, ISSN 0015-5683

Lyons, E. T., DeLong, R. L., Gulland, F. M., Melin, S. R., Tolliver, S. C. &. Spraker, T. R. (2000). Comparative Biology of Uncinaria spp. in the California Sea Lion (*Zalophus californianus*) and the Northern Fur Seal (*Callorhinus ursinus*) in California. *Journal of Parasitology*, Vol. 86, No.6, pp. 1348–1352, ISSN 0022-3395

Manter, H.W. (1940).The geographical distribution of digenetic trematodes of marine fishes in the tropical American Pacific. *Allan Hancock Foundation Publication. Allan Hancock Pacific Expeditions*, The University of Southern California Press, Vol. 2, Part 16, 1935 – 1940, pp. 531–547, Los Angeles, California

Manter, H.W. (1955). The zoogeography of trematodes of marine fishes. *Experimental Parasitology* Vol. 4, No. 1, pp. 62 – 86, ISSN 0014-4894

Manter, H.W. (1963). The zoogeographical affinities of trematodes of South American freshwater fishes. *Systematic Zoology*, Vol. 12, No. 2, pp. 45 – 70, ISSN 0039-7989

Marcogliese, D. (1995). The role of zooplankton in the transmission of helminth parasites to fish. *Reviews in Fish Biology and Fisheries*, Vol. 5, No. 3, pp. 336–371, ISSN 0960-3166

Marcogliese, D. (2002). Food webs and the transmission of parasites to marine fish. *Parasitology*, Vol. 124, No.7, pp. 83–99, ISSN 0031-1820

Martin, J. W. & Davis G. E. (2001). An updated classification of the recent crustacea. *Natural History Museum of Los Angeles County*, Science Series 39, pp. 1–124, ISSN 1-891276-27-1

Matthews, B.E. (1998). *An introduction to parasitology*. Cambridge University Press, ISBN 9780521576918, Cambridge

Möller, H. & Anders, K. (1986). Diseases and parasites of marine fishes. Verlag Heino Möller, ISBN 3923890044, Kiel

Niewiadomska, K., Pojmańska, T., Machnicka, B. & Czubaj, A. (2001). *Zarys parazytologii Ogólnej*. Wydawnictwo Naukowe PWN, ISBN 830113545X, Warszawa

Okamura, B., Curry, A., Wood, T.S. & Canning, E.U. (2002). Ultrastructure of Buddenbrockia identifies it as a myxozoan and verifies the bilaterian origin of the Myxozoa. *Parasitology*, Vol. 124, No. 2, pp. 215–223, ISSN 0031-1820

Paperna, I. (1984). Reproduction cycle and tolerance to temperature and salinity of *Amyloodinium ocellatum* (Brown, 1931) (Dinoflagellida). *Annales de Parasitologie Humaine et Comparée*, Vol. 59, No. 1, pp. 7–30, ISSN 0003-4150

Piasecki, W.& Avenant-Oldewage, A. (2008). *Diseases* caused by Crustacea, In: *Fish Diseases*. Vol 2. J.C. Eiras, H. Segner, T. Wahl and B.G. Kapoor, pp. 1115–1200 , Science Publishers, ISBN 978-1-57808-528-6, Enfield, Jersey, Plymouth

Price, P.W. 1980. Evolutionary biology of parasites. Princeton University Press. Princeton.

Polyanski, Ŭ.I. (1961). Zoogeography of parasites of the USSR marine fishes, In: *Parasitology of fishes*, Dogiel V.A., Petruševski G.K. & Polyanski Ŭ.I., pp. 230–246, Oliver and Boyd, ISBN 0876661312, Edinburgh, Scotland

Poynton, S. L., Campbell, T. W. & Palm, H. W. (1997). Skin lesions in captive lemon sharks *Negaprion brevirostris* (Carcharhinidae) associated with the monogenean *Neodermophthirius harkemai* Price, 1963 (Microbothriidae). *Diseases of Aquatic Organisms*,Vol. 31, No. 1, pp. 29–33, ISSN 0177-5103

Ravichandran, S., Rameshkumar, G. & Kumaravel, K. (2009). Variation in the Morphological Features of Isopod Fish Parasites. *World Journal of Fish and Marine Sciences*, Vol. 1, No. 2, pp. 137-140, ISSN 2078-4589

Riggs, M.R., Esch, G.W. (1987). The suprapopulation dynamics of *Bothriocephalus acheilognathi* in a North Carolina cooling reservoir: abundance, dispersion and prevalence. *Journal of Parasitology*, Vol. 73, No. 5, pp. 877-892 , ISSN 0022-3395

Rohde, K. (1981). Niche width of parasites in species-rich and species-poor communities. *Experientia*, Vol. 37, No. 4, pp. 359-361. ISSN 0014-4754

Rohde, K. (1982). *Ecology of marine parasites*. University of Queensland Press, ISBN 0-7022-1660-7, St. Lucia, Queensland

Rohde, K. (1984). Zoogeography of marine parasites. *Helgoland Marine Research*, Vol. 37, No. 1-4, pp. 35-52, ISSN 1438-387X

Rohde, K. (1994). Niche restriction in parasites: proximal and ultimate causes. *Parasitology*, Vol.109, (Suppl S), pp. 69 - 84, ISSN 0031-1820

Rohde, K. and Rohde P.P. (2005). The ecological niches of parasites, In: *Marine Parasitology*, Rohde K., pp. 286-292, Oxon, U.K., CSIRO Publ. and CAB International, ISBN 1 84593 053 3, Melbourne and Wallingford

Sclater, P.L. 1858. On the general geographical distribution of the members of the class Aves. *Journal the Proceedings the Linnean Society. Zoology*, Vol. 2, pp. 130-145

Shoop, W.L. (1991). Vertical transmission of helminthes: hypobiosis and amphiparatenesis. *Parasitology Today*, Vol. 7, No. 2, pp. 51 - 54, ISSN 0169-4758

Sousa, W.P. (1990). Spatial scale and processes structuring a guild of larval trematode parasites, In: *Parasite communities: Patterns and Processes*. Esch, G., Bush, A., & Aho, J., pp. 41-67, Chapman and Hall, ISBN 0412335409, London

Šul'man, S.S. (1961). Zoogeography of parasites of USSR freshwater fishes, In: *Parasitology of fishes*, Dogiel V.A., Petruševski G.K. & Polyanski Ŭ.I., pp. 180-229, Oliver and Boyd, ISBN 0876661312, Edinburgh, Scotland

Taraschewski, H. (2005). Acanthocephala (thorny or spiny-headed worms, In: *Marine Parasitology*, Rohde K., pp. 116 -121, Oxon, U.K., CSIRO Publ. and CAB International, ISBN 0643090258, Melbourne and Wallingford

Timi, J.T. (2007). Parasites as biological tags for stock discrimination in marine fish from South American Atlantic waters. *Journal of Helminthology*, Vol. 81, No. 2, pp. 107 - 111, ISSN 0022-149X

Whittington, I.D. (2004). The Capsalidae (Monogenea: Monopisthocotylea): a review of diversity, classificationand phylogeny with a note about species complexes. *Folia Parasitologica*, Vol. 51, pp. 109-122, ISSN 0015-5683

Williams, H. & Jones, A. (1994). *Parasitic worms of fish*. Taylor & Francis, ISBN 0-85066 425X, London and Bristol

Yamaguti, S. (1971). *Synopsis of digenetic trematodes of vertebrates*. Keigaku Publishing Co, Tokyo

Yokoyama, H. & Masuda, K. (2001). *Kudoa* sp. (Myxozoa) causing a post-mortem myoliquefaction of North-Pacific giant octopus *Paroctopus dofleini* (Cephalopoda: Octopodidae). *Bulletin of the European Assotiation of Fish Pathologists*, Vol. 21, No. 6, pp. 266-268, ISSN 01080288

Złotorzycka, J. (1998). *Słownik Parazytologiczny*, Polskie Towarzystwo Parazytologiczne, ISBN 83-901349-3-4, Warszawa

12

Connectivity as a Management Tool for Coastal Ecosystems in Changing Oceans

Nicolas Le Corre[1], Frédéric Guichard[2] and Ladd E. Johnson[1]
[1]Université Laval,
[2]McGill University,
Québec
Canada

1. Introduction

Recent theoretical management research has focused on systems from species to ecosystem at large scales (i.e., metapopulations and meta-ecosystems), and the links between habitats patches and subpopulations are of crucial importance to understand, predict, and manage resource dynamics. One of the key characteristics affecting the dynamics and demography of metapopulations is thus connectivity (Hanski, 1999; Kritzer & Sale, 2004; Moilanen & Nieminen, 2002), the exchange or flux of material between different locations (Cowen & Sponaugle, 2009). Because of its broad definition and growing relevance, "connectivity" is now employed in a number of fields, including metapopulation ecology. Consequently, several definitions of connectivity exist with the main differences between them lying in the spatial scale of study (Kadoya, 2009). In this review, we consider connectivity in its broadest sense of demographic or population connectivity: the exchange of individuals among geographically separated subpopulations in a metapopulation (Cowen & Sponaugle, 2009).

1.1 Connectivity in marine ecology

In a marine context, metapopulation structure is defined as populations occupying discrete patches, demographically connected according to a dispersal kernel (the function of propagule abundance vs. distance from the parental source) and potentially affecting the dynamic of the entire ecosystem (Kritzer & Sale, 2004). The main difference from equivalent terrestrial systems is that local extinctions rarely occur in marine systems (Kritzer & Sale, 2004) as the diverse regulation processes operating in the ocean and their inherent stochasticity lead to lower extinction rates (Hixon et al., 2002). Connectivity is one of these processes and tends to operate over larger spatial scales in marine metapopulations, due to fewer dispersal barriers and a more favorable medium for long distance movement of propagules. However, the potential for self-recruitment, i.e., the retention of propagules within a population, has recently been highlighted in many marine systems and may act as an additional mechanism to prevent extinction (Almany et al., 2007; Cowen et al., 2006; Levin, 2006).

In marine ecology, ideas have historically ranged from the extremes of demographically-open systems (fully connected) to closed populations (not connected) (Hixon et al., 2002).

For example, fisheries stock recruitment models generally ignored connectivity among populations, and local density-dependent factors were considered the most important parameter in the regulation of the populations. At the other end of the spectrum, recruitment into local populations was thought to occur from a general pool of propagules with new settlers arriving from unknown source populations. These simplifications of immigration and emigration processes were imaginable due to the spatial scales of studies that were either large enough (in the former case of some fisheries) or sufficiently small enough (open populations) to avoid dealing with the more realistic intermediate situations that characterize most marine systems. Connectivity is now, however, recognized to be a primary driver of most large-scale marine population dynamics. This is particularly true for the large number of marine species that are bentho-pelagic, with a stationary phase (e.g., sessile or sedentary juveniles and adults) and a planktonic stage (e.g., larvae, spores) during which dispersal occurs. Thus, marine ecologists have more recently focused on the dispersing agents (i.e., propagules) themselves and how they serve to connect populations. Here we focus on coastal marine invertebrate species with both a stationary and planktonic (dispersive) phase although the principles apply as well to reef fish and seaweeds.

Generally, connectivity is assumed not only to be a function of larval dispersal but also of post-larval survival (Pineda et al., 2007). It implies a large range of scales of connectivity, which are variable between and within species and locations (Cowen & Sponaugle, 2009; Kritzer & Sale, 2004). Thus, for a species in a specific area, once one knows the pattern of dispersal (i.e., dispersal kernels) and post-settlement processes, patterns of connectivity can then be derived. Unfortunately, this is often more simply said than done as dispersal parameters, such as pelagic larval duration (PLD) and post-settlement processes, that were historically considered to be stable over time (i.e., implying invariant connectivity) can be, in fact, quite variable. Indeed, several recent genetic studies have shown substantial spatial heterogeneity between life stages and temporal variability in genetic structure with metapopulations (Hogan et al., 2010; Selkoe et al., 2010). Likewise, several studies have explored hypotheses of oceanographic variability to explain fine-scale genetic patchiness (Banks et al., 2007) or chaotic genetic patchiness (Hogan et al., 2010), and certain larval transport models suggest that large variations in PLD and recruitment patterns could even be linked to hydrodynamic variability (Bolle et al., 2009; Connolly & Baird, 2010). Indeed, even knowledge of the PLD is not sufficient to predict scales of dispersal and gene flow among populations (Mitarai et al., 2009; Weersing & Toonen, 2009). Regardless, all these studies emphasize the importance of variation in larval dispersal on the resulting spatial patterns observed in different systems, and consequently, we should investigate connectivity as a varying feature of natural systems. Levels of variation (seasonality, annual variation, and periodicity) need to be examined in depth, and methods to assess connectivity should take these variations into account. Connectivity then should be thought of as the net result of all dispersal that has been observed over a given period, and the actual connectivity of the system will arise from the integration of all dispersal processes (Jacobson & Peres-Neto, 2010).

1.2 Measurement of connectivity

As the awareness of connectivity as a crucial characteristic for understanding ecosystems has emerged, a number of methods have been developed to explore and estimate connectivity within metapopulations and meta-ecosystems . These can differ, however, in their ability to assess variability in connectivity and can be further distinguished by their specificity in measuring dispersal between subpopulations and their applicability to other systems (Cowen et al., 2006; Cowen & Sponaugle, 2009; Jones et al., 2009; Levin, 2006). Other reviews about connectivity have adopted a classification based on two main categories: direct and indirect methods (Jacobson & Peres-Neto, 2010) or natural & artificial markers (Thorrold et al., 2002). Because each method targets a different goal and is often applied to a specific scale, we have chosen instead to classify methods by assessing their specificity to a species or system. The former concentrates on the connectivity of a specific species, giving information on dispersal patterns of the species in the study area. The latter focuses on the dispersal processes (e.g., hydrodynamics) and its variations in a specific study region and can thus be applied to co-occurring species having similar characteristics. Both types can include methods for assessing connectivity over multiple years (integrative) or for a single event (punctual).

Although using different methods to assess patterns of connectivity of a species in a particular system inevitably leads to different estimates, such predictions should ideally be similar. Regardless, to compare results among methods, the scale at which connectivity is evaluated ought to be the same for all methods (Palumbi, 2004; Weersing & Toonen, 2009). For example, dispersal distance of blue mussel larvae has been estimated through different methods in various systems and ranges from <5km to <100km (Gilg & Hilbish, 2003; McQuaid & Phillips, 2000; Smith et al., 2009). A part of this variability is inherent to the analysis of distinct systems with different methods; more significantly, however, these methods did not estimate the connectivity at the same spatiotemporal scale. For example, Gilg and Hilbish (2003) used a genetic method that averaged over several generations whereas Smith et al. (2009) estimated the pattern of connectivity within a single year. Consequently, when connectivity estimates are compared, attention should be given to the temporal scale employed in the method. The use of multiple methods at different temporal scales may be necessary, however, to completely understand a system, and the application of several methods in a given system should permit measures at different spatiotemporal scales and lead to a better knowledge of the crucial connections between populations.

When comparing different methods of measuring connectivity, it is important to evaluate not only differences in mean connectivity, but also how to measure variability in the pattern of connectivity within a specific system. Such variation can arise from biotic or abiotic factors and can affect the connectivity and the dynamics of the system at different scales. Depending on the system, variability of connectivity can result in periodicity, stationarity or more complex behavior of individual populations or the entire ecosystem. Therefore, the range of variation in dispersal patterns needs to be better understood to improve model predictions and management strategies, ideally using a single method over different temporal scales (e.g., day, season or year). However, because assessing connectivity employs newly developed tools, most efforts concentrate on simply evaluating the principal patterns of connectivity; only a few studies have tried to empirically estimate the variability of connectivity itself (Botsford et al., 2009; Jones et al., 2009). Moreover, in spite of the vast choice of methods, only a few are appropriate to assess potential variability of connectivity.

Beyond academic interest in ecosystem functioning, knowledge of connectivity and its variability is essential for applied environmental problems. It is particularly important for the design of marine protected areas (MPAs) to preserve biodiversity. As reserves integrate many species, it becomes important to consider the dispersal networks of all targeted species to improve coastal management. In this case, multi-scale studies are necessary because of the potential for different species to disperse at different scales, and respond differently to settlement variation.

In this chapter, we examine recent progress in our understanding of population connectivity in coastal marine systems and discuss the implications of variability of connectivity in the persistence of populations and ecosystems over large temporal and spatial scales. We hope to demonstrate how understanding connectivity and its variability can help the long-term sustainable management of entire ecosystems in a variable world. We divide our treatment into three parts. Firstly, we review the recently developed tools from different scientific disciplines concerned with connectivity and classified them as species- or system-specific as well as on their scale of applicability. Secondly, we examine evidence on the variability of observed patterns of connectivity and its causes. Finally, we discuss considerations for management and conservation of ecosystems. In particular, we review different theories and strategies related to populations and ecosystem dynamics that integrate the variability of connectivity in the context of marine protected areas.

2. Methods to assess variability in connectivity

As the interest in population connectivity has grown, so too has the number of methods to estimate connectivity patterns. Several scientific disciplines, including physics, genetics, and microchemistry, have contributed to our improved understanding of dispersal in marine systems. These approaches were originally developed for other reasons, but they can also be applied to estimate the dispersal of individuals and the flux between populations. However, as mentioned above, the diverse life histories, PLD and mobility of different species require that temporal and spatial scales are taken into account, making comparisons among methods tenuous.

The high mortality rate and high diffusion of larvae during the dispersive stage make direct measurements of larval dispersal nearly impossible. Therefore most methods measure dispersal patterns indirectly, e.g., through successful settlers (recruits). Previous reviews of the methodology used to measure connectivity distinguished between direct or indirect, or artificial or natural methods (Jacobson & Peres-Neto, 2010; Thorrold et al., 2002). Rather than following these dichotomies, we classify connectivity methods according to their applicability to different species or other systems. The first category groups methods that provide results for a particular species. The second includes techniques relevant to particular systems (e.g., bay, reef, or shoreline), and can be applied to other species. For each category, we briefly describe several methods that allow the measurement of variability in connectivity among populations, describe their scale of applicability, and discuss their potential utility. To conclude, we discuss scenarios where several complementary methods can be used within the same system.

2.1 Species-specific methods

Methods presented here have been developed recently to evaluate the dispersal kernel of individual key or representative species. They can be applicable to other species, but require further development to fit the species of interest. Both of the main methodological approaches rely on sampling individuals for genetic or geochemical markers.

2.1.1 Parentage analysis and assignment tests

Population genetics is the most widely used approach for making inferences about dispersal and connectivity in marine organisms (Hellberg, 2009). Traditionally, spatial variation in frequencies of alleles and genotypes (F_{st} and G_{st}) was the most common indirect method to genetically assess genetic divergence and long-term connectivity among populations (Hedgecock, 2010). However, limitations in the resolution of temporal scales, especially the inherent integration of dispersal over multiple generations, made it impossible to assess connectivity patterns over shorter ecological scales (Hedgecock et al., 2007). Recently new more direct genetic methods such as population and parentage analysis have been developed to more precisely estimate connectivity among populations (Christie et al., 2010b; Hedgecock, 2010; Manel et al., 2005). These methods are based on the multilocus genotype of individuals at different locations (Manel et al., 2005). Assignment tests provide the probability that an individual originated from one of a number of different known source populations. However, precise assignment of a given individual to a population requires that populations are genetically distinct and is unsuccessful when populations are too similar (Christie et al., 2010a; Saenz, 2009). Parentage analysis is a particular type of assignment test used to determine the parents of an individual or group of individuals based on shared alleles between individuals (Manel et al., 2005). As populations of marine invertebrates are usually comprised of large numbers of individuals with possibly long dispersal phases, the fraction of sampled individuals is usually too small for precise parental assignment (Hedgecock et al., 2007), and a persistent challenge associated with these techniques is the necessity for genotyping many individuals, both adults and recruits, from all of the different populations within the metapopulation. However, a promising new technique of parentage assignment (Christie et al., 2010a) requires fewer individuals from a given population than previous techniques. Using a Bayesian classification approach for the kind of organisms, this approach has been used successfully to document connectivity patterns of marine organisms with long PLD (Christie et al., 2010b; Richards et al., 2007; Underwood et al., 2007).

The spatial scale over which these methods can be used depends on the characteristics of the species (e.g., PLD, larval behaviour) and of the system (e.g., currents, topography) in question. In addition, the temporal scale of the sampling will depend on the frequency of reproduction of the species and on the variability of the oceanographic conditions encountered by the larvae. Consequently, the assessment of the variability of connectivity pattern necessitates an extensive sampling of all the potentially connected populations over different cohorts. Despite the high costs of these methods, they offer very precise techniques to measure connectivity pattern. Unfortunately, estimating the variability of connectivity requires multi-year studies.

2.1.2 Geochemical signatures in calcified structures

While genetic assignment tests measure connectivity by determining the natal origin of juveniles that are collected from different sites within a region, calcified structures (e.g., otoliths, statoliths or shells) can retain chemical traces of the environment (due to spatial and temporal variations of seawater) encountered by individuals during their entire life. Researchers are using such chemical signatures (e.g., isotope ratios, trace elements) of calcified structures formed during early development to identify the region or site of origin of individuals (Thorrold et al., 2002; Zacherl, 2005). These structures are either naturally marked by the environment or artificially "tagged" by transgenerational isotope labelling (TRAIL) at their origin. Natural markers can be found in the otoliths of fish, the statoliths or shells of molluscs. This process is usually bipartite – first the microchemistry of the calcareous parts corresponding to early life is analysed to define the trace elemental profile of a location of interest. Then, the trace elemental fingerprint of post-dispersal individuals is compared with the elemental profiles of individuals from which the original location is known (Becker et al., 2005; Becker et al., 2007). The source identification is obviously more reliable when differences in elemental composition are great among possible source locations (Thorrold et al., 2007). However, this method necessitates identifying the elemental profiles from all potential sources (Berumen et al., 2010), and moreover, it can be variable in time (seasonally, yearly) (Cook, 2011; Fodrie et al., 2011; Walther & Thorrold, 2009). Even though some statistical methods can be used to increase the precision of assignments (White et al., 2008), some limitations of this method appear for marine organisms because chemical distinctions among origin areas are sometimes too small to enable accurate assignment of individuals (Berumen et al., 2010), and the processes of integration of these trace elements in the hard parts of these organisms is not fully understood (Thorrold et al., 2007; Warner et al., 2005). The applicability of these techniques over multiple years has been shown recently and has provided new insights on the variability in connectivity. In particular, multi-year studies on fish otoliths (Clarke et al., 2010), oyster and mussel shells (Carson, 2010) have revealed seasonally and yearly variations in connectivity and the importance of self-recruitment in different systems. This variability underscores the need to identify the source elemental profiles over appropriate temporal scales if needed.

The second approach involves directly tagging individuals with enriched isotopes at possible source populations (Thorrold et al., 2006). Stable isotopes at concentrations an order of magnitude higher than those found in nature are injected into gravid females and are subsequently incorporated into internally developing embryos, thereby acting as unequivocal tags (Thorrold et al., 2006). This method permits the marking of many individuals at one time, and at low doses does not alter larval and juvenile behaviour (Williamson et al., 2009). TRAIL has been used mostly to assess the self-recruitment hypothesis in cephalopods (Pecl et al., 2010) and reef fishes (Almany et al., 2007; Planes et al., 2009). However, because of high mortality rates during the larval dispersal stage, this technique cannot be employed to assess connectivity at large scales. Moreover, as a large part of the population needs to be marked at one time, it is almost impossible to use this method with benthic invertebrates or large populations of reef fishes. Finally, the use of this method is limited by the different markers available and questions regarding the incorporation process of markers (Pecl et al., 2010). Thus, this method appears to be most useful in assessing variability of connectivity at limited spatial scales of dispersal in system where larval retention and self-recruitment are thought to be important.

2.1.3 Invasive species

The establishment and subsequent spread of non-indigenous species ("invasive species") is an emerging environmental problem of global extent, but a "silver lining" of biological invasions is a relative easy opportunity to examine rates and patterns of dispersal (Johnson & Padilla, 1996). Estimates of rates of spread can be made from sequential observations at the edge of the range (Grosholz, 1996; Lyons & Scheibling, 2009; McQuaid & Phillips, 2000) assuming that sampling efforts are reasonably constant over time. Such information has already been used in the planning for MPA (Shanks et al., 2003) for comparison with range shifts associated with climate change (Sorte et al., 2010). Unfortunately, the monitoring of most past invasions has been a rather piecemeal affair with different observers using different techniques and/or efforts to document the distribution of the invasive species over time. Current interest in aquatic invasive species has, however, provided better information and interest in documenting the secondary spread of established invaders.

Measures of the rates of spread of invasive species are not, however, exact equivalents of either dispersal or connectivity in metapopulations. First, invasive species often spread by both natural (e.g., currents) and human-mediated (boat hull fouling) vectors (Goldstien et al., 2010), and the latter can give artificially high estimates of dispersal. Second, population densities at the edge of an invader's range are likely to be lower than in sites where the species is well established. This attribute may reduce the propagule supply available for dispersal and lead to underestimates of the normal dispersal that occurs within a fully developed metapopulation. Finally, given the number of propagules that are likely to settle will diminish with increasing distance from the edge of an invader's range, there are likely to be sites where populations will not become established due to demographic limitations such as "Allee effects" (Leung et al., 2004) which would not exist if dispersal was simply occurring between populations within a metapopulation. Thus, again estimates of range expansion for either invasive or native species are likely to underestimate dispersal distances. Such measures can serve, however, as first approximations for dispersal within metapopulation of similar species and provide information on pathway of transport within coastal ecosystems.

2.2 System specific methods

System-specific approaches depend primarily on properties of the physical system to assess connectivity patterns in the study area. These methods are developed to fit to a specific system but can be adapted to other systems and to a range of species. Two such methods are particularly useful in evaluating variability in connectivity: geostatistics and biophysical models.

2.2.1 Geostatistics

Geostatistics, the statistical analysis of spatially-referenced data over large spatial extents, represents a powerful new tool to assess connectivity in marine ecosystems. These analyses are based on estimates of the adult abundance and the number of recruits at different sites within the study area. Significant coupling between sites at a particular spatial scale gives an estimate of the distance over which a given source population has an impact on recruitment in a recipient population (Fig 1.). Appropriately oriented and homogeneous coastal systems (e.g., estuaries, rivers, straight shoreline) are currently the preferred systems for the application of this method because such coastal configurations facilitate the detection of significant signals between adult populations and their impact on juvenile recruitment.

This method has already been used to estimate the distance of demographic coupling of blue mussels in the St. Lawrence estuary (Fig. 2; Smith et al., 2009). Depending on the different dispersal properties of the species, distinct signals can be distinguished, ranging from no association at all to significant relationships at a given distance between adults and recruits. Different theories can be tested (e.g., post-recruitment effects, supply-side limitations, demographic coupling), and the covariogram developed from the data provides a system specific estimation of the dispersal characteristics of the species. Also, this approach can avoid difficulties in separating different kind of variability (e.g., environmental variations among sites versus pure connectivity variations) by using detrended data to partition the variance (Le Corre et al., unpublished).

1. Three alternative hypotheses (A, B and C):

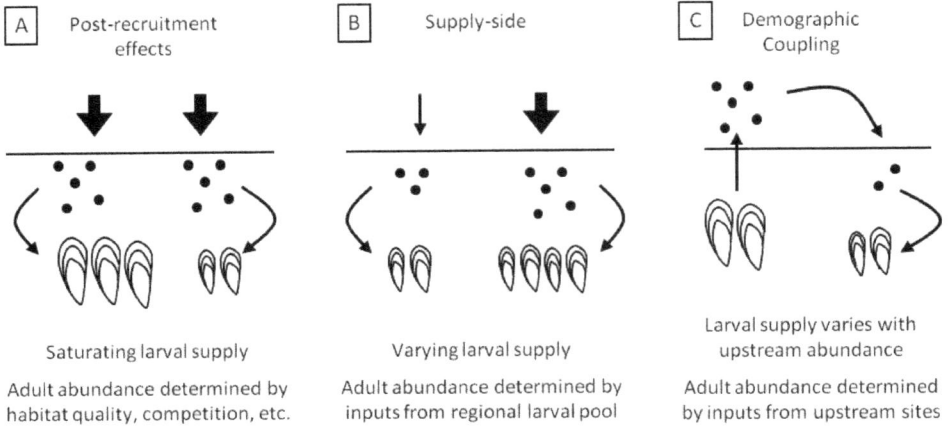

A Post-recruitment effects	B Supply-side	C Demographic Coupling

Saturating larval supply

Varying larval supply

Larval supply varies with upstream abundance

Adult abundance determined by habitat quality, competition, etc.

Adult abundance determined by inputs from regional larval pool

Adult abundance determined by inputs from upstream sites

2. Expected plots of covariance (cross-covariance) against lag distance for two variables (adult abundance and recruits) at multiple sites:

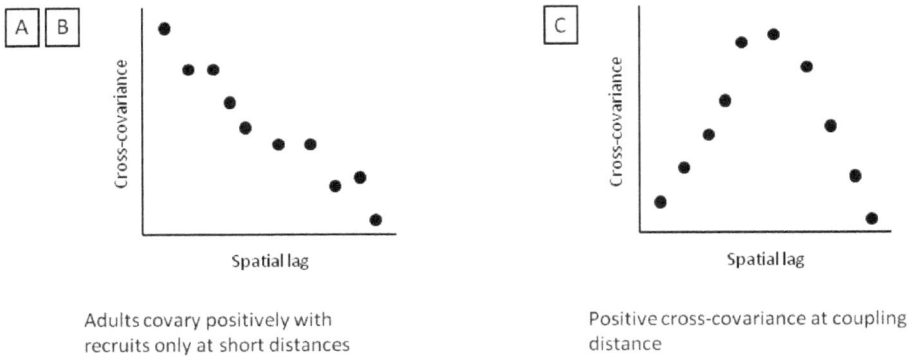

Adults covary positively with recruits only at short distances

Positive cross-covariance at coupling distance

Fig. 1. (1) Different scenarios of mussel recruitment (A, B, and C) that can be tested using cross-covariance geostatistics. (2) Expected cross-covariograms for each scenarios.

Although system-specific by nature, this method is less expensive as the data are relatively easy to collect and analyse. Consequently, it permits the repeated sampling necessary for evaluating the temporal variability of the connectivity pattern of the study species. Depending on the frequency of reproductive/dispersal events, the analysis can be repeated yearly or even seasonally to estimate the effective scale of connectivity, to infer dispersal patterns, and to capture the temporal variability of connectivity. This method is particularly useful for estimating the scale of connectivity and the variability in dispersal pattern, and be used to complement other methods, such as genetic analyses that provide longer spatial and temporal scale information.

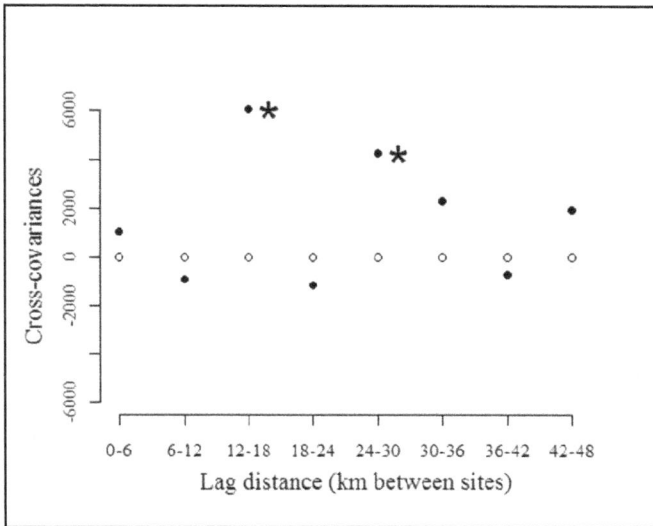

Fig. 2. Example of demographic connectivity in marine mussel metapopulations: significant cross-covariances (filled circles with asterisk) were observed between upstream adult mussels and downstream recruitment, indicating spatial connectivity between sites at 12-18km and 24-30km apart from one another. Open circles indicate mean values of a random process in that system. This figure is modified from Smith et al. 2009 (Copyright (2009) by the Association for the Sciences of Limnology and Oceanography, Inc.).

2.2.2 Coupled biophysical models

With advances in computational abilities, many three-dimensional hydrodynamic models have been developed to better understand geochemical processes, hydrology, and sedimentology. In recent years, ecologists have started exploring the power of these tools to better comprehend dispersal of larvae which are almost impossible to track directly (Cowen et al., 2007; Thorrold et al., 2007; Werner et al., 2007). These powerful tools can model entire coastal ecosystems, incorporating diverse attributes such as coastal geometry and the influence of wind. The spatial scale of the model is a key parameter for coastal marine species because of the important interaction between dispersing larvae and of nearshore physical processes, but depending on the complexity required, their spatial resolution can be adjusted accordingly (Greenberg et al., 2007). For example, models with high resolution are preferred to simulate coastal processes like eddies or waves.

The best way to model the dispersal of larvae appears to be the Lagrangian dispersal process (Siegel et al., 2003), particularly through the use of Individual Based Modeling (IBM) (Werner et al., 2007). At its simplest, Lagrangian dispersal assumes that larvae are transported advectively as passive particles (Mitarai et al., 2009; Siegel et al., 2003), and thus the process consists of following a parcel of water that is characterized by a particular set of conditions (e.g., initial density, PLD). Movements of the Lagrangian particles are then viewed as proxies for passively dispersing larvae and the analysis of the trajectories of several particles (used to create probability density functions) allows the estimation of dispersal kernels, the connectivity matrix, or potential connectivity (Cowen et al., 2006;

Siegel et al., 2003; Watson et al., 2010). Recently, IBMs have also been able to integrate behavioural traits of organisms (e.g., diel migration, mortality, feeding).

In a context of marine coastal species, the most important factors influencing dispersal, in addition to physical currents, are species properties such as timing of spawning, PLD, and competence, i.e., development to a stage able to settle back into the benthic environment (Mitarai et al., 2009). Integration of the interaction between individuals and their physical environment into physical oceanographic models, including the addition of specific larval behaviours, will certainly increase the precision of estimates of dispersal kernels and connectivity. IBMs have already permitted researchers to explore and integrate the role of specific behaviours during the dispersal phase and better explain diverse phenomena such as self-recruitment and limited dispersal (Werner et al., 2007). Depending on the species, the incorporation of processes such as the relationship between growth and water temperature (O'Connor et al., 2007), diel migration (Ayata et al., 2010; Cowen et al., 2006), chemical cues (Gerlach et al., 2007), and attraction by turbulence and waves (Fuchs et al., 2007) can have large impacts on resulting dispersal patterns. The added complexity due to the integration of such processes can be mitigated by the judicious identification and selection of behaviours that are found to be most critical in the dispersal of the individual species in question.

The development of 3-D hydrodynamic models necessitates high level programming capacities, good calibration before and after development of the model, and extensive validation before they are used by ecologists to run simulations. Also larval behaviours require considerable effort to be properly integrated into the model and a strong understanding by the programmer of the processes most critical for accurately describing the behaviour. Consequently, the development of a good hydrodynamic model requires substantial time and associated costs. However, once appropriately developed and validated, these models offer a powerful tool to explore a given system and enable researchers to test hypotheses with increasing realism. In particular, the low costs associated with exploring different environmental scenarios provide a fantastic tool to assess population connectivity and its variability across different spatiotemporal scales. The modeled patterns can then be used to generate testable hypotheses regarding connectivity which can, with the use of targeted experiments, provide data critical for model refinement, increasing our ability to understand the mechanisms driving patterns of connectivity. Ultimately, such coupled biophysical models will permit ecologists to test implications of different scenarios of climate change for population connectivity and persistence.

2.3 Use of multiple methods

Given the inherent limitations of any given approach, attempts have been made to compare different methodologies. In particular, several studies have confirmed predictions of biophysical models with genetic analysis (Galindo et al., 2010; White et al., 2010) or the inverse (Gilg & Hilbish, 2003), and the comparison of these tools has permitted researchers to verify predictions and to identify the main genetic processes involved in marine dispersal. Perhaps more importantly, mismatches between theoretical predictions and empirical data have directed researchers to explore further the mechanisms involved (Galindo et al., 2010). A good example is that of Gilg and Hilbish (2003) who combined simulated hydrodynamic data (2D) and allele frequencies in a region with strong differentiation among populations to estimate the geographic scale of larval dispersal. The use of 3 years of averaged simulation data in combination with the genetic allele frequencies data enabled them to define dispersal distance. They were not, however, able to assess the temporal variability in their estimate because the temporal resolution of connectivity assessed by these methods was too different. In the future, the use of multiple methodologies should provide more robust estimates of connectivity patterns by incorporating newly developed methods. As previously mentioned, it remains important to use multiple complementary methods with similar temporal resolution, especially when assessing the variability of connectivity patterns.

3. Variability of connectivity for populations: Causes and consequences

As discussed above, variability in estimates of connectivity can be attributed to methodology alone, i.e., due simply to technical artifacts. However, there are many natural processes that result in connectivity being truly and inherently variable. At larger spatial and temporal scales, this variability has important impacts on different characteristics (e.g., demography, genetics) of populations, communities, and ecosystems. In this section we review the primary processes affecting and causing variability in connectivity and then discuss the implications of these variations in dispersal patterns over larger spatiotemporal scales.

3.1 Main factors causing variability of connectivity

Dispersal, which underlies connectivity, involves three distinct sequential steps – it begins with the release of propagules (e.g., gametes, spores or larvae), is followed by a pelagic dispersal phase and ends with settlement to an appropriate habitat (Cowen & Sponaugle, 2009; Pineda et al., 2010). Each of these phases represents a potential source of variability because of the distinct suite of biological, physical, and biophysical processes involved in each period (Fig. 3). Depending on species, region, and timing of a particular study, these phases of connectivity may differentially impact dispersal and consequently produce variability in connectivity patterns.

3.1.1 Spawning

The first phase, release of propagules, can be affected by variation in the abiotic environment (e.g., temperature), maternal condition, food availability, and local oceanographic conditions. All of these parameters can influence the timing and location of spawning (Levitan, 2005). The importance of spawning, and its influence on connectivity depends greatly on the study species and their various modes of reproduction. Species-specific variation in offspring size reflects evolutionary strategies to deal with unpredictable variation of the environment (Marshall et al., 2008) and represents a compromise between quantity and quality as reflected by the number, size, and feeding strategy (e.g., lecithotrophic vs. planktotrophic) of propagules produced. Maternal condition can also be an important factor in terms of brood quality and is likely linked to the food available to parents prior to the spawning event. Spawning is influenced by both intrinsic (e.g., gonadal condition) and extrinsic (e.g., spawning cues) properties of the system (Starr et al., 1990), and as local oceanographic features are highly variable at small spatial and temporal scales (i.e., due to topography, bathymetry, waves, and tides), the timing of spawning will be a key factor influencing fertilization in broadcast spawners and larval dispersal trajectories of planktotrophic and lecithotrophic species (Largier, 2003).

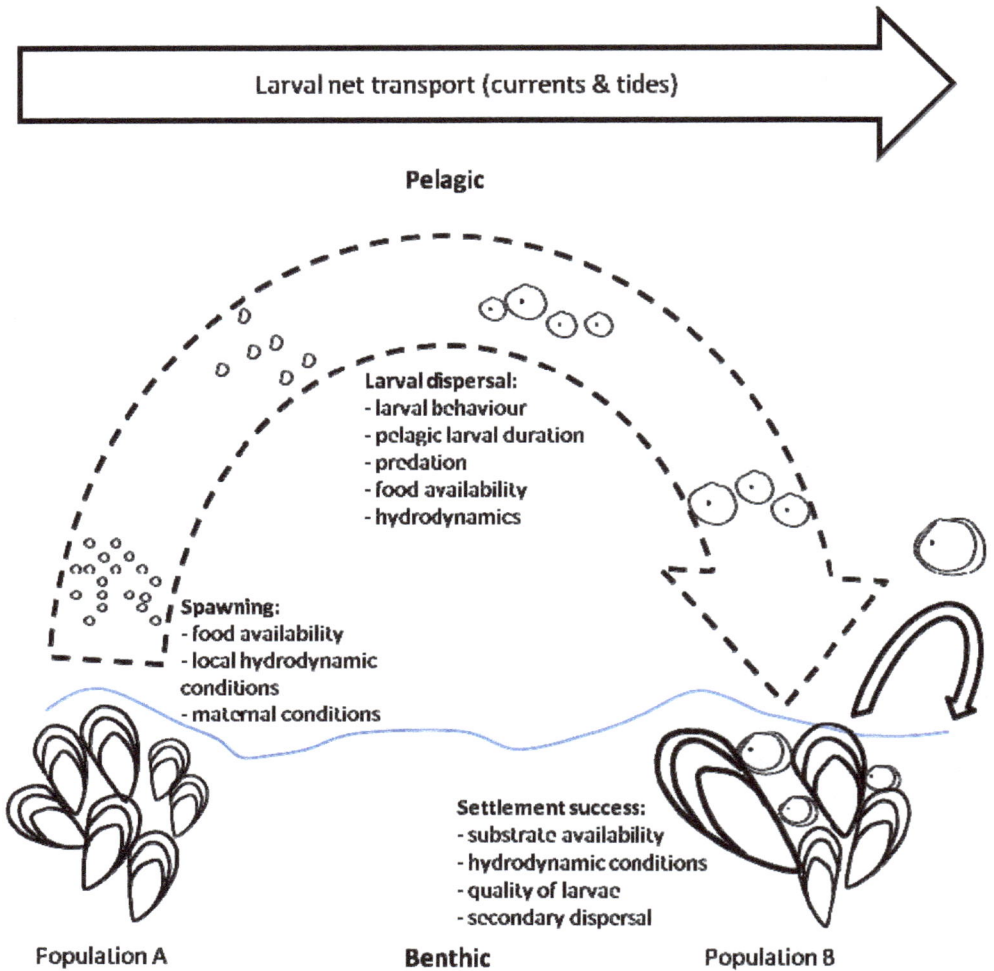

Fig. 3. Overview of the different processes contributing to the variability in connectivity among populations (e.g., marine mussels).

3.1.2 Larval dispersal

Once the planktonic phase begins, the spatial and temporal scales over which connectivity varies increase relative to the spawning phase. During this phase, many features influence larval dispersal: hydrodynamics of the system, larval behavior, prey availability, PLD and predation (Levin, 2006). In the simplest case, many larvae can be thought of as passively dispersing particles subject to oceanographic dynamics (see above), but these complex processes vary enormously over all spatial and temporal scales. Certain hydrodynamic conditions may increase larval dispersal distances (e.g., alongshore currents, wind driven surface current), while others may act to reduce dispersal distance (e.g., eddies, waves, tides) (Largier, 2003; McQuaid & Phillips, 2000; Mitarai et al., 2009; Sponaugle et al., 2005). More realistically, larval behaviour, especially vertical swimming, can influence dispersal. Indeed, the swimming ability of many pelagic larvae often increases with ontogeny enabling complex swimming behavior in both horizontal and vertical directions; which in turn has associated impacts on dispersal trajectories (Cowen et al., 2006; Fuchs et al., 2007; Gerlach et al., 2007). Likewise, because many species have planktotrophic larvae which must feed and develop during the dispersive phase, there is an obligatory pelagic period (the PLD) of days to weeks during which dispersal is occurring. PLD can vary greatly both within and among species (Kinlan & Gaines, 2003; Kinlan et al., 2005; Shanks & Eckert, 2005), depending on abiotic and biotic conditions. In particular, temperature and prey availability play key roles in larval development due to their high spatiotemporal variability, resulting in large variation in mean PLD and mortality among cohorts (O'Connor et al., 2007). In addition, if a suitable substratum is not available when larvae become competent to settle down to benthic habitat, organisms can even delay metamorphosis, so that the PLD can extend significantly beyond the mean pelagic larval duration (Pechenik et al., 1990). Consequently, the dispersal phase can last a relatively long time with concomitant increases in possible dispersal trajectories. All of these factors, in addition to direct predation on larvae, result in very high mortality during the pelagic larval phase (Houde, 1997).

3.1.3 Settlement
Even if larvae do survive the challenges of the pelagic realm and find suitable substrata, post-settlement processes can dramatically influence the chances of these "recruits" becoming part of the local population. First, post-settlement survivorship is closely tied to larval quality (Pechenik et al., 1998), and thus the diverse factors influencing larval life (see above) can produce variation as well in the quality of settlers (e.g., size, physiological conditions). This variability in quality has important demographic impacts, particularly on the growth and survival rate of individuals (Pechenik et al., 1998; Phillips, 2002). Various selection processes occur early in the development of juveniles because high quality metamorphosed larvae have a higher probability of recruiting to the adult population (Cowan & Shaw, 1988). Nevertheless, favorable larval traits do not necessary produce successful juveniles because the benthic environment can require different traits for survival. Quality of settlers then, as with quantity and quality of larvae, appears to be highly variable over spatial and temporal scales and consequently contribute to produce heterogeneous and highly variable recruitment patterns across locations (e.g., sink locations, areas of low recruitment). At smaller spatial scales, some species aggregate during the dispersal phase because of currents and differences among water masses (Natunewicz et al., 2001); larval delivery appears to be highly variable at spatial scales less than one hundred meters and necessitates additional sampling effort to characterize the population (Pineda et al., 2010; Siegel et al., 2008). Even if most long distance dispersal occurs during the pelagic larval stage of invertebrates, dispersal has been observed as post-larvae, juveniles and adults, which also contributes to variability exhibited across the three stages of the connectivity process (Bayne, 1964; Petrovic & Guichard, 2008).

3.2 Ecological implications of variability in connectivity.
All the phenomena described above can produce variability in connectivity. Therefore, the ecological consequences at larger spatiotemporal scales (metapopulation and ecosystem) are

diverse. Only a few theoretical studies have been conducted to assess these impacts (e.g., variable dispersal kernels). In this section, we discuss the general implications of considering variable connectivity patterns on large scale demography and genetics.

As previously emphasized, marine systems are characterized by variability in environmental conditions. Stochasticity of dispersal kernels or the connectivity matrix is often used to incorporate this variability in models. When stochasticity of connectivity is increased, large increases occur in the mean abundance of individuals in the metapopulation as well as in its variance (Aiken & Navarrete, 2011). Moreover, persistence of the metapopulation is logically enhanced by adding dispersal variability corresponding to what is observed in empirical studies (Aiken & Navarrete, 2011). In practice, strong variations in recruitment occur at a regional scale due to variation in bathymetry and winds. Because of this, some areas can be identified as "hot spots" where recruitment is always higher than the regional average (Siegel et al., 2008), in spite of the large variation in recruitment observed among years (Shima et al., 2010). Since some species are reproducing all year round, important variations can be observed in the number of recruitment events. Depending on seasonality, certain patterns of dispersal at different scales can emerge during a given year (Siegel et al., 2008). Some systems are strongly oriented by ocean currents, causing asymmetrical dispersal, and higher sensitivity to climate change has been observed in simulations (Aiken & Navarrete, 2011). The metapopulation system can then shift continually between stable and unstable states, according to variability in recruitment (Aiken & Navarrete, 2011). However, long-term empirical evidence for variability in connectivity is rare, so the long-term analysis of its impact on the demography of populations remains unknown. However, some methods developed recently (e.g., biophysical model, geostatistics; see above) should allow a better assessment of connectivity patterns with previously collected data and permit better validation of theoretical work.

Variability in dispersal distance also leads to different levels of gene flow between populations; parent populations of different cohorts of recruits at a given location are therefore variable and gene flow may occur, in time, over various distances (O'Connor et al., 2007). High variability in connectivity can lead to unstable genetic structure at seasonal and annual temporal scales in adults and juveniles (chaotic genetic patchiness), each site consisting of an admixture of cohorts from multiple sources. Chaotic genetic patchiness provides stability to the entire metapopulation and operates as a buffer against strong fluctuations in population size (Hogan et al., 2010).

4. Implications for management and conservation of biodiversity and ecosystems

Management of biodiversity and conservation of ecosystems, often through the establishment of marine protected areas (MPAs), has been well studied in past decades, and agreement has been reached on the necessity of a solid knowledge of population size, genetic diversity, representativeness to the entire system, and connectivity pattern across the area for effective management (O'Connor et al., 2007; Sundblad et al., 2011). However, different goals exist in species management, so benefits are different depending on the adopted strategy (e.g., specific fisheries protection or biodiversity conservation), opportunities, budget, and number and types of species targeted by the MPA network (e.g., homing or sessile vs. migratory species) (Kritzer & Sale, 2004; McCook et al., 2010). Goals of MPA networks are evaluated and chosen by policy decisions and are often established according to the opportunities present in the region, rather than following strict ecological analysis (Kritzer & Sale, 2004; Sundblad et al., 2011). In spite of the recent interest and extensive research on fisheries management on an ecosystem level, the effectiveness of MPAs to protect ecosystems and their biodiversity has unfortunately received relatively little attention (Jones et al., 2007).

4.1 Design of MPA networks

MPAs are generally established to help a given metapopulation or an ecosystem persist demographically. When defining a network of protected areas, the determination of the size and structure (e.g., a single or several areas, spacing, location) requires a good knowledge of the scale of dispersal and the size of discrete local populations (Jones et al., 2007; Kritzer & Sale, 2004). In MPAs where the goal is to preserve biodiversity, several areas are normally required; if the goal is to protect a single species (e.g., fishery), fewer and larger areas are generally used (Fogarty & Botsford, 2007; Jones et al., 2007). Size of reserves also depends on the effective number of individuals surviving to the next generation (Almany et al., 2009). With regard to the distance among reserves, the greater the dispersal range, the larger the protected areas should be and the longer the distances among them can be (Jones et al., 2007). For example, if the goal is the persistence of species, Moffitt et al. (2011) recommended increasing the size of MPAs and diminishing distances among them to allow a higher number of species to persist via network connectivity rather than self-replenishment, particularly species with long dispersal. Additionally, the spacing among reserves to protect a given species is variable among regions because of local differences in larval development time. For example, because water is warmer in tropical regions relative to temperate ones, PLD is generally shorter and thus spacing among reserves in networks should be smaller in tropics to ensure connectivity (O'Connor et al., 2007).

Because low levels of larval exchange limit success of MPAs (Bell, 2008), networks of protected areas should be designed as a function of observed connectivity patterns, but can also include potential connectivity among areas that can be linked and where suitable substrata exist. Consequently, some of the methods described above (section 2) to assess variability in connectivity can be used in determining locations that are potentially connected and can reinforce the MPA network. Moreover, a good description of the demography of the population through statistical methods (Aiken & Navarette 2011) or source/sink population analysis (Almany et al., 2009) will help to define sectors of the coast that would have a greater impact if protected. For example, isolated populations or retention areas, which have high conservation values, should be preferred because they ensure the persistence of the metapopulation (Almany et al., 2009; White et al., 2010). However, in retention zones, connectivity is less important, but sensitivity to stochastic disturbances might be higher. Therefore, such populations could go extinct, endangering the whole network because of its weaker connectivity (White et al., 2010). Generally, Jones (2007) suggests protecting source populations, isolated populations and spawning aggregation sites.

4.2 Management of biodiversity

When applied to whole ecosystems, management becomes more complex and reserves should be designed differently. Metacommunity levels should be considered in spatially-explicit models to manage effectively reserves in MPAs network (Guichard et al., 2004).

More recently, MPAs have been designed to be large enough to protect a suite of populations and have emphasized the importance of protecting different functional groups in ecosystems (McLeod et al., 2009). Also, the protection of vulnerable or fragile species may necessitate focussing on the other species upon which they depend, perhaps at multiple spatial and temporal scales (Almany et al., 2009). As dispersal distances occur at different scales among species, variability in spacing between reserves is desirable to reduce dependence of the system to a specific distance and better protect diverse groups of species (Kaplan, 2006).

When a MPA network is designed, population genetics also need to be considered. Because panmictic populations are rare, it is important to study direction and strength of the gene flow (von der Heyden, 2009). Gene flow should be maintained by frequent, medium and rare (long distance) dispersal of individuals among populations and its inclusion in MPA network designs is highly recommended (O'Connor et al., 2007). Also, von der Heyden (2009) recommends favoring multiple MPAs to avoid excessive population genetic structuring and population isolation; spacing of reserves should be designed to ensure adequate demographic connectivity and maintenance of genetic diversity (Almany et al., 2009). Even if substantial self-recruitment has been observed in dispersal analyses and may permit the persistence of specific populations, the exchange of individuals among populations remains crucial from a perspective of genetic diversity.

Because high environmental variability is an inherent part of marine systems, conservation strategies have to be developed to reduce its impacts on biodiversity and ecosystems. For example, in a context of global climate change, McLeod et al. (2009) suggested "spreading the risk" to avoid coral reef extinction by protecting several replicates of all kinds of habitats. To identify potential habitats, they proposed to use past incidents of coral bleaching and sea surface temperature. Another risk-spreading strategy that limits the impact of variability of connectivity patterns and strong fluctuations in MPAs networks involves using more, but smaller reserves (Almany et al., 2009; Hogan et al., 2010). Also, under high environmental variability, Baeza & Estades (2010) have shown that enhancement of the habitat quality in small reserves has better effects on surrounding landscapes than large and costly enhancement of large reserves.

5. Conclusion

Limited connectivity and the resulting metapopulation dynamics are now recognized features of coastal ecosystems. This overview of connectivity has revealed both the inherent shortcomings and future potential of applying this approach to the understanding and management of coastal ecosystems. There is clearly an emerging set of techniques that can now be applied to estimate and document dispersal between populations and the concomitant effects on metapopulation connectivity. There are, however, biases in these techniques in terms of the temporal and spatial scales over which they can be applied, and future effort will need to strive for the integration of these different approaches to better understand the role of connectivity in maintaining demographic stability and genetic diversity within metapopulations across scales. Moreover, connectivity can no longer be considered a static, invariant property of metapopulations. It too is inherently variable, subject to intrinsic and extrinsic factors that can affect the dispersal and survival of propagules. The importance of documenting and incorporating this variability in our theoretical and empirical understanding of metapopulation dynamics and ecosystem function is a new challenge, but one that must be met to address the environmental challenges associated with the sustainable management of ecosystems threatened by overexploitation and climate change.

6. Acknowledgments

We thank G. Cook, C. Leroux, and E. Pedersen for critical reading of the manuscript. We acknowledge and appreciate the financial support provided by the Canadian National Science and Engineering Research Council (NSERC Strategic Grant Project 336324).

7. References

Aiken, C. M. & Navarrete, S. A. (2011). Environmental fluctuations and asymmetrical dispersal: generalized stability theory for studying metapopulation persistence and marine protected areas. *Marine Ecology-Progress Series*, 428, pp. 77-88, 0171-8630

Almany, G. R.; Berumen, M. L.; Thorrold, S. R.; Planes, S. & Jones, G. P. (2007). Local replenishment of coral reef fish populations in a marine reserve. *Science*, 316, 5825, (May), pp. 742-744, 0036-8075

Almany, G. R.; Connolly, S. R.; Heath, D. D.; Hogan, J. D.; Jones, G. P.; McCook, L. J.; Mills, M.; Pressey, R. L. & Williamson, D. H. (2009). Connectivity, biodiversity conservation and the design of marine reserve networks for coral reefs. *Coral Reefs*, 28, 2, (Jun), pp. 339-351, 0722-4028

Ayata, S. D.; Lazure, P. & Thiebaut, E. (2010). How does the connectivity between populations mediate range limits of marine invertebrates? A case study of larval dispersal between the Bay of Biscay and the English Channel (North-East Atlantic). *Progress in Oceanography*, 87, 1-4, (Oct-Dec), pp. 18-36, 0079-6611

Baeza, A. & Estades, C. F. (2010). Effect of the landscape context on the density and persistence of a predator population in a protected area subject to environmental variability. *Biological Conservation*, 143, 1, (Jan), pp. 94-101, 0006-3207

Banks, S. C.; Piggott, M. P.; Williamson, J. E.; Bove, U.; Holbrook, N. J. & Beheregaray, L. B. (2007). Oceanic variability and coastal topography shape genetic structure in a long-dispersing sea urchin. *Ecology*, 88, 12, (Dec), pp. 3055-3064, 0012-9658

Bayne, B. L. (1964). Primary and secondary settlement in *Mytilus edulis* L. (Mollusca). *Journal of Animal Ecology*, 33, 3, pp. 513-523, 0021-8790

Becker, B. J.; Fodrie, F. J.; McMillan, P. A. & Levin, L. A. (2005). Spatial and temporal variation in trace elemental fingerprints of mytilid mussel shells: a precursor to invertebrate larval tracking. *Limnology and Oceanography*, 50, 1, (Jan), pp. 48-61, 0024-3590

Becker, B. J.; Levin, L. A.; Fodrie, F. J. & McMillan, P. A. (2007). Complex larval connectivity patterns among marine invertebrate populations. *Proceedings of the National Academy of Sciences of the United States of America*, 104, 9, (Feb), pp. 3267-3272, 0027-8424

Bell, J. J. (2008). Connectivity between island marine protected areas and the mainland. *Biological Conservation*, 141, 11, (Nov), pp. 2807-2820, 0006-3207

Berumen, M. L.; Walsh, H. J.; Raventos, N.; Planes, S.; Jones, G. P.; Starczak, V. & Thorrold, S. R. (2010). Otolith geochemistry does not reflect dispersal history of clownfish larvae. *Coral Reefs*, 29, 4, (Dec), pp. 883-891, 0722-4028

Bolle, L. J.; Dickey-Collas, M.; van Beek, J. K. L.; Erftemeijer, P. L. A.; Witte, J. I. J.; van der Veer, H. W. & Rijinsdorp, A. D. (2009). Variability in transport of fish eggs and larvae. Iii. Effects of hydrodynamics and larval behaviour on recruitment in plaice. *Marine Ecology-Progress Series*, 390, pp. 195-211, 0171-8630

Botsford, L. W.; White, J. W.; Coffroth, M. A.; Paris, C. B.; Planes, S.; Shearer, T. L.; Thorrold, S. R. & Jones, G. P. (2009). Connectivity and resilience of coral reef metapopulations in marine protected areas: matching empirical efforts to predictive needs. *Coral Reefs*, 28, 2, (Jun), pp. 327-337, 0722-4028

Carson, H. S. (2010). Population connectivity of the Olympia oyster in southern California. *Limnology and Oceanography*, 55, 1, (Jan), pp. 134-148, 0024-3590

Christie, M. R.; Johnson, D. W.; Stallings, C. D. & Hixon, M. A. (2010a). Self-recruitment and sweepstakes reproduction amid extensive gene flow in a coral-reef fish. *Molecular Ecology*, 19, 5, (Mar 10), pp. 1042-1057, 0962-1083

Christie, M. R.; Tissot, B. N.; Albins, M. A.; Beets, J. P.; Jia, Y.; Ortiz, D. M.; Thompson, S. E. & Hixon, M. A. (2010b). Larval connectivity in an effective network of marine protected areas. *Plos One*, 5, 12, (Dec 21), pp. 1932-6203

Clarke, L. M.; Munch, S. B.; Thorrold, S. R. & Conover, D. O. (2010). High connectivity among locally adapted populations of a marine fish (*Menidia menidia*). *Ecology*, 91, 12, (Dec), pp. 3526-3537, 0012-9658

Connolly, S. R. & Baird, A. H. (2010). Estimating dispersal potential for marine larvae: dynamic models applied to scleractinian corals. *Ecology*, 91, 12, (Dec), pp. 3572-3583, 0012-9658

Cook, G. S. (2011). Changes in otolith microchemistry over a protracted spawning season influence assignment of natal origin. *Marine Ecology-Progress Series*, 423, pp. 197-209, 0171-8630

Cowan, J. H. & Shaw, R. F. (1988). The distribution, abundance, and transport of larval sciaenids collected during winter and early spring from the continental-shelf waters off west Louisiana. *Fishery Bulletin*, 86, 1, (Jan), pp. 129-142, 0090-0656

Cowen, R. K.; Paris, C. B. & Srinivasan, A. (2006). Scaling of connectivity in marine populations. *Science*, 311, 5760, (Jan 27), pp. 522-527, 0036-8075

Cowen, R. K.; Gawarkiewic, G.; Pineda, J.; Thorrold, S. R. & Werner, F. E. (2007). Population connectivity in marine systems: an overview. *Oceanography*, 20, 3, (Sep), pp. 14-21, 1042-8275

Cowen, R. K. & Sponaugle, S. (2009). Larval dispersal and marine population connectivity. *Annual Review of Marine Science*, 1, pp. 443-466, 1941-1405

Fodrie, F. J.; Becker, B. J.; Levin, L. A.; Gruenthal, K. & McMillan, P. A. (2011). Connectivity clues from short-term variability in settlement and geochemical tags of mytilid mussels. *Journal of Sea Research*, 65, 1, (Jan), pp. 141-150, 1385-1101

Fogarty, M. J. & Botsford, L. W. (2007). Population connectivity and spatial management of marine fisheries. *Oceanography*, 20, 3, (Sep), pp. 112-123, 1042-8275

Fuchs, H. L.; Neubert, M. G. & Mullineaux, L. S. (2007). Effects of turbulence-mediated larval behavior on larval supply and settlement in tidal currents. *Limnology and Oceanography*, 52, 3, (May), pp. 1156-1165, 0024-3590

Galindo, H. M.; Pfeiffer-Herbert, A. S.; McManus, M. A.; Chao, Y.; Chai, F. & Palumbi, S. R. (2010). Seascape genetics along a steep cline: using genetic patterns to test predictions of marine larval dispersal. *Molecular Ecology*, 19, 17, (Sep), pp. 3692-3707, 0962-1083

Gerlach, G.; Atema, J.; Kingsford, M. J.; Black, K. P. & Miller-Sims, V. (2007). Smelling home can prevent dispersal of reef fish larvae. *Proceedings of the National Academy of Sciences of the United States of America*, 104, 3, (Jan), pp. 858-863, 0027-8424

Gilg, M. R. & Hilbish, T. J. (2003). Patterns of larval dispersal and their effect on the maintenance of a blue mussel hybrid zone in southwestern England. *Evolution*, 57, 5, (May), pp. 1061-1077, 0014-3820

Goldstien, S. J.; Schiel, D. R. & Gemmell, N. J. (2010). Regional connectivity and coastal expansion: differentiating pre-border and post-border vectors for the invasive tunicate *Styela clava*. *Molecular Ecology*, 19, 5, (Mar), pp. 874-885, 0962-1083

Greenberg, D. A.; Dupont, F.; Lyard, F. H.; Lynch, D. R. & Werner, F. E. (2007). Resolution issues in numerical models of oceanic and coastal circulation. *Continental Shelf Research*, 27, 9, (May), pp. 1317-1343, 0278-4343

Grosholz, E. D. (1996). Contrasting rates of spread for introduced species in terrestrial and marine systems. *Ecology*, 77, 6, (Sep), pp. 1680-1686, 0012-9658

Guichard, F.; Levin, S. A.; Hastings, A. & Siegel, D. (2004). Toward a dynamic metacommunity approach to marine reserve theory. *Bioscience*, 54, 11, (Nov), pp. 1003-1011, 0006-3568

Hanski, I. (1999). Habitat connectivity, habitat continuity, and metapopulations in dynamic landscapes. *Oikos*, 87, 2, (Nov), pp. 209-219, 0030-1299

Hedgecock, D.; Barber, P. H. & Edmands, S. (2007). Genetic approaches to measuring connectivity. *Oceanography*, 20, 3, (Sep), pp. 70-79, 1042-8275

Hedgecock, D. (2010). Determining parentage and relatedness from genetic markers sheds light on patterns of marine larval dispersal. *Molecular Ecology*, 19, 5, (Mar), pp. 845-847, 0962-1083

Hellberg, M. E. (2009). Gene flow and isolation among populations of marine animals. *Annual Review of Ecology Evolution and Systematics*, 40, pp. 291-310, 1543-592X

Hixon, M. A.; Pacala, S. W. & Sandin, S. A. (2002). Population regulation: historical context and contemporary challenges of open vs. closed systems. *Ecology*, 83, 6, (Jun), pp. 1490-1508, 0012-9658

Hogan, J. D.; Thiessen, R. J. & Heath, D. D. (2010). Variability in connectivity indicated by chaotic genetic patchiness within and among populations of a marine fish. *Marine Ecology-Progress Series*, 417, pp. 263-U289, 0171-8630

Houde, E. D. (1997). Patterns and trends in larval-stage growth and mortality of teleost fish. *Journal of Fish Biology*, 51, (Dec), pp. 52-83, 0022-1112

Jacobson, B. & Peres-Neto, P. R. (2010). Quantifying and disentangling dispersal in metacommunities: How close have we come? How far is there to go? *Landscape Ecology*, 25, 4, (Apr), pp. 495-507, 0921-2973

Johnson, L. E. & Padilla, D. K. (1996). Geographic spread of exotic species: ecological lessons and opportunities from the invasion of the zebra mussel *Dreissena polymorpha*. *Biological Conservation*, 78, 1-2, (Oct-Nov), pp. 23-33, 0006-3207

Jones, G. P.; Srinivasan, M. & Almany, G. R. (2007). Population connectivity and conservation of marine biodiversity. *Oceanography*, 20, 3, (Sep), pp. 100-111, 1042-8275

Jones, G. P.; Almany, G. R.; Russ, G. R.; Sale, P. F.; Steneck, R. S.; van Oppen, M. J. H. & Willis, B. L. (2009). Larval retention and connectivity among populations of corals and reef fishes: history, advances and challenges. *Coral Reefs*, 28, 2, (Jun), pp. 307-325, 0722-4028

Kadoya, T. (2009). Assessing functional connectivity using empirical data. *Population Ecology*, 51, 1, (Jan), pp. 5-15, 1438-3896

Kaplan, D. M. (2006). Alongshore advection and marine reserves: consequences for modeling and management. *Marine Ecology-Progress Series*, 309, pp. 11-24, 0171-8630

Kinlan, B. P. & Gaines, S. D. (2003). Propagule dispersal in marine and terrestrial environments: a community perspective. *Ecology*, 84, 8, (Aug), pp. 2007-2020, 0012-9658

Kinlan, B. P.; Gaines, S. D. & Lester, S. E. (2005). Propagule dispersal and the scales of marine community process. *Diversity and Distributions*, 11, 2, (Mar), pp. 139-148, 1366-9516

Kritzer, J. P. & Sale, P. F. (2004). Metapopulation ecology in the sea: from Levins' model to marine ecology and fisheries science. *Fish and Fisheries*, 5, 2, (Jun), pp. 131-140, 1467-2960

Largier, J. L. (2003). Considerations in estimating larval dispersal distances from oceanographic data. *Ecological Applications*, 13, 1, (Feb), pp. S71-S89, 1051-0761

Leung, B.; Drake, J. M. & Lodge, D. M. (2004). Predicting invasions: propagule pressure and the gravity of allee effects. *Ecology*, 85, 6, (Jun), pp. 1651-1660, 0012-9658

Levin, L. A. (2006). Recent progress in understanding larval dispersal: new directions and digressions. *Integrative and Comparative Biology*, 46, 3, (Jun), pp. 282-297, 1540-7063

Levitan, D. R. (2005). Sex-specific spawning behavior and its consequences in an external fertilizer. *American Naturalist*, 165, 6, (Jun), pp. 682-694, 0003-0147

Lyons, D. A. & Scheibling, R. E. (2009). Range expansion by invasive marine algae: rates and patterns of spread at a regional scale. *Diversity and Distributions*, 15, 5, (Sep), pp. 762-775, 1366-9516

Manel, S.; Gaggiotti, O. E. & Waples, R. S. (2005). Assignment methods: matching biological questions techniques with appropriate. *Trends in Ecology & Evolution*, 20, 3, (Mar), pp. 136-142, 0169-5347

Marshall, D. J.; Bonduriansky, R. & Bussiere, L. F. (2008). Offspring size variation within broods as a bet-hedging strategy in unpredictable environments. *Ecology*, 89, 9, (Sep), pp. 2506-2517, 0012-9658

McCook, L. J.; Ayling, T.; Cappo, M.; Choat, J. H.; Evans, R. D.; De Freitas, D. M.; Heupel, M.; Hughes, T. P.; Jones, G. P.; Mapstone, B.; Marsh, H.; Mills, M.; Molloy, F. J.; Pitcher, C. R.; Pressey, R. L.; Russ, G. R.; Sutton, S.; Sweatman, H.; Tobin, R.; Wachenfeld, D. R. & Williamson, D. H. (2010). Adaptive management of the Great Barrier Reef: a globally significant demonstration of the benefits of networks of marine reserves. *Proceedings of the National Academy of Sciences of the United States of America*, 107, 43, (Oct), pp. 18278-18285, 0027-8424

McLeod, E.; Salm, R.; Green, A. & Almany, J. (2009). Designing marine protected area networks to address the impacts of climate change. *Frontiers in Ecology and the Environment*, 7, 7, (Sep), pp. 362-370, 1540-9295

McQuaid, C. D. & Phillips, T. E. (2000). Limited wind-driven dispersal of intertidal mussel larvae: in situ evidence from the plankton and the spread of the invasive species *Mytilus galloprovincialis* in South Africa.a. *Marine Ecology-Progress Series*, 201, pp. 211-220, 0171-8630

Mitarai, S.; Siegel, D. A.; Watson, J. R.; Dong, C. & McWilliams, J. C. (2009). Quantifying connectivity in the coastal ocean with application to the Southern California Bight. *Journal of Geophysical Research-Oceans*, 114, (Oct), pp. 0148-0227

Moffitt, E. A.; White, J. W. & Botsford, L. W. (2011). The utility and limitations of size and spacing guidelines for designing marine protected area (MPA) networks. *Biological Conservation*, 144, 1, (Jan), pp. 306-318, 0006-3207

Moilanen, A. & Nieminen, M. (2002). Simple connectivity measures in spatial ecology. *Ecology*, 83, 4, (Apr), pp. 1131-1145, 0012-9658

Natunewicz, C. C.; Epifanio, C. E. & Garvine, R. W. (2001). Transport of crab larval patches in the coastal ocean. *Marine Ecology-Progress Series*, 222, pp. 143-154, 0171-8630

O'Connor, M. I.; Bruno, J. F.; Gaines, S. D.; Halpern, B. S.; Lester, S. E.; Kinlan, B. P. & Weiss, J. M. (2007). Temperature control of larval dispersal and the implications for marine ecology, evolution, and conservation. *Proceedings of the National Academy of Sciences of the United States of America*, 104, 4, (Jan), pp. 1266-1271, 0027-8424

Palumbi, S. R. (2004). Marine reserves and ocean neighborhoods: the spatial scale of marine populations and their management. *Annual Review of Environment and Resources*, 29, pp. 31-68, 1543-5938

Pechenik, J. A.; Eyster, L. S.; Widdows, J. & Bayne, B. L. (1990). The influence of food concentration and temperature on growth and morphological differentiation of blue mussel *Mytilus edulis* L. larvae. *Journal of Experimental Marine Biology and Ecology*, 136, 1, pp. 47-64, 0022-0981

Pechenik, J. A.; Wendt, D. E. & Jarrett, J. N. (1998). Metamorphosis is not a new beginning. *Bioscience*, 48, 11, (Nov), pp. 901-910, 0006-3568

Pecl, G. T.; Doubleday, Z. A.; Danyushevsky, L.; Gilbert, S. & Moltschaniwskyj, N. A. (2010). Transgenerational marking of cephalopods with an enriched barium isotope: a promising tool for empirically estimating post-hatching movement and population connectivity. *ICES Journal of Marine Science*, 67, 7, (Oct), pp. 1372-1380, 1054-3139

Petrovic, F. & Guichard, F. (2008). Scales of *Mytilus* spp. Population dynamics: importance of adult displacement and aggregation. *Marine Ecology-Progress Series*, 356, pp. 203-214, 0171-8630

Phillips, N. E. (2002). Effects of nutrition-mediated larval condition on juvenile performance in a marine mussel. *Ecology*, 83, 9, (Sep), pp. 2562-2574, 0012-9658

Pineda, J.; Hare, J. A. & Sponaugle, S. (2007). Larval transport and dispersal in the coastal ocean and consequences for population connectivity. *Oceanography*, 20, 3, (Sep), pp. 22-39, 1042-8275

Pineda, J.; Porri, F.; Starczak, V. & Blythe, J. (2010). Causes of decoupling between larval supply and settlement and consequences for understanding recruitment and population connectivity. *Journal of Experimental Marine Biology and Ecology*, 392, 1-2, (Aug), pp. 9-21, 0022-0981

Planes, S.; Jones, G. P. & Thorrold, S. R. (2009). Larval dispersal connects fish populations in a network of marine protected areas. *Proceedings of the National Academy of Sciences of the United States of America*, 106, 14, (Apr), pp. 5693-5697, 0027-8424

Richards, V. P.; Thomas, J. D.; Stanhope, M. J. & Shivji, M. S. (2007). Genetic connectivity in the Florida reef system: comparative phylogeography of commensal invertebrates with contrasting reproductive strategies. *Molecular Ecology*, 16, 1, (Jan), pp. 139-157, 0962-1083

Saenz, A. (2009). Estimating connectivity in marine populations: an empirical evaluation of assignment tests and parentage analysis under different gene flow scenarios. (vol 18, pg 1765, 2009). *Molecular Ecology*, 18, 11, (Jun), pp. 2543-2543, 0962-1083

Selkoe, K. A.; Watson, J. R.; White, C.; Ben Horin, T.; Iacchei, M.; Mitarai, S.; Siegel, D. A.; Gaines, S. D. & Toonen, R. J. (2010). Taking the chaos out of genetic patchiness: seascape genetics reveals ecological and oceanographic drivers of genetic patterns in three temperate reef species. *Molecular Ecology*, 19, 17, (Sep), pp. 3708-3726, 0962-1083

Shanks, A. L.; Grantham, B. A. & Carr, M. H. (2003). Propagule dispersal distance and the size and spacing of marine reserves. *Ecological Applications*, 13, 1, (Feb), pp. S159-S169, 1051-0761

Shanks, A. L. & Eckert, G. L. (2005). Population persistence of california current fishes and benthic crustaceans: a marine drift paradox. *Ecological Monographs*, 75, 4, (Nov), pp. 505-524, 0012-9615

Shima, J. S.; Noonburg, E. G. & Phillips, N. E. (2010). Life history and matrix heterogeneity interact to shape metapopulation connectivity in spatially structured environments. *Ecology*, 91, 4, (Apr), pp. 1215-1224, 0012-9658

Siegel, D. A.; Kinlan, B. P.; Gaylord, B. & Gaines, S. D. (2003). Lagrangian descriptions of marine larval dispersion. *Marine Ecology-Progress Series*, 260, pp. 83-96, 0171-8630

Siegel, D. A.; Mitarai, S.; Costello, C. J.; Gaines, S. D.; Kendall, B. E.; Warner, R. R. & Winters, K. B. (2008). The stochastic nature of larval connectivity among nearshore marine populations. *Proceedings of the National Academy of Sciences of the United States of America*, 105, 26, (Jul), pp. 8974-8979, 0027-8424

Smith, G. K.; Guichard, F.; Petrovic, F. & McKindsey, C. (2009). Using spatial statistics to infer scales of demographic connectivity between populations of the blue mussel, *Mytilus* spp. *Limnology and Oceanography*, sous presse, pp.

Sorte, C. J. B.; Williams, S. L. & Carlton, J. T. (2010). Marine range shifts and species introductions: comparative spread rates and community impacts. *Global Ecology and Biogeography*, 19, 3, (May), pp. 303-316, 1466-822X

Sponaugle, S.; Lee, T.; Kourafalou, V. & Pinkard, D. (2005). Florida current frontal eddies and the settlement of coral reef fishes. *Limnology and Oceanography*, 50, 4, (Jul), pp. 1033-1048, 0024-3590

Starr, M.; Himmelman, J. H. & Therriault, J. C. (1990). Direct coupling of marine invertebrate spawning with phytoplankton blooms. *Science*, 247, 4946, (Mar), pp. 1071-1074, 0036-8075

Sundblad, G.; Bergstrom, U. & Sandstrom, A. (2011). Ecological coherence of marine protected area networks: a spatial assessment using species distribution models. *Journal of Applied Ecology*, 48, 1, (Feb), pp. 112-120, 0021-8901

Thorrold, S. R.; Jones, G. P.; Hellberg, M. E.; Burton, R. S.; Swearer, S. E.; Neigel, J. E.; Morgan, S. G. & Warner, R. R. (2002). Quantifying larval retention and connectivity in marine populations with artificial and natural markers. *Bulletin of Marine Science*, 70, 1, (Jan), pp. 291-308, 0007-4977

Thorrold, S. R.; Jones, G. P.; Planes, S. & Hare, J. A. (2006). Transgenerational marking of embryonic otoliths in marine fishes using barium stable isotopes. *Canadian Journal of Fisheries and Aquatic Sciences*, 63, 6, (Jun), pp. 1193-1197, 0706-652X

Thorrold, S. R.; Zacherl, D. C. & Levin, L. A. (2007). Population connectivity and larval dispersal using geochemical signatures in calcified structures. *Oceanography*, 20, 3, (Sep), pp. 80-89, 1042-8275

Underwood, J. N.; Smith, L. D.; Van Oppen, M. J. H. & Gilmour, J. P. (2007). Multiple scales of genetic connectivity in a brooding coral on isolated reefs following catastrophic bleaching. *Molecular Ecology*, 16, 4, (Feb), pp. 771-784, 0962-1083

von der Heyden, S. (2009). Why do we need to integrate population genetics into South African marine protected area planning? *African Journal of Marine Science*, 31, 2, (Sep), pp. 263-269, 1814-232X

Walther, B. D. & Thorrold, S. R. (2009). Inter-annual variability in isotope and elemental ratios recorded in otoliths of an anadromous fish. *Journal of Geochemical Exploration*, 102, 3, (Sep), pp. 181-186, 0375-6742

Warner, R. R.; Swearer, S. E.; Caselle, J. E.; Sheehy, M. & Paradis, G. (2005). Natal trace-elemental signatures in the otoliths of an open-coast fish. *Limnology and Oceanography*, 50, 5, (Sep), pp. 1529-1542, 0024-3590

Watson, J. R.; Mitarai, S.; Siegel, D. A.; Caselle, J. E.; Dong, C. & McWilliams, J. C. (2010). Realized and potential larval connectivity in the Southern California Bight. *Marine Ecology-Progress Series*, 401, pp. 31-48, 0171-8630

Weersing, K. & Toonen, R. J. (2009). Population genetics, larval dispersal, and connectivity in marine systems. *Marine Ecology-Progress Series*, 393, pp. 1-12, 0171-8630

Werner, F. E.; Cowen, R. K. & Paris, C. B. (2007). Coupled biological and physical models present capabilities and necessary developments for future studies of population connectivity. *Oceanography*, 20, 3, (Sep), pp. 54-69, 1042-8275

White, C.; Selkoe, K. A.; Watson, J.; Siegel, D. A.; Zacherl, D. C. & Toonen, R. J. (2010). Ocean currents help explain population genetic structure. *Proceedings of the Royal Society B-Biological Sciences*, 277, 1688, (Jun), pp. 1685-1694, 0962-8452

White, J. W.; Standish, J. D.; Thorrold, S. R. & Warner, R. R. (2008). Markov Chain Monte Carlo methods for assigning larvae to natal sites using natural geochemical tags. *Ecological Applications*, 18, 8, (Dec), pp. 1901-1913, 1051-0761

Williamson, D. H.; Jones, G. P. & Thorrold, S. R. (2009). An experimental evaluation of transgenerational isotope labelling in a coral reef grouper. *Marine Biology*, 156, 12, (Nov), pp. 2517-2525, 0025-3162

Zacherl, D. C. (2005). Spatial and temporal variation in statolith and protoconch trace elements as natural tags to track larval dispersal. *Marine Ecology-Progress Series*, 290, pp. 145-163, 0171-8630

13

Geochemical Changes in Aquatic Environment Caused by Deep Dredging – A Case Study: The Puck Bay (Baltic Sea)

Bożena Graca[1], Katarzyna Łukawska-Matuszewska[1],
Dorota Burska[1], Leszek Łęczyński[2] and Jerzy Bolałek[1]
[1]Institute of Oceanography, University of Gdańsk, Gdynia,
[2]Maritime Institute in Gdańsk, Gdańsk,
Poland

1. Introduction

Dredge activities is very widespread antropogenic seabed disturbance. It is used to replenish sand on beaches, to create and maintain harbor, berth, waterways, may also be used for underwater mining activities and as a technique for fishing certain species of crabs or edible clams. Dredging has many deleterious environmental effects (Johnston 1981). Changes in bottom topography due to dredging can influence water dynamic and in consequence sediment transport (Maa et al., 2004; Work at al. 2004). Dredge pits and deep furrows can create a sink for fine-grained sediments, organic matter and contaminants (Desprez 2000) and result in hypoxic and anoxic conditions, as well as sulfate reduction in sediment (Bolałek et al., 1996, Flocks&Franze 2002, Graca at al. 2004) (Fig. 1). Geochemical changes resulted in deep dredging can affect benthic organisms. Limited recolonization of dredge pits was observed (Palmer et al. 2008, Szymelfenig et al. 2006). Such condition can influence nutrients dynamic and potentially stimulate eutrophication (Graca et al., 2004).

Impact of dredging depends on its intensity and the type of used method, as well as the environmental condition in dredge area (Boyd et al. 2005, Robinson et al. 2005). Puck Bay is a small water body located on the Polish Baltic coast. Deep dredging works carried in this reservoir, creates great opportunity to study the impact of deep dredging in areas with different water dynamics. From the north the bay is restricted from the open Baltic by the Hel Peninsula. It is 34 km long, and its width varies from 0.2 km to 2.9 km. A change of peninsula land cover during the last few decades, especially the construction of the new harbor, enhanced erosion processes. At present, the Hel Peninsula requires intensive reinforcement (Urbański & Solanowska, 2009). Eighty-two percent (c.a. 6,98 mln m[3]) of sands for peninsula's bank protection was gained from the bottom of the Puck Bay. First large beach nourishment was carried out by the end of 80's in last century. As an effect, in the years 1989-95, five the dredge pits were created in the bottom of the Puck Bay along the Hel Peninsula (Fig.2). The depth of the pits reaches 7-14 m, while natural depth in the surrounding area does not exceed 2 m.

a)

b)

Fig. 1. Bottom at the deep part of the dredge pit in the Puck Bay. Photos made in
Władysławowo dredge pit (a) in full growing season and (b) at the end of growing season.

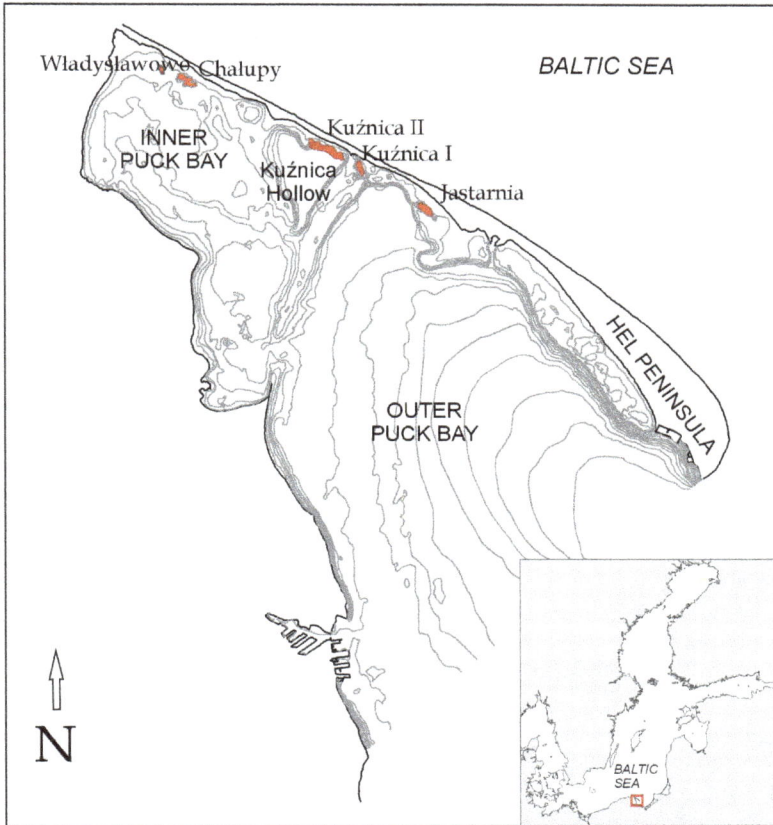

Fig. 2. Location of dredge pits in the Puck Bay.

Dredging for marine minerals has occurred in Poland waters for many years, in response to the need for sand and gravel used as construction aggregate and for beach replenishment. The first attempt to take account of environmental aspects of sands excavation from the bottom of the Puck Bay has been taken in 1992. Ciszewski and Kruk-Dowgiałło (1992) in their evaluation including some biological (phyto- and zoobentos), microbiological (bacterioplancton) and chemical (oxygen conditions) aspects suggested that, dredging should not be deeper then 3 m. It is necessary in order to reproduce disturbed biocenosis. But this suggestion was not obligatory. Currently, based on legal requirements the Environmental Impact Assessment is required (including the development of Environmental Impact Report) in the case of projects with potentially significant impacts on natural environment.

The Puck Bay is a protected area, in the frame of the European Ecological Natura 2000 Network (birds directive and habitat directive). Its northern part, containing the dredge pits, is included in the Seaside Landscape Park – a national protected area. According to EU Water Framework Directive, Poland should provide an improvement of quality of sea waters, especially in the protected areas.

Despite the research conducted so far in two of the dredge pits (Bolałek et al. 1996, Graca et al. 2004, Graca & Dudkowiak 2006, Szymelfenig et al. 2006), it is not known, which pits in the Puck Bay have undergone the largest changes in respect to natural conditions, what is the season influence on environmental conditions in the pits, and what is the scale of the problem created by the geochemical changes in bottom sediments of pits? An aim of this work is answering these questions. This is necessary to decide about the eventual recultivation of the dredge pits, and to choose a way to conduct it. Results of this study can be additionally used during the planning stage of environment-friendly dredging work.

2. Materials and methods

Samples of sediments, nearbottom and interstitial waters were collected by means of scuba diving in November-December 2007, in March, May-June and in August 2008.

Nearbottom waters were collected to glass and polyethylene bottles. Interstitial waters from the surface sediment layer (c. 4-5 cm) were obtained *in situ* by means of syringe tipped with pumice stone. This method detailed description has been given in Graca et al. (2004). Sediments (3 cores at each station) were sampled with Plexiglas pipes, 30 cm long and 3.6 cm in diameter (Fig. 3)

In each of the five studied the dredge pits (Fig. 2) samples were collected in three locations: in shallow parts of the slopes (stations S1), in deep parts of the slopes (stations S2) and at the bottom of the pit (stations D) (Fig. 4). Additionally samples were collected at two stations located on the outskirts (stations labeled O). One of those stations was situated between the dredge pit and the peninsula (labeled OP), and the other between the pit and the bay (OZ) (Fig. 4).

Ammonia, phosphates and hydrogen sulfide were analyzed in nearbottom and interstitial waters (Grasshoff et al. 1983). Oxygen in the nearbottom water has been determined by titration (Strickland & Parsons, 1972).

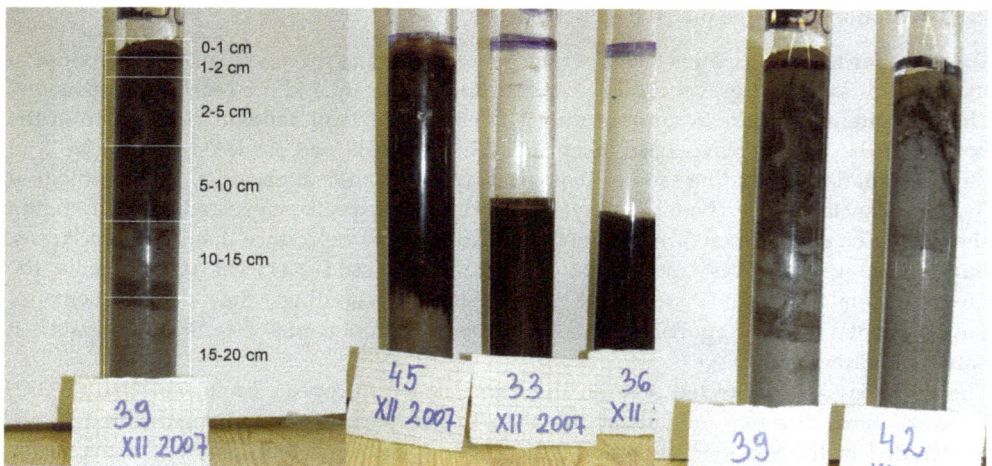

Fig. 3. Sediment cores collected in dredge pits (scheme of core division).

Sediment cores of maximum length of 20 cm were divided into six layers: 0-1cm, 1-2 cm, 2-5 cm, 5-10 cm, 10-15 cm and 15-20 cm (Fig. 3). In those layers moisture content (W) and loss on ignition (LOI) has been determined by drying the sample to constant mass. Samples were dried in first case at 105°C, in the latter in 550°C.

Organic carbon (C_{org}) and total nitrogen (N_{tot}) contents were measured using the Perkin Elmer 2400 CHNS/O analyzer. Sediment samples were dried at 60° C until reaching a constant mass. Samples for organic carbon determination were treated with 1M HCl in order to remove carbonates (Hedges& Stern, 1984, Burska, 2010).

Total phosphorus (P_{tot}) has been determined by spectrophotometry using acid-molybdate method after previous ignition (24 h in 483°C) and leaching with 1 M hydrochloric acid. Inorganic phosphorus (P_{inorg}) has been analyzed in similar way, but without ignition. Organic phosphorus (P_{org}) concentration has been calculated as a difference between P_{tot} and P_{inorg} (Froelich et al., 1988).

Chlorophyll *a* (chl *a*) and pheophytin (pheo) contents in sediment were determined by applying the method developed by Plante-Cuny (1974). Lyophilized sediment samples were homogenized and later extracted with 90% acetone for 8 hrs. Spectrophotometric measurements of extinction in extract samples were performed before and after acidification with 1M HCl. Redox potential (Eh) and pH were measured by means of electrodes.

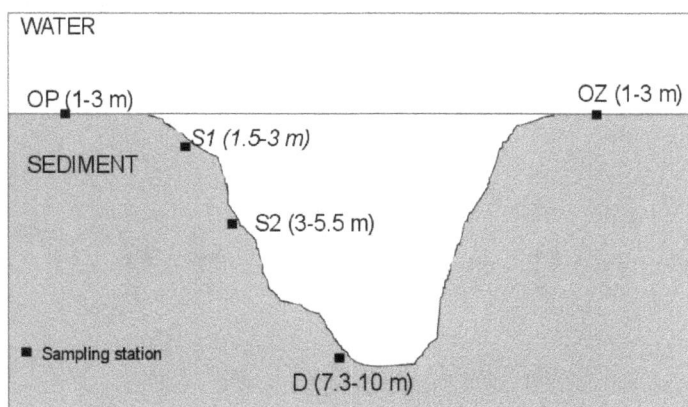

Fig. 4. Sampling stations arrangement in the dredge pits and at their outskirts.

3. Results

3.1 Nutrients, oxygen and hydrogen sulphide in nearbottom and interstitial waters

Nearbottom waters in the study area were characterized with good oxygen conditions. The lowest oxygen concentrations (4,1 cm^3 dm^{-3}) were observed at the outskirts of the Władysławowo pit in December. Decreased oxygen concentrations were also noticed in spring in the deepest parts of the Władysławowo (4,8 cm^3 dm^{-3}) and Chałupy (5,8 cm^3 dm^{-3}) pits.

Concentrations of phosphates (0,0-3,9 µmol dm^{-3}) and ammonia (0-4,1 µmol dm^{-3}) in the nearbottom waters at the outskirts and inside the pits did not differ significantly and were similar to those observed in the Puck Bay (Bolałek et al. 1993).

Concentration of hydrogen sulfide in nearbottom water varied from 0 to 2,3 µmol dm^{-3}. Only 8% of the data exceeded 0,2 µmol dm^{-3}. Those values were observed in the deepwater parts of the Władysławowo pit (November-December and August), Chałupy pit (November-December and May), Kuźnica I (November-December, May-June) and Kuźnica II pit (November-December), and also on the shallow slope of the Jastarnia pit (November-December).

In the interstitial waters hydrogen sulfide concentration varied in range 0-88,7 μmol dm^{-3} (Fig. 5a). The highest values occurred in the deepwater part of the dredge pits. 79% of the data did not exceed 10 μmol dm^{-3}. Concentrations higher than 10 μmol dm^{-3} were observed in August, and additionally in the Władysławowo pit also in May-June (stations O, S2 and D), and in March (station D).

Phosphate concentrations in the interstitial waters ranged from 0,2 to 58,3 μmol dm^{-3} (Fig. 5b). In most cases (83% of data) they did not exceed 10 μmol dm^{-3}. Values greater than 10 μmol dm^{-3} were observed in August in the deepwater part of all the dredge pits (stations D and S2) and additionally at the shallow slope (stations S1) of the Władysławowo, Kuźnica I and Jastarnia pits. In the interstitial waters from the bottom (station D) of the Władysławowo pit, phosphate concentrations exceeding 10 μmol dm^{-3} occurred also in March and in May-June.

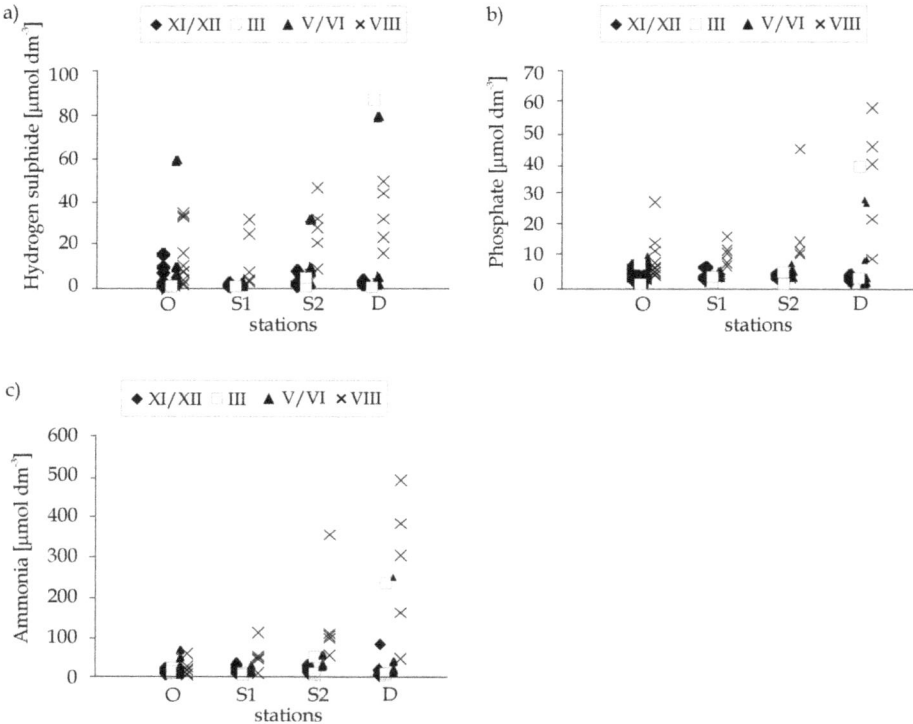

Fig. 5. Concentrations of (a) hydrogen sulfide, (b) phosphates and (c) ammonia in the interstitial waters observed in particular sampling seasons at the outskirts of the dredge pits (O), at their shallow (S1) and deeper (S2) slopes, and at the bottom (D).

Ammonia concentrations in the interstitial waters varied from 3,5 to 389,1 µmol dm⁻³ (Fig. 5b). Majority of the results (89% of data) did not exceed 100 µmol dm⁻³. Concentrations higher than 100 µmol dm⁻³ were observed in August in the deepwater parts of all the dredge pits (stations S2 and D), and in the Władysławowo pit additionally at the shallow slope (station S1). At the center (station D) of the Władysławowo pit ammonia concentrations exceeding 100 µmol dm⁻³ were also noticed in March and May-June

3.2 Sediments
3.2.1 Grain size distribution, W, LOI, Eh and phosphorus
An increase of the fine sediment fractions has been observed (<125 µm) in the dredge pits sediments in comparison to their outskirts (Fig. 6). This increase was most pronounced in the deepest parts of the dredge pits (stations D). Except the Jastarnia and Chałupy pits, it was also visible on the slopes. It has to be noted that content of fractions 125 µm and 63 µm in the sediments of the Władysławowo pit's outskirts were clearly higher than this found at the other outskirts.

Three quarters of the moisture content results in the sediments of the outskirts did not exceed 20% (Tab. 1). Moisture content has increased in the dredge pit areas. Sediments in their deepest parts (station D) were most hydrated (Fig. 7). Only quarter of the moisture content results in these areas were below 26%. LOI variability has been similar (Tab. 1, Fig. 8).

Outskirts (stations O) and slopes (stations S1 and S2) of the dredge pits did not differ significantly in regard to phosphorus content in the sediments (Tab. 1). Clearly higher concentrations were observed at the bottom (stations D) of the pits.

Redox potential of the sediments varied from (-495) to 489 mV (Tab. 1). In the deepwater part of the pits (stations S2 and D), in approximately half of the cases, sediments were reductive (Eh<100 mV) from the first analyzed layer. Reductive conditions intensified with depth below the sediment surface (Fig. 9). Below the topmost centimeter of sediment, 82% of the results were lower than 100 mV.

At the outskirts (stations O) and at the shallow slope of the dredge pits (stations S1) reductive conditions occurred in 15% of the cases. In half of the cases they appeared in the sediment layer below the 0-2 cm.

Outskirts (stations O), slopes (stations S1 and S2) and the centers of the dredge pits (stations D) were similar in regard to pH in the sediments (ANOVA Kruskall-Wallis test, p=0,61, Tab. 1)

3.2.2 Organic carbon, total nitrogen and photosynthetic pigments
Organic carbon contents in the analyzed sediment ranged from 0,021 to 168,50 mg g⁻¹d.w., while the median (Md) value was 1,52 mg g⁻¹d.w. (Tab. 2). Most of C_{org} concentration values (75%) were lower than 3,42 mg g⁻¹d.w. Total nitrogen concentration values were ca. 10 times lower and ranged from 0,01 to 22,11 mg g⁻¹d.w. (Tab. 2).

Similarly to C_{org} concentration values, high values of N_{tot}, i.e. >0,45 mg g⁻¹ d.w. constituted only 25% of all measurements.

Chlorophyll a concentrations varied from the value equal to the limit of detection (LD) to 124,12 µg g⁻¹ d.w., while the median value was 0,35 µg g⁻¹ d.w. Most of chl a concentrations (75%) were lower than 1,42 µg g⁻¹ d.w. Phaeopigment concentrations were on average six times higher than the concentrations of chlorophyll a (Tab. 2).

Władysławowo

Chałupy

Kuźnica II

Kuźnica I

Jastarnia

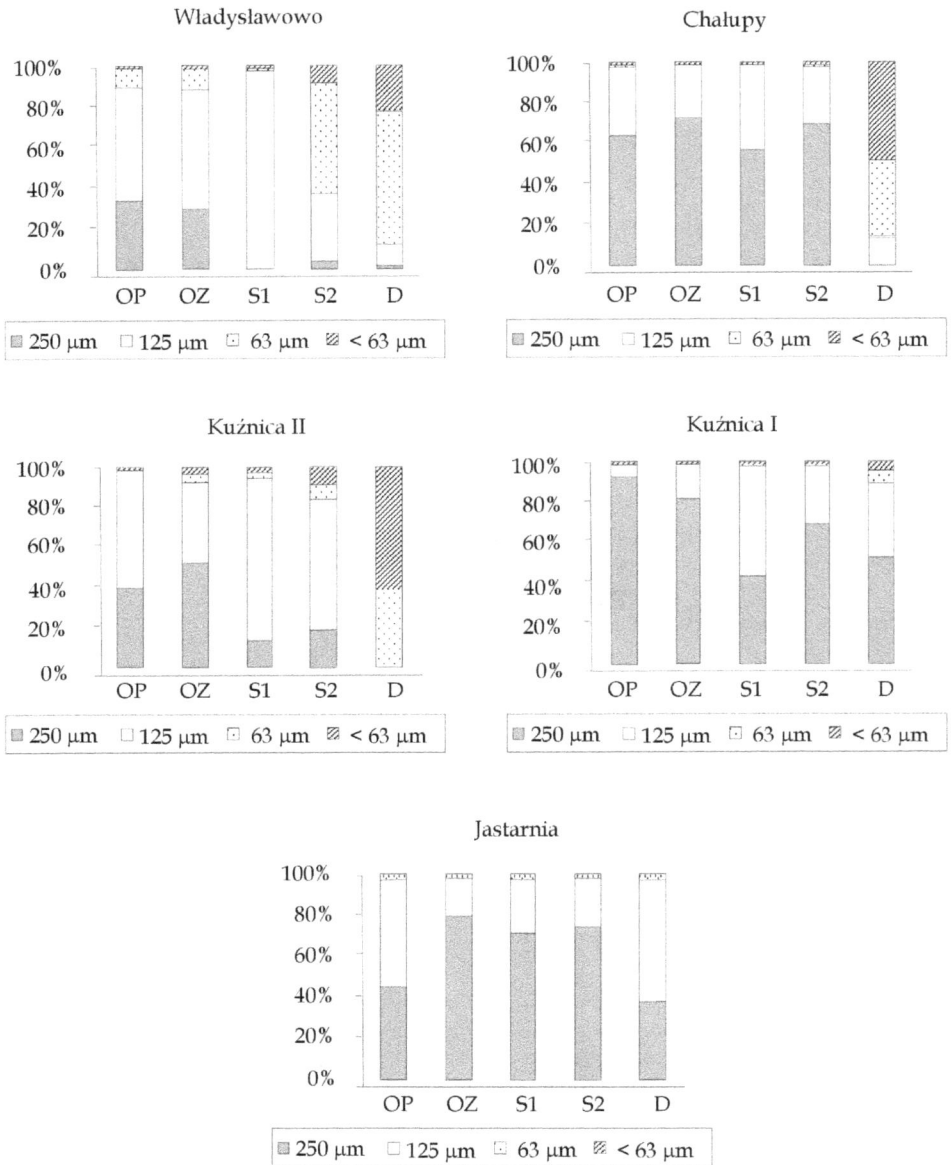

Fig. 6. Contribution of different sediment fractions (in the topmost layer - 0-5 cm) in the dredge pits and at their outskirts (OP – outskirt of the pit, peninsula side, OZ – outskirt of the pit, bay side, S1 – shallow slope, S2 - deep slope, D – bottom of the pit).

a)

WŁADYSŁAWOWO

Median; Box: 25%-75%; Whisker: Non-Outlier Range

b)

JASTARNIA

Median; Box: 25%-75%; Whisker: Non-Outlier Range

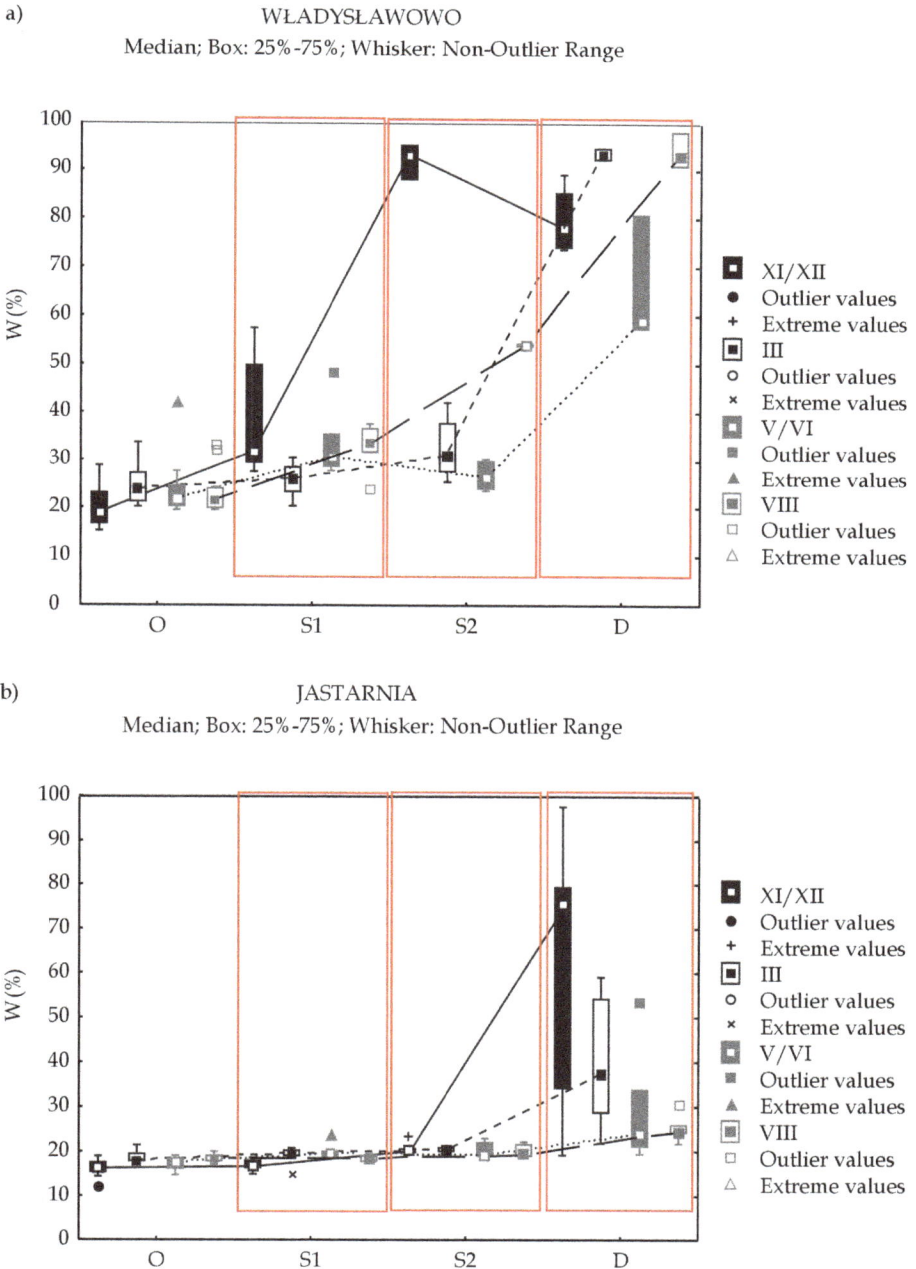

Fig. 7. Seasonal variation in sediment moisture content (W) on the basis of
(a) Władysławowo pit and its outskirts, and (b) Jastarnia pit and its outskirts
(O- pit's outskirt, S1-pit's shallow slopes, S2- pit's deep slopes, D- bottom of the pit).

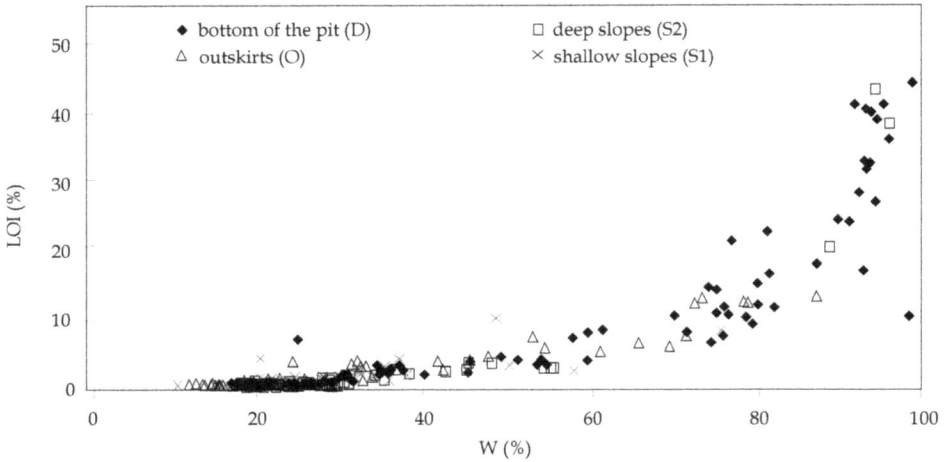

Fig. 8. Relationship between moisture content (W) and lost of ignition (LOI) in sediment.

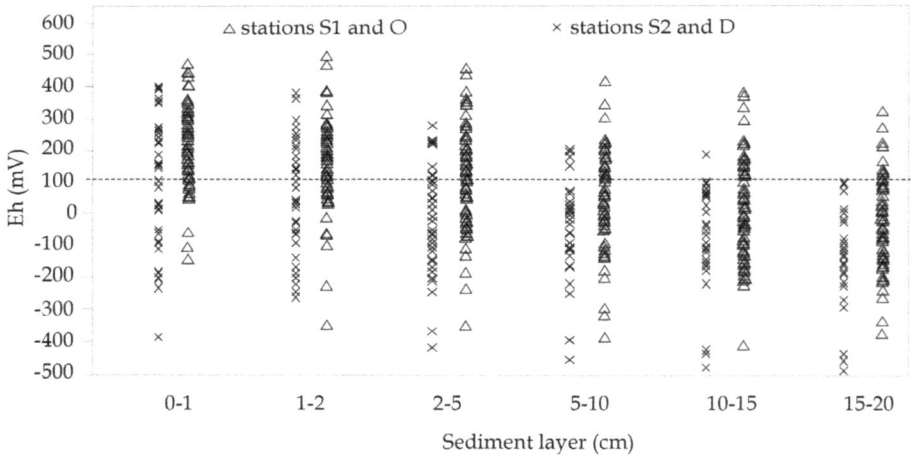

Fig. 9. Redox potential in particular sediment layers in the deepwater part of the dredge pits (stations S2 and D) and at the outskirts of the pits and shallow part of their slopes (stations O and S1) (dashed line marks the Eh value of 100mV, which is a threshold of reductive conditions (Boyd, 1995).

Generally sediments at outskirts of the pits were characterized by low carbon and total nitrogen content as well as low chl a and pheophytin concentration. Obtained values did not differ from those reported in literature for similar sediment types (Burska et al., 1999).
Among outskirts of all investigated pits, the outskirts of the Władysławowo and Kuźnica II dredge pits were characterized by the higher contents of organic carbon, total nitrogen and photosynthetic pigments (ANOVA Kruskal-Wallis test, p<0.05). In the case of sediment

from the Kuźnica II pit, this finding is connected to the location of its outskirt within the natural pit Kuźnica Hollow (Fig. 2). The increased values of chemical parameters measured in sediment in this location are connected to the natural sedimentation processes and the accumulation of organic matter.

Values of C_{org}/N_{tot}, chl a/(chl a+pheo), and C_{org}/(chl a+pheo) ratios indicate the differences in quality of organic matter in sediments in the study area (Tab. 2). Strongly decomposed organic matter occurred in the deepest part of the dredging pits (stations D). It was manifestated mainly by chl a/(chl a+pheo) ratio.

Organic matter in sediments from the outskirts of the Władysławowo dredge pit displayed statistically significant higher molar C_{org}/N_{tot} ratio (ANOVA Kruskal-Wallis test, $p<0,05$) and higher value of C_{org}/(chl a+pheo) ratio (ANOVA Kruskal-Wallis test, $p<0,05$) in comparison to the outskirts of other pits. In the case of sediments sampled from the deepest part of dredge pits difference in organic matter quality was demonstrated by C_{org}/(chl a+pheo) ratio only. The highest value of this ratio was noted in Władysławowo pit.

Organic carbon and total nitrogen concentration was decreasing with the increasing sediment depth (Fig. 11). Chl a and pheophytin concentrations reached the highest values in samples of surficial sediments (0-2 cm). In deeper sediment layers the content of chl a and pheo decreased in a stepwise fashion to reach a stable level in the 5-10 cm layer.

4. Discussion

4.1 An extent of deep dredging impact

It has been assumed, that conditions at the outskirts of each the dredge pit represent natural environment, and the differences between pit and its outskirts are an effect of deep dredging. In order to assess the extent of geochemical changes in each pit, the Mann-Whitney U test ($p<0,05$) has been used to compare selected geochemical parameters (W, LOI Eh, pH and phosphorus content in sediment) between the dredge pits and their outskirts. Then, statistically significant cases were summed up for each the dredge pit. Similar procedure was used to appraise the season influence on the extent of conditions changes in the pits due to deep dredging (cases with significant changes were summed up for each season).

Significant differences between the outskirts and subsequently: shallow slope, deep slope and deepest part of the dredge pit were observed in 23, 31 and 61 % of the cases, respectively. This indicates that the most vulnerable for geochemical condition changes resulting from deep dredging was the deepwater part of the dredge pits. This confirms previous findings (Bolałek et al. 1996).

Most prevalent changes between the pits and their outskirts were observed in the Władysławowo area (Fig. 12a). Number of changes in the case of Chałupy and Kuźnica I areas was similar, and greater than in the Kuźnica II and Jastarnia regions (Fig. 12a). Among analyzed parameters, the most frequently impacted by deep dredging was moisture content and loss on ignition (Fig. 12a). These parameters reflects the water dynamic and consequently conditions of sediment deposition (Håkanson et al. 2003, Jönsson et al. 2005). High moisture content and loss on ignition, observed in deep part of the dredge pits (Tab. 1, Fig. 7), are characteristic for areas of low water dynamics, where deposition of fine particles takes place, while low values are characteristic for areas where high water dynamics prevents or inhibits settling of those particles. Obtained results indicate, that deep dredging has weakened water dynamics in the area of the dredge pits which enabled the deposition of fine sediments in their deepwater parts (Fig. 6) and favored organic matter accumulation (Tab. 2).

		n^1	Min.[2]	Max.[3]	Mean	Md.[4]	Q_L^5	Q_U^6
Outskirts of the dredge pits (stations O)	W	215	11	42	20	19	18	20
	LOI	215	0,03	6,,03	0,42	0,27	0,19	0,42
	P_{tot}	216	1,5	13,3	7,9	7,9	6,9	9,1
	P_{inorg}	214	0,1	8,3	5,7	5,7	5,1	6,4
	P_{org}	215	0,2	6,3	2,3	2,2	1,7	2,8
	Eh	211	-415	489	104	112	-15	234
	pH	206	5,04	8,22	6,92	6,83	6,67	7,20
Shallow slope of the dredge pits (stations S1)	W	118	10	75	23	21	19	23
	LOI	118	0,08	9,87	0,81	0,37	0,26	0,79
	P_{tot}	118	5,1	21,7	8,5	8,3	7,1	9,6
	P_{inorg}	118	3,2	10,4	5,7	5,7	5,1	6,4
	P_{org}	118	0,0	11,3	2,8	2,6	2,0	3,1
	Eh	118	-344	432	62	94	-46	173
	pH	117	5,01	8,16	6,87	6,93	6,55	7,21
Deeper slope of the dredge pits (stations S2)	W	114	18	95	26	22	20	25
	LOI	114	0,10	42,95	1,58	0,52	0,27	0,85
	P_{tot}	111	4,0	87,0	10,3	8,4	7,1	9,6
	P_{inorg}	111	3,9	48,4	6,6	5,9	5,1	6,5
	P_{org}	112	0,0	58,8	3,7	2,6	2,0	3,2
	Eh	112	-495	385	2	10	-115	136
	pH	108	5,95	8,41	6,96	6,95	6,66	7,29
The deepest part of the dredge pits (stations D)	W	92	18	98	54	50	26	81
	LOI	92	0,09	44,00	10,08	3,67	0,91	15,50
	P_{tot}	90	1,7	71,9	22,1	13,8	8,7	33,0
	P_{inorg}	90	1,3	42,9	13,3	8,9	5,7	18,1
	P_{org}	90	0,0	44,7	8,8	3,7	2,5	13,4
	Eh	88	-465	393	-55	-82	-192	79
	pH	86	5,23	8,07	6,85	6,88	6,45	7,26

[1] number of observation; [2] minimum; [3] maximum; [4] median; [5] lower quartile, [6] upper quartile

Table 1. Outcome of statistical analysis of geochemical parameters in sediment (0-20 cm)(W-%-moisture content, LOI -% – loss on ignition, P_{tot}, P_{inorg}, P_{org}- µmol g^{-1} d.w.,- total, inorganic and organic phosphorus content, Eh-mV – redox condition and pH) in the study area.

		n^1	Min.[2]	Max.[3]	Mean	Md.[4]	Q_L^5	Q_U^6
Outskirts of the dredge pits (stations O)	C_{org}	221	0,21	51,56	2,39	1,05	0,59	1,87
	N_{tot}	221	0,01	5,73	0,29	0,14	0,08	0,21
	chl a	230	0,03	6,07	0,64	0,21	0,09	0,48
	pheo	230	0,03	120,82	2,75	1,11	0,60	1,95
	C_{org}/N_{tot}	215	3,3	48,0	10,9	9,3	7,2	12,9
	chl a/(chl a+ pheo)	220	1,2	90,7	23,0	17,6	9,5	30,4
	C_{org} /(chl a+ pheo)	222	86	6836	1059	741	362	1286
Shallow slope of the dredge pits (stations S1)	C_{org}	110	0,38	5,46	1,54	1,20	0,84	2,09
	N_{tot}	110	0,05	0,61	0,21	0,18	0,11	0,27
	chl a	113	0,04	5,47	0,68	0,26	0,14	0,64
	pheo	113	0,06	15,44	2,42	1,92	1,20	2,93
	C_{org}/N_{tot}	107	3,0	37,4	9,84	8,78	6,6	12,2
	chl a/(chl a+ pheo)	113	1,3	65,7	18,5	14,4	7,9	25,6
	C_{org} /(chl a+ pheo)	110	72	14545	856	536	310	821
Deeper slope of the dredge pits (stations S2)	C_{org}	105	0,31	105,92	6,10	1,65	0,96	3,21
	N_{tot}	105	0,02	12,18	0,75	0,23	0,14	0,45
	chl a	105	0,00	35,25	1,66	0,38	0,17	1,23
	pheo	105	0,11	184,79	8,37	3,00	1,83	4,50
	C_{org}/N_{tot}	101	3,0	34,7	9,7	8,5	6,9	10,6
	chl a/(chl a+ pheo)	102	0,8	47,5	16,4	14,4	6,6	22,0
	C_{org} /(chl a+ pheo)	96	48	6905	632	475	270	681
The deepest part of the dredge pits (stations D)	C_{org}	95	0,21	168,50	35,12	19,98	2,42	51,02
	N_{tot}	95	0,03	22,11	4,51	1,72	0,39	6,62
	chl a	95	0,00	124,12	7,88	1,84	0,44	7,92
	pheo	95	0,11	257,13	52,39	28,31	4,16	77,85
	C_{org}/N_{tot}	95	4,5	23,1	9,8	9,2	7,7	11,1
	chl a/(chl a+ pheo)	88	0,3	43,5	12,0	11,1	7,3	14,9
	C_{org} /(chl a+ pheo)	98	22	8155	908	561	201	930

1 – 6 as in Table 1

Table 2. Outcome of statistical analysis of selected organic matter components and indicators of organic matter quality: organic carbon (C_{org} - mg g^{-1} d.w.), total nitrogen (N_{tot} - mg g^{-1} d.w.), chlorophyll a (chl a - µg g^{-1} d.w.), pheophytin (pheo - µg g^{-1} d.w.), C_{org}/N_{tot}, C_{org} /(chl a+pheo) (%), chl a/(chl a+pheo) (%)) in sediment (0-20 cm).

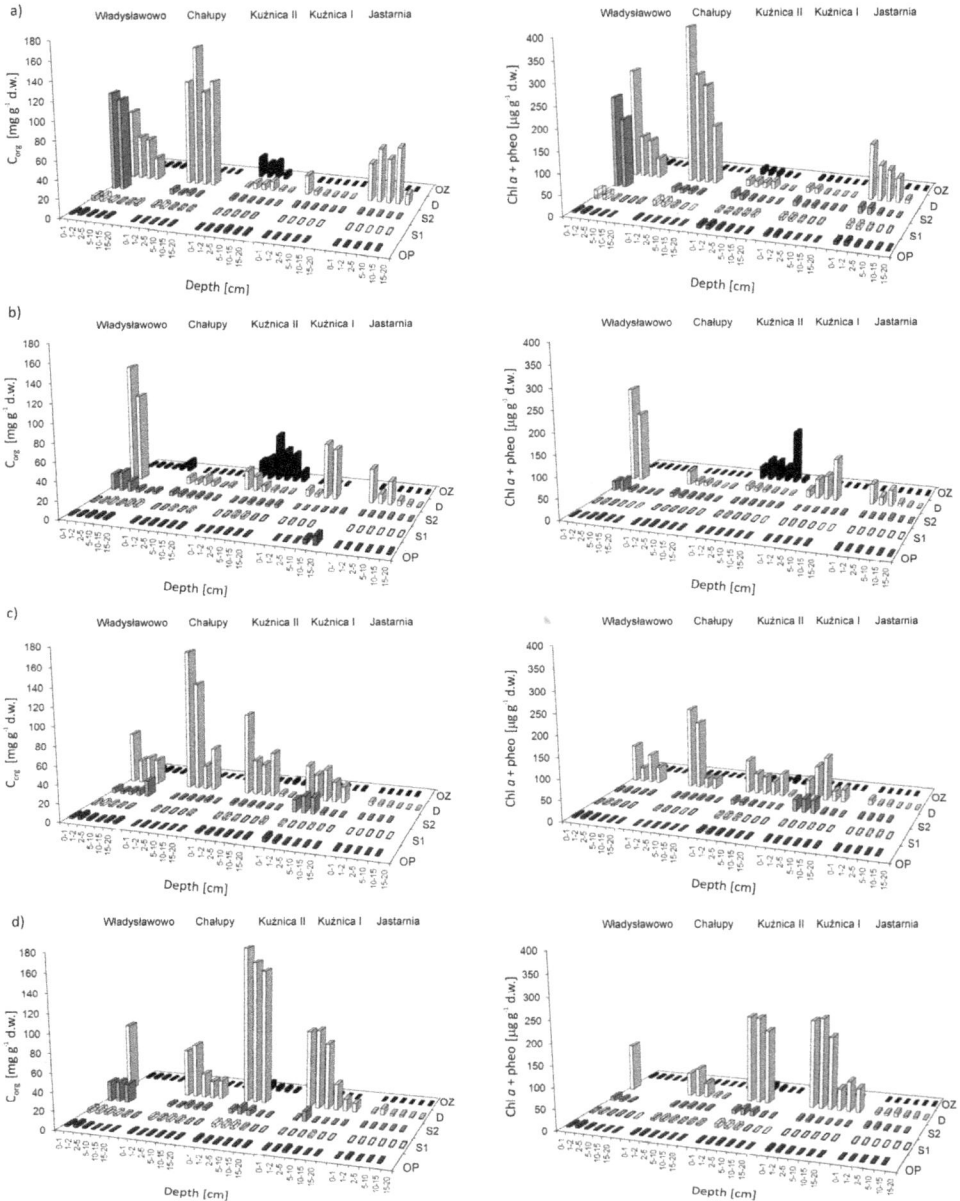

Fig. 10. Organic carbon content (Corg - mg g^{-1} d.w.) and the sum of photosynthetic pigments (chl a+pheo - μ g g^{-1} d.w.) in sediments from the dredge pits and their outskirts in a) November-December 2007, b) March 2008, c) May-June 2008 and d) August 2008.
(OP - outskirt of the pit, peninsula side, OZ - outskirt of the pit, bay side, S1 - shallow slope of pit, S2 - deep slope of pit, D - bottom of the pit).

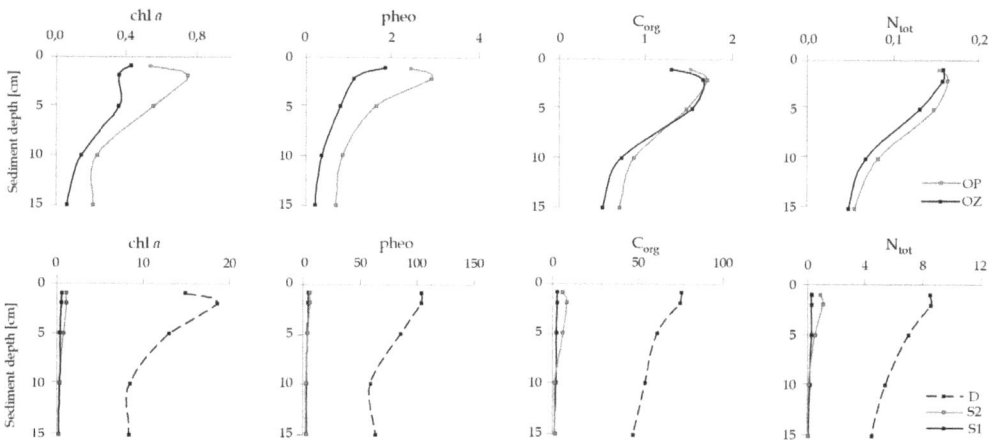

Fig. 11. Sediment profiles of mean concentrations of chlorophyll a (chl a - μ g g^{-1}d.w.), pheophytin (pheo - μ g g^{-1}d.w.), organic carbon (C_{org} - mg g^{-1}d.w.) and total nitrogen (N_{tot} - mg g^{-1}d.w.) in the dredge pits (stations: S1-pit's shallow slopes, S2- pit's deep slopes, D- bottom of the pit) and their outskirts (stations OP and OZ) in August 2008.

This caused sulfate reduction, and in consequence hydrogen sulfide availability, as well as an increase of ammonia and phosphate concentration in the interstitial waters (Fig. 5). Elevated levels of organic matter associated with dredged sediments and deposition of fine-grained sediment in dredge pits have been documented by researchers elsewhere (Johnnston 1981, Nayar et al. 2007, Palmer et al, 2008). In very deep dredge pits such as located in Lake Pontchartrain's, sulfate reduction was dominant process of organic matter degradation (Flocks&Franze 2002). Also previous studies conducted in Kuźnica II and Władysławowo pits indicated anoxic degradation of organic matter. They also indicated temporal oxygen deficits in the nearbottom waters of the Władysławowo pit (Bolałek et al. 1996, Graca, 2009). During presented study, high concentrations of hydrogen sulfide temporally observed in the dredge pits (Fig. 5) so suggest oxygen deficits in the nearbottom waters. Oxygen was determined in samples of the water collected c. 0.5 m over bottom, so probably oxygen deficits closer to the sediment surface.

Impact of deep dredging decreased seawards (Fig. 12a) and has been strongest in the Władysławowo pit area characterized with the weakest water dynamics. Weak water dynamics at the Władysławowo sampling site is indicated by a relatively high content of fine sediment fractions at the outskirts of the dredge pit. (Fig. 6). It also corresponds to the highest content of organic carbon and total nitrogen in sediments collected from the pit and its outskirts (Fig. 10). Moreover, high values of C_{org}/N_{tot} and $C_{org}/(chl/a/+pheo)$ ratio, which have been observed in this area (Tab. 2), indicate the larger share of terrigenic matter in sediments from Władysławowo as compared to those from other sampling sites.

Factor diminishing the impact of deep dredging in the Kuźnica II pit, could be the vicinity of a natural bottom depression, the Kuźnicka Hollow and the fact that the dredge pit is partially connected with it – which is proved by an up-to-date bathymetry (Fig. 2).

Deep dredging impact on geochemical condition was not clearly seasonally dependent in Jastarnia and Kuźnica I pits. (Fig. 12b). Jastarnia pit is located in the Outer Puck Bay. Kuźnica I pit is situated in the areas characterized with intensive currents, through which the water exchange between Inner and Outer Puck Bay occurs. In other dredge impacted regions (Władysławowo, Chałupy, Kuźnica II), located in the Inner Puck Bay, differences were rarely significant in March. In August and November/December an increase of significant differences frequency has been visible in Kuźnica II region (Fig. 12b).

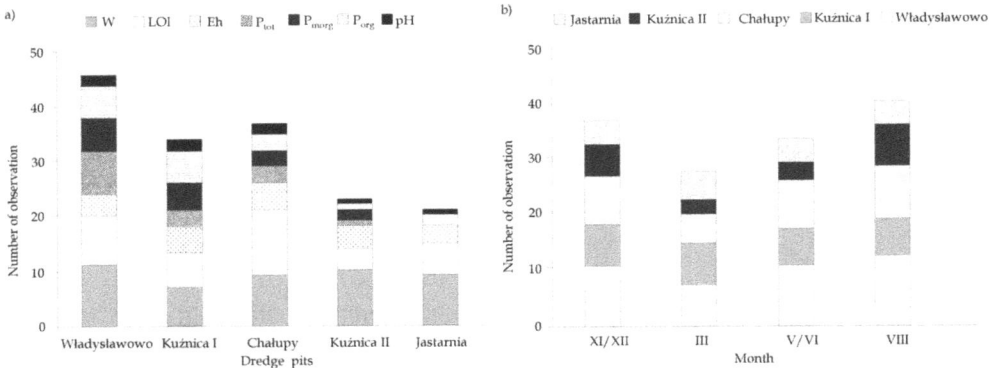

Fig. 12. Number of statistically significant differences in geochemical conditions (Mann-Whitney U test, $p<0.05$) between (a) particular pits and their outskirts (b) dredge pits and their outskirts in particular sampling seasons.

It seems that season influence on the extent of dredging impact depended on pit location and associated water exchange and circulation. The Inner Puck Bay, where season influence was clearly visible, is a shallower, isolated part of the Puck Bay, with hindered water exchange, when compared to the Outer Puck Bay (Nowacki 1993).

The complex interaction of organic matter degradation and increased water dynamics during storms could contribute to the decrease of differences between dredge pits and their outskirts in the cold half year. In consequence, the end of winter is probable a good time to monitor the geochemical conditions in the deep dredging impacted area, because represents "permanent" changes resulting from dredging.

4.2 Deep dredging impact on nitrogen and phosphorus internal loading

A complex interplay of high supply of organic matter to the sediment, hydrogen sulfide production and lost of benthic organisms diminish sediment ability to phosphorus storage (Gächter&Müller 2003, Gunnars at al. 2002, Karlson et al. 2007), decrease coupled nitrification/denitryfication (Sørensen et al., 1980, Karlson et al. 2005, 2007) and increase ammonia regeneration (Kelly et al. 1985; Kemp et al. 1990). As a result deep dredging potentially influences nitrogen and phosphorus dynamic and increases their internal loading in the dredge pit areas. Previous studies, conducted in the Puck Bay, indicated a decrease of phosphorus accumulation to its removal ratio from 9,3 in natural bottom to 2,2 in the Kuźnica II dredge pit, and a decrease of denitrification to ammonification ratio from 6,3 in natural bottom to 0,2 in the dredge pit area (Graca et al. 2004).

Ammonia and phosphate are main components of nitrogen and phosphorus benthic fluxes in the study area, as well as in other eutroficated regions of the Baltic Sea (Conley et al. 1997; Graca et al. 2004, Pitkänen et al. 2001). In order to assess potential enhancement of eutrofication due to dredge pits impact on nutrient dynamic, their benthic fluxes have been estimated in and around the dredge pits basing on Fick's first law. Calculation's details were the same, as in previous study conducted in Kuźnica II dredge pit (Graca et al. 2004). Nutrients loadings released from the sediments were calculated, considering the surface of each depth in the dredge pits (calculated by using Surfer 8) and fluxes seasonal variability. Both actual ($P_{ac.}$, $N_{ac.}$) and potential ($P_{pot.}$, $N_{pot.}$) loads have been calculated, where potential load denotes the load in situation, when the conditions in the dredge pits would be identical to that found in their outskirts (Tab. 3). From the difference between the actual and potential loads, a surplus of N and P removed to water due to deep dredging has been calculated. This is only a rough estimation, because Fick's first law allows only diffusive fluxes. Not include mineralization of organic matter on the surface of sediment and influences of water dynamics and benthic organisms on the benthic fluxes (Forja & Gómez-Parra, 1998; Val Klump & Martens, 1983; Zabel et al., 2000). Therefore fluxes at the outskirts and shallow pit's slopes are probable underestimated. In deepwater part of dredge pits, effects of benthic organisms and turbulent diffusion are potentially limited. As the result, fluxes due to molecular diffusion are probably the major component of the total nutrient exchange in this area. However, when sediments are rich in organic matter and the nearbottom water contain oxygen, as in the dredge pits deepest area, fluxes of phosphate and ammonia calculated based on Fick's first law could have been overestimated. The reason is phosphate precipitation and/or adsorption, as well as coupled nitrification/denitrification at the sediment surface. On the other hand, permanent presence of hydrogen sulfide in deepwater part of dredge pits, potentially inhibit nitrification and in turn, hamper denitrification (Dollhopf et al. 2005, Herbert 1999). Moreover, hydrogen sulfate availability limits the sediment phosphorus sorption capacity. Formations of FeS and FeS2 results in removal of the phosphates previously bound to iron oxyhydroxide despite oxygenated nearbottom water (e.g. Gähter&Müler, 2003). Though presented estimations are rough, they indicate the scale of the problem. Increase of internal N and P loadings has been observed in all dredge pits (Tab. 3). Surplus of N and P removed to water column has been the highest in the area of the biggest pit - Kuźnica II. However, the ratio of actual to potential loadings indicate, that the Władysławowo pit was the area, where internal loading has been intensified the most (Tab. 3).

dredge pit	area (m²)	load (kg y⁻¹)						loads ratio	
		$P_{ac.}$	$P_{pot.}$	$\Delta P^{1)}$	$N_{ac.}$	$N_{pot.}$	$\Delta N^{2)}$	$P_{ac.}/P_{pot.}$	$N_{ac.}/N_{pot.}$
Władysławowo	6519	0,4	0,03	0,4	4,1	0,1	4,0	13,3	40,1
Chałupy	205 068	5,3	0,9	4,4	54,3	4,7	49,6	5,8	11,6
Kźnica II	873 874	23,3	7,0	16,3	262,1	56,3	205,8	3,3	4,6
Kuźnicy I	187 389	2,7	2,0	0,7	25,5	3,8	21,7	1,3	6,7
Jastarnia	220 315	2,2	1,6	0,6	12,7	11,1	1,6	1,4	1,1
sum	1 493 165	33,9	11,5	22,3	358,8	76,0	282,8		

1) - $\Delta P = P_{ac.} - P_{pot.}$ 2) - $\Delta N = N_{ac.} - N_{pot.}$

Table 3. Phosphorus ($P_{ac.}$) and nitrogen ($N_{ac.}$) loadings released from the sediments of particular dredge pits and loadings of those elements which would have been released ($P_{pot.}$, $N_{pot.}$), if the conditions in the dredge pits would be identical to those observed at their outskirts.

In the contrary, Jastarnia pit was the area, where internal loading has been intensified insignificantly. From sediments of all the dredge pits a surplus of c. 22 kg of P and c. 238 kg of N was removed annually into the water column. Such values are trivial in respect to for example external loadings of N and P into the Puck Bay (16,1 t P a^{-1} and 18 t N a^{-1}, Pempkowiak, 1997). This is connected to relatively small dredge pit's area (1,49 km^2). Therefore dredge pits create an important problem, but rather in scale of coastal areas of the Hel Peninsula where pits are located, than in scale of the whole Puck Bay.

Water residence time is important factor influencing susceptibility to eutrophication in aquatic ecosystems (Wulff & Stigebrandt 1989, Nedwell et al. 1999, Dettman 2001, Savchuk & Wulff 2007). Nitrogen and phosphorus loadings released from sediments probable can affect the environment more in the Inner Puck Bay, because of longer water residence time than in the Outer Puck Bay (Nowacki 1993) and consequently, longer N and P residence times in this area. It is worth mentioning, that hydrodynamic studies conducted in the frame of current project (not published) showed that the dredge pits additionally slightly hinder the water exchange between Inner and Outer Puck Bay.

5. Conclusions

All the dredge pits in the Puck Bay create sink for organic matter and fine-grained sediments. As the result hydrogen sulfide production has been stimulated, especially in deepwater parts of pits. Important implication of hydrogen sulfide production could be restricted pit's recolonization by benthos organisms. Moreover, the complex interplay of hydrogen sulfide availability and lost of benthic organisms could probable influence nitrogen and phosphorus dynamic. Rough estimations showed significant increase of internal N and P loadings in some of the dredge pits. Due to their relatively small area (1,49 km^2), surplus of N and P annually released from the pit's sediments (several hundred kilos of nitrogen and a dozen or so kilos of phosphorus) create problem rather in scale of coastal areas of the Hel Peninsula where pits are located, than in scale of the whole Puck Bay.

Deep dredging impact on geochemical condition was the most pronounced and clearly seasonal dependent in a shallower, isolated part of the Puck Bay. In this region of the bay organic matter degradation in sediment and increased water dynamics during storms, decrease differences in geochemical conditions between dredge pits and their outskirts in the cold half year.

6. Acknowledgement

Research was conducted within the project entitled "Creation of recultivation program for post-dredging area in Puck Bay (R 1404203) financed by the polish Ministry of Science and Higher Education and realized by Maritime Institute in Gdańsk.

7. References

Bolałek J., Falkowska L. & Korzeniewski K. (1993). Hydrochemia Zatoki, In: *Zatoka Pucka*, K. Korzeniewski, pp. 222-281, Fundacja Rozwoju Uniwersytetu Gdańskiego, Gdańsk

Bolałek J., Jankowska H., Łęczyński L., Frankowski L. & Podgórska B. (1996). Geological, geochemical and bacteriological conditions in the postdredging pit in Puck Bay. (Southern Baltic, Poland). *Oceanological Studies*, XXV (3), pp. 111-122

Boyd, C.E. (1995). *Bottoms soils, sediment and pond aquaculture*. Chapman&Hall, New York

Bradtke K. & Urbański, J. (2008). (http://pim.iopan.gda.pl/ZatokaPucka).

Burska D., Frankowski L. & Bolałek J. (1999). Temporal variability in the chemical composition of bottom sediments in the Pomeranian Bay (Southern Baltic). *Oceanology*, 41, pp. 445-459

Burska D. (2010). Analiza elementarna (CHNS), In: *Fizyczne, biologiczne i chemiczne badania morskich osadów dennych*, J. Bolałek, pp. 289-295, Wydawnictwo Uniwersytetu Gdańskiego, Gdańsk

Ciszewski P. & Kruk-Dowgiałło L. (1992). Ecological requirements for conducting the dredging works in the region of Chałupy and Kuźnica on the open sea side of the Hel Peninsula. Typescript (in Polish)

Conley, D.J., Stockenberg, A., Carman, R., Johnstone, R.W., Rahm, L. & Wulff, F. (1997). Sediment-water nutrient fluxes in the Gulf of Finland, Baltic Sea. *Estuarine, Coastal and Shelf Science*, 45, pp. 591-598

Desprez, M. (2000). Physical and biological impact of marine aggregate extraction along the French coast of the Eastern English Channel: short- and long-term post-dredging restoration. *ICES Journal of Marine Science*, 57, pp. 1428-1438

Dettmann, E.H. (2001). Effect of water residence time on annual export and denitrification of nitrogen in estuaries: A model analysis. *Estuaries*, 24(4), pp. 481–490

Dollhopf, S.L., Hyun, J., Smith, A.C., Adams, H.J., O'Brien, S. & Kostka, J.E. (2005). Quantification of ammonia-oxidizing bacteria and factors controlling nitrification in salt marsh sediments. *Applied and Environmental Microbiology*, 71(1), pp. 240–246

Flocks & Franze (2002). Environmental Issues-Dredge Pit Characterization, Environmental Atlas of the Lake Pontchartrain Basin, (http://pubs.usgs.gov/of/2002/of02-206/env-issues/pit-characterization.html)

Forja J.M. & Gómez-Parra A. (1998). Measuring nutrient fluxes across the sediment-water interface using benthic chambers. *Marine Ecology and Progress Series*, 164, pp. 95-105

Froelich P.N., Arthur M.A., Burnett W.C., Deakin M., Hensley V. Jahnke R., Kaul L., Kim K.-H., Roe K., Soutar A. & Vathakanon C. (1988). Early diagenesis of organic matter in Peru continental margin sediments: phosphorite precipitation. *Marine Geology*, 80, pp. 309-343

Gächter R. & Müller B. (2003) Why the phosphorus retention of lakes does not necessarily depend on the oxygen supply to their sediment surface. *Limnology and Oceanography*, 48(2), pp. 929-933

Golterman H.L. (1975). *Physiological limnology*. Elsevier Scientific Pub. Co., N.Y.

Graca B., Burska D., Dudkowiak M., Jędrasik J., Kowalewski M., Łęczyński L., Łukawska K., Szymelfenig M. & Zarychta A. (2003). *Ocena wpływu wyrobisk porefulacyjnych na zmiany ekosystemu Zatoki Puckiej*. RAPORT Projektu KBN 6PO4E 01709 (in Polish)

Graca B., Burska D. & Matuszewska K. (2004). The impact of dredging deep pits on organic matter degradation in sediments. *Water Air and soil Polution*, 158, pp. 237-259

Graca B., Witek, Z., Burska, D., Białkowska, I., Pawelec, A., & J. Bolałek (2006). Porewater's nutrients (phosphate, ammonia and silicate) in southern Baltic Sea. *Oceanological and Hydrobiological Studies*, XXXV (3), pp. 237-25

Graca B. & Dudkowiak M. (2007). Microbiological changes in environment caused by deep dredging. A case study: post-dredging pit Kuźnica II. *Oceanological and Hydrobioogical Studies*, XXXVI (1), pp. 17-27

Graca, B. (2009). *The dynamics of nitrogen and phosphorus transformations at the sediment-water interface in the Gulf of Gdańsk*, Wydawnictwo UG, Gdańsk (in polish with English abstract)

Grasshoff K., Ehrhardt M. & Kremling K. (1983). *Methods of sea water analysis*, Verlag Chem., Weinheim

Gunnars, A. Blomqvist, S., Johansson, P. & Andersson, C. (2002). Formation of Fe(III) oxyhydroxide colloides in freshwater and brackish seawater, with incorporation of phosphate and calcium. *Geochimica et Cosmochimica Acta*, 66, pp. 745-758

Håkanson L., Lundin L.-C., Savchuk O., Ionov V., Musielak S. & Furmańczyk K. (2003). The Baltic Sea, In: L. Ryden, P. Migula, M. Andersson, pp. 120-147, *Environmental Science*, The Baltic University Press, Uppsala

Hedges J.I. & Stern J.H. (1984). Carbon and nitrogen determinations of carbonate containing solids. *Limnol. Oceanogr.*, 29, pp. 657-663

Herbert, R. A. (1999). Nitrogen cycling in coastal marine ecosystems. *FEM Microbiol. Rev*, 23, pp. 563–590

Johnnston S.A. Jr (1981). Estuarine dredge and fill activities: A review of impacts. *Environmental Management*, 5 (5), pp. 427–440

Jönsson A., Danielsson Å. & Rahm L. (2005). Bottom type distribution based on wave friction velocity in the Baltic Sea. *Continental Shelf Research*, 25, pp. 419–435

Karlson, K., Hulth, S. & Ringdahl, K., Rosenberg, R. (2005). Experimental recolonisation of Baltic Sea reduced sediments: survival of benthic macrofauna and effects on N and P biogeochemistry. *Marine Ecology Progress Series*, 294, pp. 35–49

Karlson K., Bonsdorff E. & Rosenberg R. (2007). The impact of benthic macrofauna for nutrient fluxes from Baltic Sea sediments. *Ambio*, 36 (2/3), pp. 1-7

Kemp, W.M., Sampou, P., Caffrey, J. & Mayer, M. (1990). Ammonium recycling versus denitrification in Chesapeak Bay Sediments. *Limnology and Oceanography*, 35 (7), pp. 1545-1563

Kelly J.R., Berounsky V.M., Nixon S.W. & Oviatt C.A. (1985). Benthic-pelagic coupling and nutrient cycling across an experimental eutrophication gradient. *Marine Ecology Progress Series*, 26, pp. 207-219

Maa, J. P.Y., Hobbs, C.H., Kim S.C. & Wei, E. (2004). Possible impacts on physical oceanpgraphic processes by cumulative sand mining offshore of Maryland and Delaware. *Journal of Coastal Research*, 20 (1), pp. 44-60

Maksymowska D., *Degradacja materii organicznej w toni wodnej i osadach dennych Zatoki Gdańskiej*, Instytut Oceanografii UG, Gdynia (in Polsih)

Nayar S., · Miller D.J., Hunt A. & · Goh B.P.L. Chou L.M. (2007). Environmental effects of dredging on sediment nutrients, carbon and granulometry in a tropical estuary. *Environ Monit Assess*, 127, pp. 1–13

Nedwell D.B., Jickells T.D., Trimmer M. & Sanders, R. (1999). Nutrient in estuaries. *Advances in Ecological Research*, 29, pp. 43-92

Nowacki, J. (1993). Cyrkulacja i wymiana wód, In: *Zatoka Pucka*, K. Korzeniewski, pp. 181-206, Fundacja Rozwoju Uniwersytetu Gdańskiego, Gdańsk

Palmer T.A., Paul A. Montagna P:A. & Nairn R.B. (2008). The effects of a dredge excavation pit on benthic macrofauna in offshore Louisiana. *Environmental Management*, 41, pp. 573–583

Pempkowiak J. (1994). *Przyczyny katastrofy ekologicznej w rejonie Zatoki Puckiej. Zanieczyszczenie i odnowa Zatoki Gdańskiej*, Gdynia, Polska

Pitkänen, H., Lehtoranta J. & Räike, A. (2001). Internal nutrient fluxes counteract decreases in external load: The case of the estuarial eastern Gulf of Finland, Baltic Sea. *Ambio*, 30 (4-5), pp. 195-201

Plante-Cuny M. (1974). Evaluation par spectrophotometrie des teneurs en chlorophylle a fonctionelle et en pheopigments des substrats meubles marins. Mission Orstom Nosy-Be, Madagascar, 45, pp. 1-76

Savchuk O. & Wulff F. (2007). Modeling the Baltic Sea eutrophication in a decision support system. *Ambio*, 36 (2), pp. 141-148

Sørensen J., Tiedje J.M. & Fiestone R.B. (1980). Inhibition by sulfide of nitric and nitrous oxide reduction by denitrifying Pseudomonas fluorescens. *Applied and Environmental Microbiology*, 39, pp. 105-108

Strickland J.D.H. & ParsonsT.R. (1972). A practical handbook of seawater analysis. *Bulletin of the Fisheries Research Board of Canada*, 167

Szymelfenig M., Kotwicki L. & Graca B. (2006). Benthic re-colonization in post-dredging pits in the Puck Bay (the southern Baltic Sea). *Estuarine and Coastal Shelf Science*, 68, pp. 489-498

Val Klump, J. & Martens, C.S. (1983). Benthic nitrogen regeneration, In: *Nitrogen in the marine environment*, E.J., Carpenter, D., Capone, pp. 411-459Academic Press, London

Schindler D.W. & Vallentyne J.R. (2008). *Algal bowl: overfertilization of the world's freshwaters and estuaries*, Michigan State Univ Pr

Work P.A., Fehrenbacher F., & Voulgaris G. (2004). Nearshore impacts of dredging for beach nourishment. *Port, Coast., and Oc. Engrg.*,130(6), pp. 303-311

Wulff F. & Stigebrandt A. (1989). A time-dependent budget model for nutrients in the Baltic Sea. *Global Biogeochemical Cycle*, 3, pp. 63-78

Zabel M., Hensen C. & Schlüter M. (2000). Back to the ocean cycles: benthic fluxes and their distribution patterns, In: *Marine Geochemistry*, H.D., Schulz, M. & Zabel, pp. 373-394, Springer Verlag Berlin-Heidelberg

Urbański & Solanowska (2009).
    http://www.ocean.univ.gda.pl/~oceju/zmiany_polwyspu_helskiego.pdf

14

On the Chemical Profile of Marine Organisms from Coastal Subtropical Environments: Gross Composition and Nitrogen-to-Protein Conversion Factors

Graciela S. Diniz, Elisabete Barbarino and Sergio O. Lourenço
Universidade Federal Fluminense, Departamento de Biologia Marinha,
Niterói, RJ
Brazil

1. Introduction

After more than a century of predominance of more traditional studies in marine biology (e.g. ecology, reproduction, distribution, feeding, taxonomy), in the last decades the scientific community has been diversifying tremendously research approaches (Duarte et al., 2011). In this context, studies on the chemical composition of marine organisms have been becoming more common in the world (e.g. Amsler, 2009; Hedges et al., 2002; Janecki & Rakusa-Suszczewski, 2004). Information on chemical characteristics can offer important subsidies for the study of physiology, biochemistry, ecology, and conservation of marine species, for example (Barbarino & Lourenço, 2009), as well as in applied science.

Despite the recognition of the importance of data on the chemical composition of organisms in marine biology, there are still significant constraints in the basic knowledge of species, especially in tropical and subtropical environments. Studies in this field are very necessary to fulfill the existing gap. Data on the chemical components of the primary metabolism (such as nitrogen, phosphorus, protein, carbohydrate and lipid) are not available for many species, especially for non-edible organisms (Barbarino & Lourenço, 2009).

Nitrogen is an essential element which is incorporated into fundamental macromolecules of the primary metabolism, such as proteins, peptides, nucleic acids, pigments and amino acids (Karl et al., 2002). Phosphorus is a key component of the ATP and many molecules, such as phospholipids. Information on the tissue concentrations of nitrogen and phosphorus allows interpretations on the availability of dissolved inorganic nutrients in the environment, especially when primary producers are analyzed (Lobban & Harrison, 1994).

Proteins exert fundamental roles in all biological processes: enzymatic catalysts, transport and storage of substances, movements, mechanical support, immunological protection, generation and transmission of nerve impulses, control of growth and cell differentiation (Zaia et al., 1998). The amino acids are the basic structural units of the proteins. Carbohydrates exert a structural role in the cells and act as a reserve of chemical energy. They are particularly abundant in plant materials. Lipids have important functions in the cells, as structural components of biological membranes, reserve of chemical energy, and diverse metabolic activities, such as activation of enzymes and participation in the transport of electrons, for example (Carvalho & Recco-Pimentel, 2007).

Moreover, measurements of proteins, carbohydrates and lipids have important applications in to characterize organisms which serve as food (Tacon et al., 2009). Chemical characterization of macromolecules of marine organisms are also very relevant for other applied uses, such as in industry, agriculture, aquaculture and health care.

The need for data on the chemical composition of marine organism is evident, but some methodological constraints still limit the progress in this field. In general, many basic chemical analyses must be adapted to the characteristics of the marine organisms, regarding the effects of moisture, salt content and differences in algae cell wall composition, for example (Barbarino & Lourenço, 2009). This is particularly relevant in protein analysis, because differences procedures used for protein extraction establish remarkable influence on final results.

Some of the most common methods used for protein determination in marine organisms, Lowry's method (Lowry et al., 1951) and Bradford method (Bradford, 1976), are subject to interferences from many factors. The interferences are a consequence of the effects of some substances on specific amino acids, since that the chemical reactions which produce the protein quantification depends on the reactivity of the amino acid side groups (Legler et al., 1985). By contrast, the total nitrogen analysis is relatively simple and easy to perform, and nitrogen-to-protein conversion factors (N-Prot factors) can be used to estimate the crude protein content. The use of N-Prot factors has some important advantages if compared to other methodologies: accuracy, low cost and no need of a laborious protein extraction before the analysis.

A few studies on the protein content of marine organisms used N-Prot factors. In addition, the factor predominantly used in these studies was the traditional factor 6.25, calculated by Jones (1931) from bovine muscles. The use of the factor 6.25 is based on two assumptions: i) that the average protein contains about 16 percent nitrogen by weight ($6.25 \times 16 = 100$) (Tacon et al., 2009); and ii) that an insignificant amount of non-protein nitrogen (NPN) is present in the samples (Coklin-Brittain et al., 1999). However, the percentage of nitrogen in protein depend on the amino acid composition, which is variable from one protein source to another, and in practice a variation of between 12 and 19 percent nitrogen is possible between individual proteins (Tacon et al., 2009). Moreover, several organisms contain high concentrations of other nitrogenous compounds, such as nucleic acids, amines, urea, inorganic intracellular nitrogen, vitamins and alkaloids (Lourenço et al., 2004). The contribution of non-protein nitrogenous compounds to the total crude protein content of different kinds of marine biological samples ranges from 10% to 40%, according to Venugopal & Shahidi (1996).

The use of specific N-Prot factors is widely recommended in order to get more accurate estimates of protein content (Sosulski & Imafidon, 1990). The nitrogen:protein ratio does vary according to the source considered (Mariotti et al., 2008). The use of N-Prot factors is particularly wide in food science, such as certain cereals (e.g 5.26 for rise, 5.47 for wheat; Fujihara et al., 2008), legumes (e.g. 4.75-5.87 for cassava root; Yeoh & Truong, 1996), mushroom (4.70; Mattila et al., 2002), and fish and fish products (4.94, Salo-Väänänen & Koivistoinen, 1996), among other products. The use of N-Prot factors is still unusual in sea science, possibly because most of the scientific community ignores this methodological alternative.

The use of the factor 6.25, in case of NPN-rich samples, tends to overestimate the protein
data. Despite this, several authors continue to use the factor 6.25 to estimate the protein
content of aquatic organisms (e.g. Nurnadia et al., 2011; Undeland et al., 1999; Zaboukas et
al., 2006). Except for a short list of specific N-Prot factors available for marine organisms, the
general factor 6.25 is still used for most plant and animal sources from the sea.

The purpose of this study was to determine specific N-Prot factors for 23 species of marine
organisms from coastal areas of Rio de Janeiro State, Brazil, based on the ratio of amino acid
composition to total nitrogen (TN) content. In addition, we also characterized and compared
the organisms regarding their carbohydrate, lipid, nitrogen and phosphorus contents.

2. Material and methods

2.1 Organisms and sampling

The organisms studied in this work were collected at different locations in the State of Rio
de Janeiro, distributed by the municipalities of Arraial do Cabo, Armação dos Búzios,
Niteroi, Rio de Janeiro and Angra dos Reis (Figure 1). The organisms were chosen regarding
their local abundance and taxonomic diversity, including species that live in the transition
between land and sea. Seaweeds, spermatophytes, invertebrates and fishes were assessed in
this study, totalizing 23 species (Table 1).

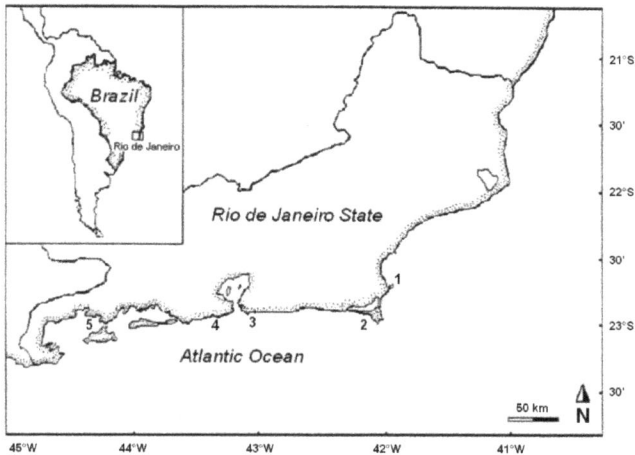

Fig. 1. Sampling sites in Rio de Janeiro State, Brazil: Armação dos Búzios (1), Arraial do
Cabo (2), Niterói (3), Rio de Janeiro (4) and Angra dos Reis (5).

Red (*Asparagopsis taxiformis* and *Centroceras clavulatum*), green (*Chaetomorpha aerea*) and brown
(*Sargassum filipendula*) algae were analyzed. Whole thalli of adult seaweeds were collected and
washed in the field with local seawater in order to remove epiphytes, sediment and organic
matter. Plants were packed in plastic bags and kept on ice until returned to the laboratory,
where the samples were gently brushed under running seawater, rinsed with distilled water,
dried with paper tissue and frozen at -18°C. Subsequently, the samples were freeze dried in a
Terroni Fauvel, model LB1500TT device. Statistically, each seaweed sample resulted of more
than a single specimen, because of the small size of the individuals.

Organisms	Family	Sampling sites
Seaweeds		
Asparagopsis taxiformis (Delile) Trevisan de Saint-Léon, 1845	Bonnemaisoniaceae	Angra dos Reis
Centroceras clavulatum (C. Agardh) Montagne, 1846	Ceramiaceae	Arraial do Cabo
Chaetomorpha aerea (Dillwyn) Kützing, 1849	Cladophoraceae	Arraial do Cabo
Sargassum filipendula (Agardh, 1824)	Sargassaceae	Arraial do Cabo
Spermatophytes		
Acrostichum aureum (Linnaeus, 1753)	Polypodiaceae	Rio de Janeiro
Avicennia schaueriana (Stapft & Leechman) Moldenke, 1939	Avicenniaceae	Rio de Janeiro
Blutaparon portulacoides (St. -Hill) Mears, 1982	Amaranthaceae	Rio de Janeiro
Halodule wrightii (Ascherson, 1868)	Cymodoceaceae	Arm. Búzios
Laguncularia racemosa (Linnaeus) Gaertner, 1807	Combretaceae	Rio de Janeiro
Ruppia maritima (Linnaeus, 1753)	Ruppiaceae	Rio de Janeiro
Spartina alterniflora (Loiseleur-Deslongchamps, 1807)	Poaceae	Rio de Janeiro
Typha domingensis (Persoon) Steudel, 1807	Typhaceae	Rio de Janeiro
Invertebrates		
Aplysia brasiliana (Rang, 1828)	Aplysiidae	Arm. Búzios
Bunodossoma caissarum (Correa, 1964)	Actiniidae	Arraial do Cabo
Desmapsamma anchorata (Carter, 1882)	Desmacidadae	Angra dos Reis
Echinaster (Othilia) brasiliensis (Muller & Troschel, 1842)	Echinasteridae	Angra dos Reis
Echinometra lucunter (Linnaeus,1758)	Echinometridae	Arraial do Cabo
Eledone massyae (Voss, 1964)	Octopodidae	Arraial do Cabo
Fishes		
Dactylopterus volitans (Linnaeus, 1758)	Dactylopteridae	Niterói
Genypterus brasiliensis (Regan, 1903)	Ophidiidae	Arraial do Cabo
Mullus argentinae (Hubbs & Marini, 1933)	Mullidae	Arraial do Cabo
Rhizoprionodon lalandii (Muller & Henle, 1839)	Carcharhinidae	Niterói
Rhizoprionodon porosus (Poey, 1861)	Carcharhinidae	Niterói

Table 1. Organisms studied, with indication of taxonomic details and sampling sites in Rio de Janeiro State, Brazil. Arm. Búzios = Armação dos Búzios.

The spermatophytes analyzed here included two mangrove plants (*Laguncularia racemosa* and *Avicennia schaueriana*), one salt marsh plant (*Spartina alterniflora*), two wetland plants (*Acrostichum aureum* and *Typha domingensis*), a sand dune plant (*Blutaparon portulacoides*), and two seagrasses (*Halodule wrightii* and *Ruppia maritima*). Only mature and healthy leaves were used in this study: both young and senescent leaves were not collected. Leaves were collected and prepared similarly to the description presented for seaweeds. Each statistical sample resulted from leaves collected from a single plant.

Six species of marine invertebrates belonging to four phyla were analyzed: *Desmapsamma anchorata* (Porifera), *Bunodossoma caissarum* (Cnidaria), *Aplysia brasiliana, Eledone massyae* (Mollusca), *Echinaster brasiliensis* and *Echinometra lucunter* (Echinodermata). The animals were collected in the inter-tidal region or by free diving in small depths (< 5.0 m). The animals were washed in the field with seawater, packed in plastic bags and kept on ice until returned to the laboratory. In the laboratory samples were washed with distilled water, cut, frozen and freeze-dried. Internal soft parts of the bodies were used for echinoderms, and for mollusks only the mantle was used. For other invertebrates (e.g. sponge) whole bodies were processed. Each statistical sample resulted from a single individual.

Five species of not-sexed adult fishes (with similar body sizes) were collected with bottom trawls and washed and transported as described for the invertebrates. In the laboratory samples were cut and only the muscles were used. Each statistical sample resulted from a single individual.

For all samples, the freeze-dried material was powdered manually using a mortar and pestle, and it was kept in desiccators containing silica-gel, under vacuum and at room temperature, until the chemical analyses were carried out. Four statistical samples were prepared for each species assessed in this study.

2.2 Chemical analyses

Total nitrogen and phosphorus were determined after peroxymonosulphuric acid digestion, using a Hach digestor (Digesdhal®, Hach Co.). Total N and P contents were determined spectrophotometrically, following analytical details provided by Lourenço et al. (2005).

The Lowry's method (Lowry et al., 1951) was used to analyze hydrosoluble protein in the samples, using bovine serum albumin as a protein standard. Spectrophotometric determinations were performed at 750 nm, 35 min after the start of the chemical reaction. Total carbohydrate was extracted with 80% H_2SO_4, according to Myklestad & Haug (1972). The carbohydrate concentration was determined spectrophotometrically at 485 nm, 30 min after the start of the chemical reaction, by the phenol-sulfuric acid method (Dubois et al., 1956), using glucose as a standard. Total lipid was extracted according to Folch et al. (1957), and determined gravimetrically after solvent (chloroform) evaporation.

Total amino acid (AA) was determined by high performance liquid chromatography with pre-column derivatization with AccQ.Fluor® reagent (6-aminoquinolyl-N-hydroxysuccinimidyl carbamate), reverse phase column C_{18} AccQ.Tag® Nova-Pak (150x3.9 mm; 4 µm), ternary mobile phase in gradient elution composed by sodium acetate 140 mM + TEA 17 mM pH 5.05 (solvent A), acetonitrile (solvent B) and water (solvent C), flow 1 ml.min^{-1} (Cohen & De Antonis, 1994). A Waters, model Alliance 2695 chromatograph was used, equipped with a fluorescence detector Waters® 2475 ($\mu_{ex.}$ 250 nm, $\mu_{em.}$ 395 nm). Analytical conditions were suitable to determine all amino acids, except tryptophan, cysteine + cistine and methionine. The percent of nitrogen in each amino acid was used to calculate nitrogen recovered from total amino acid analysis. Aspartic acid, threonine, serine, glutamic acid, proline, glycine, alanine, valine, isoleucine, leucine, tyrosine, phenylalanine, histidine, lysine, and arginine contents were multiplied by 0.106, 0.118, 0.134, 0.096, 0.123, 0.188, 0.158, 0.120, 0.108, 0.108, 0.078, 0.085, 0.271, 0.193, and 0.322, respectively (Diniz et al., 2011). The sum of amino acid residues was used to estimate total protein in the samples (Fujihara et al., 2008; Yeoh & Truong, 1996).

2.3 Calculation of N-Prot factors

N-Prot factors were determined for each species by the ratio of amino acid residues (AA-Res) to total nitrogen (TN) of the sample: N-Prot factor = AA-Res / TN. Thus, for a 100 g (dry weight) sample having 16.21 g of amino acid residues and 3.48 g of TN, a N-Prot factor of 4.66 is calculated.

The amino acid residues of the samples was calculated by summing up the amino acid masses retrieved after acid hydrolysis (total amino acids), minus the water mass (18 H_2O/mol of amino acid) incorporated into each amino acid after the disruption of the peptide bonds (Fujihara et al., 2008).

2.4 Statistical analysis

The results were analyzed by one-way analysis of variance (ANOVA) with significance level $\alpha = 0.05$, followed, where applicable, with a Tukey's multiple comparison test (Zar, 1996). The statistical analysis was applied within a given category of organisms (ex.: invertebrates), thus comparisons of chemical composition between seaweeds and fishes, for instance, were not performed.

3. Results

3.1 Nitrogen and phosphorus

The percentage of nitrogen showed wide variations among seaweeds, ranging from 1.75% (S. filipendula, a brown alga) to 5.56% (A. taxiformis) of the dry weight. Red algae showed higher total nitrogen concentrations in the thalli, with significant differences ($p < 0.001$) to the other algae (Table 2). The green alga C. aerea showed an intermediate concentration of N in comparison to red and brown algae. Spermatophytes also showed a great variation in the nitrogen concentration among the species. Values ranged from 0.99% (L. racemosa) to 3.83% (R. maritima). The highest concentrations of nitrogen were observed in seagrasses, followed by the wetland plant T. domingensis (Table 2).

The animals present more nitrogen concentration in their tissues than the photosynthetic organisms. In addition, remarkable variations in N concentrations were observed among the invertebrate species, ranging from 3.79% (D. anchorata, sponge) to 12.7% (E. massyae, an octopus). Fishes tended to show higher values of total N, ranging from 11.6% (M. argetinae) to 14.9% (R. porosus) of the d.w., with minor variations among the fish species (Table 2).

Wide concentrations of phosphorus occurred in seaweeds. C. clavulatum showed the highest value (0.54%, $p < 0.001$) and S. filipendula and A. taxiformis showed the lowest concentrations (0.27% and 0.30% respectively). The lowest values for phosphorus were observed in spermatophytes, especially in the sand dune plant B. portulacoides (0.13%). The seagrasses showed the highest phosphorus concentrations among the spermatophytes, circa 0.58%.

The concentrations of phosphorus in animals were typically higher than in algae and plants. Values found for invertebrates showed wide variations, ranging from 0.41% in the sponge D. anchorata to 1.16% in the octopus E. massyae. Narrower variations in P concentrations were found in fishes, ranging from 0.94% (D. volitans) to 1.29% (G. brasiliensis).

3.2 Hydrosoluble protein, carbohydrate and lipid

The hydrosoluble protein contents in seaweeds ranged from 8.72% (S. filipendula) to 16.1% (C. aerea) of the d.w. with intermediate and similar concentrations ($p > 0.05$) in the red algae.

On the Chemical Profile of Marine Organisms from Coastal Subtropical Environments: Gross Composition and
Nitrogen-to-Protein Conversion Factors

315

The concentrations of hydrosoluble protein also varied widely among spermatophytes. Peaks were found in *T. domingensis* (21.4%, $p < 0.001$) and the seagrass *H. wrightii* showed the lowest concentrations (4.94%). Wide variations in protein content were recorded for the invertebrates, with lower values in the sponge *D. anchorata* (12.2%) and highest percentages in the octopus *E. massyae* (47.6%). Contents of hydrosoluble protein were similar among the fish species, always higher than 45% of the dry weight (Table 2).

Groups / species	Total Nitrogen	Total Phosphorus	Hydrosoluble protein	Total Carbohydrate	Total lipid
Seaweeds	***	***	***	***	***
A. taxiformis	5.56 ± 0.29a	0.30 ± 0.04c	11.7 ± 0.58b	22.9 ± 1.35b	4.80 ± 0.24b
C. clavulatum	4.63 ± 0.15b	0.54 ± 0.03a	11.3 ± 0.64b	27.1 ± 1.73a	2.78 ± 0.23c
C. aerea	2.56 ± 0.13c	0.43 ± 0.04b	16.1 ± 0.25a	29.4 ± 0.78a	5.49 ± 0.09a
S. filipendula	1.75 ± 0.03d	0.27 ± 0.02c	8.72 ± 0.54c	16.8 ± 0.97c	2.92 ± 0.13c
Spermatophytes	***	***	***	***	***
H. wrightii	2.72 ± 0.09b	0.57 ± 0.18a	4.94 ± 0.29f	19.0 ± 1.17d	4.14 ± 0.33c
R. maritima	3.83 ± 0.04a	0.59 ± 0.02a	6.44 ± 0.30e	26.8 ± 0.71b	6.11 ± 0.53a
A. aureum	1.73 ± 0.06c	0.21 ± 0.01b	12.7 ± 0.52c	24.2 ± 0.99c	4.69 ± 0.39c
B. portulacoides	1.75 ± 0.10c	0.13 ± 0.01c	7.73 ± 0.17e	20.3 ± 0.58d	3.71 ± 0.22d
T. domingensis	2.51 ± 0.10b	0.16 ± 0.01c	21.4 ± 1.05a	35.8 ± 1.81a	5.34 ± 0.54b
A. schaueriana	1.95 ± 0.07c	0.14 ± 0.01c	9.53 ± 0.34d	19.1 ± 1.21d	6.76 ± 0.54a
L. racemosa	0.99 ± 0.19d	0.14 ± 0.05c	14.7 ± 0.69c	9.68 ± 0.70e	6.07 ± 0.51a
S. alterniflora	1.94 ± 0.07c	0.24 ± 0.01b	17.0 ± 1.00b	22.4 ± 1.74cd	4.88 ± 0.46c
Invertebrates	***	***	***	***	***
D. anchorata	3.79 ± 0.32e	0.41 ± 0.03c	12.2 ± 0.69e	4.67 ± 0.26c	4.92 ± 0.29c
B. caissarum	9.71 ± 0.36b	1.00 ± 0.08a	29.6 ± 0.72b	6.88 ± 0.61b	8.78 ± 0.66b
A. brasiliana	7.53 ± 0.31c	0.84 ± 0.05b	25.4 ± 1.35c	18.4 ± 1.83a	5.02 ± 0.40c
E. massyae	12.7 ± 0.50a	1.16 ± 0.03a	47.6 ± 3.12a	1.85 ± 0.15d	3.80 ± 0.15d
E. brasiliensis	7.62 ± 0.15c	0.86 ± 0.04b	24.7 ± 1.57cd	3.70 ± 0.25c	25.3 ± 1.09a
E. lucunter	5.42 ± 0.32d	0.82 ± 0.23b	22.5 ± 2.69d	7.22 ± 0.12b	8.59 ± 0.92b
Fishes	***	***	***	***	***
D. volitans	11.7 ± 0.23b	0.94 ± 0.05d	47.1 ± 1.32b	3.45 ± 0.25a	7.81 ± 0.84b
G. brasiliensis	14.0 ± 0.67a	1.29 ± 0.08ab	56.0 ± 1.79a	1.11 ± 0.03de	8.28 ± 0.72b
M. argentinae	11.6 ± 0.72b	1.12 ± 0.08c	47.7 ± 2.24b	1.03 ± 0.10e	16.1 ± 3.68a
R. lalandii	14.4 ± 0.41a	1.15 ± 0.04bc	48.2 ± 1.39b	1.40 ± 0.09bc	4.40 ± 0.37c
R. porosus	14.9 ± 0.28a	1.10 ± 0.01cd	45.7 ± 0.30b	1.17 ± 0.04cde	5.28 ± 0.33bc

Table 2. Gross chemical composition of 23 species of marine organisms sampled in subtropical sites of Brazil. Values are expressed as percentage of the dry weight and represent the mean of four replicates ± standard deviation (n = 4). Mean values significantly different: ***$p < 0.001$, a > b > c > d > e. Identical superscript letters (a, a; b, b; c, c) indicate that mean values are not significantly different.

Total carbohydrate was abundant in all photosynthetic organisms, typically achieving concentrations higher than 19% (except in the mangrove plant *L. racemosa*, with only 9.68%), with a peak of 35.8% in the wetland plant *T. domingensis*. The seaweeds showed values ranging from 16.8% (*S. filipendula*) to 29.4% (*C. aerea*) of the d.w. Carbohydrate was the less abundant organic substance measured in all animals, except for sea slug *A. brasiliana*, which showed a high content of carbohydrate (18.4%). In the other invertebrates, the values of carbohydrates ranged from 1.85% (*E. massyae*, octopus) to 7.22% (*E. lucunter*, sea urchin). In fish species, the low carbohydrate content ranged from 1.03% (*M. argentinae*) to 3.45% (*D. volitans*) of the d.w.

Algae and plants exhibited low concentrations of lipid. In seaweeds, the peak was recorded in *C. aerea* (5.49%, d.w.) and the lowest value occurred in *C. clavulatum* (2.78%) ($p < 0.001$). In spermatophytes the values ranged from 3.71% (*B. portulacoides*) to 6.76% (*A. schaueriana*).

The concentration of lipid varied widely among the animals. In invertebrates, the values ranged from 3.8% in the octopus *E. massyae* to 25.3% in the starfish *E. brasiliensis*. In fishes, the lowest lipid concentrations were registered in elasmobranch fishes (*R. porosus* and *R. lalandii*), with less than 5.5% of the d.w. *M. argentinae* showed a significantly higher lipid concentration (16.1%) than the other species ($p < 0.001$).

3.3 Amino acid composition

In order to save space, results for the amino acid profiles presented in the Table 3 involve some selected organisms only.

In seaweeds, the highest concentration of glutamic acid (16.3% of total amino acids) was found in *S. filipendula*, while *A. taxiformis* had the lowest (10.3%). Aspartic acid was the second most abundant AA, varying from 9.59% (*A. taxiformis*) to 12.7% (*C. aerea*). Histidine was the less abundant amino acid in all species, and only *S. filipendula* achieved values close to 2%. *A. taxiformis* showed higher concentrations of valine, phenylalanine and arginine than the other seaweeds. Peaks of tyrosine were observed in *C. clavulatum*, with values close to 5%. Percentages of leucine and threonine were similar among all seaweeds.

In spermatophytes, the highest concentration of glutamic acid (12.2% of total amino acids) was found in the seagrass *R. maritima*, while the seagrass *H. wrightii* had the lowest (9.26%) concentrations. Aspartic acid and leucine were other abundant amino acids. Histidine and tyrosine were the less abundant amino acids among the plant species. The concentration of proline was also high in the seagrass *H. wrightii* (12.8%)

The amino acid profiles were remarkably different among invertebrates. For the anemone *B. caissarum*, the octopus *E. massyae* and the starfish *E. brasiliensis* the major amino acid was arginine, with values ranging from 13.3% to 20.6%. For the sea urchin *E. lucunter*, the sponge *D. anchorata* and the sea slug *A. brasiliana* the most abundant AA was glycine, with values ranging from 11.9% to 17.9%. Glutamic acid was the second most abundant amino acid in all species and histidine and tyrosine were the less abundant.

Fish species showed similar amino acid profiles, with glutamic acid as the most abundant amino acid, which fluctuate from 12.5% (*M. argentinae*) and 14.8% (*R. porosus*). Fishes were also rich in lysine, with values fluctuating from 9.63% (*D. volitans*) to 10.7% (*R. porosus*). The percentage of histidine was the lowest in all species, with an overall average value of 2.12%. *D. volitans* showed the higher concentration of arginine (11.0%) and the percentages of the other amino acids were typically similar among all species.

3.4 Total protein and nitrogen-to-protein conversion factors

The total protein content of the samples is showed in Table 4 as the sum of the total amino acid residues. Values of total protein were higher than hydrosoluble protein for most of the species analyzed (all animal samples), except for two seaweeds and five spermatophytes (Tables 2 and 4). The seaweeds showed wide variations in total protein concentrations, varying from 8.62% (*S. filipendula*) to 25.1% (*A. taxiformis*) of the d.w. Peaks of total protein in seaweeds were found in rhodophytes. For spermatophytes, concentrations of total protein varied from 5.56% (*L. racemosa*) to 19.2% (*R. maritima*).

The animals presented higher concentrations of total protein. For invertebrates, the values ranged from 19.4 % (*D. anchorata*, sponge) to 66.7% (*E. massyae*, octopus). Fishes showed higher total protein concentrations than invertebrates, with values varying from 66.2% (*M. argentinae*) to 81.5% (*G. brasiliensis*) of the d.w.

Amino acids	C. clavulatum	R. maritima	A. schaueriana	S alterniflora	E. massyae	E. brasiliensis	M. argentinae	R. lalandii
Aspartic acid	11.1 ± 0.49	12.8 ± 1.17	8.62 ± 0.13	9.68 ± 0.15	6.27 ± 1.02	7.97 ± 0.49	9.20 ± 0.45	8.96 ± 0.12
Threonine	5.24 ± 0.11	4.95 ± 0.14	5.35 ± 0.04	5.07 ± 0.17	4.39 ± 0.51	4.79 ± 0.37	4.44 ± 0.13	4.72 ± 0.05
Serine	5.15 ± 0.06	4.91 ± 0.36	4.48 ± 0.19	5.03 ± 0.30	6.04 ± 1.33	5.20 ± 0.95	4.32 ± 0.08	4.59 ± 0.05
Glutamic acid	11.8 ± 0.39	12.2 ± 0.73	9.78 ± 0.14	11.1 ± 0.12	9.34 ± 0.97	10.2 ± 0.34	12.5 ± 0.17	13.3 ± 0.27
Proline	4.95 ± 0.18	5.93 ± 0.33	5.40 ± 0.09	5.00 ± 0.27	5.54 ± 0.44	4.20 ± 0.50	3.63 ± 0.08	3.54 ± 0.03
Glycine	5.18 ± 0.09	6.60 ± 1.75	8.93 ± 0.27	6.42 ± 1.52	7.59 ± 0.10	5.40 ± 1.98	4.77 ± 0.25	5.14 ± 0.16
Alanine	6.76 ± 0.14	6.16 ± 0.16	6.30 ± 0.03	6.74 ± 0.92	5.55 ± 0.33	4.41 ± 0.13	6.14 ± 0.02	5.69 ± 0.07
Valine	6.18 ± 0.17	6.00 ± 0.11	6.84 ± 0.03	6.56 ± 0.32	5.11 ± 0.18	5.58 ±0.09	5.68 ± 0.08	5.50 ± 0.13
Isoleucine	5.46 ± 0.14	5.10 ± 0.41	5.50 ± 0.02	5.33 ± 0.16	5.74 ± 0.26	4.90 ± 0.15	5.41 ± 0.10	5.76 ± 0.09
Leucine	7.39 ± 0.18	8.87 ± 0.12	9.84 ± 0.05	9.33 ± 0.34	8.77 ± 0.35	7.70 ± 0.15	8.74 ± 0.15	8.78 ± 0.13
Tyrosine	4.96 ± 0.14	3.69 ± 0.58	3.77 ± 0.19	3.80 ± 0.58	3.12 ± 0.18	4.10 ± 0.35	3.53 ± 0.12	2.45 ± 0.08
Phenylalanine	5.02 ± 0.09	6.13 ± 0.48	6.09 ± 0.12	6.04 ± 0.11	5.09 ± 0.47	5.38 ± 0.41	5.12 ± 0.06	5.73 ± 0.01
Histidine	1.84 ± 0.17	1.90 ± 0.26	1.94 ± 0.03	1.73 ± 1.06	1.05 ± 1.00	1.62 ± 1.40	2.48 ± 0.07	2.47 ± 0.01
Lysine	6.80 ± 0.10	6.15 ± 0.16	5.45 ± 0.30	6.22 ± 0.94	8.36 ± 0.35	7.68 ± 0.63	10.4 ± 0.15	10.6 ± 0.14
Arginine	7.31 ± 0.09	6.28 ± 0.15	7.17 ± 0.20	6.66 ± 0.17	17.9 ± 4.68	20.6 ± 3.27	10.2 ± 0.08	8.16 ± 0.05
Total	95.1 ± 2.52	97.6 ± 6.93	95.5 ± 1.81	94.7 ± 7.12	99.9 ± 12.2	99.8 ± 11.2	96.6 ± 1.99	95.4 ± 1.39

Table 3. Total amino acid composition of eight organisms from coastal environments of Rio de Janeiro State, Brazil. Results are expressed as grams of amino acid measured in 100 g of protein and represent the actual recovery of amino acids after acid hydrolysis. Values are the mean of three replicates ± SD (n = 3).

The Table 4 also presents the nitrogen mass within total amino acid (Amino acid–N), the percentage of N present in protein (protein-N) and the N-Prot factors calculated for each species. The relative percentage of protein nitrogen was estimated as the ratio of nitrogen recovered from amino acid (Table 4) to total nitrogen (Table 2). From the ratio of the mass of amino acid residues to total nitrogen specific N-Prot factors were calculated.

Nitrogen mass within total amino acid in seaweeds ranged from 1.36% (*S. filipendula*) to 4.14% (*A. taxiformis*). Protein nitrogen ranged from 69.5% (*C. clavulatum*) to 78% (*S. filipendula*) and the red algae tended to show higher percentages of non-proteinaceus nitrogen (NPN). The N-Prot factors calculated for seaweeds ranged between 4.51 (*A. taxiformis*) to 4.98 (*C. clavulatum*). An overall average N-Prot factor of 4.78 was calculated for all seaweed species.

For spermatophytes, amino acid-N ranged from 0.92% (*L. racemosa*) to 3.08% (*R. maritima*). The percentage of protein-N varied widely among the species, from 63.8% (*T. domingensis*) to 97% (*A. aureum*). The species that presented more protein-N showed the highest N-Prot factors; conversely, the species with lower protein-N showed the smaller values for N-Prot factors. The N-Prot factors ranged from 3.97 (*T. domingensis*) to 6.00 (*A. aureum*). An overall average N-Prot factor of 4.82 was calculated from the data for all spermatophytes.

Groups/species	Amino acid residues	Amino acid-N	Protein-N	N-Prot factor
Seaweeds				
Asparagopsis taxiformis	25.1 ± 2.07	4.14 ± 0.34	74.5 ± 6.14	4.51 ± 0.37
Centroceras clavulatum	23.0 ± 0.68	3.22 ± 0.09	69.5 ± 2.04	4.98 ± 0.15
Chaetomorpha aerea	12.01 ± 0.82	1.94 ± 0.13	75.8 ± 5.20	4.69 ± 0.32
Sargassum filipendula	8.62 ± 0.33	1.36 ± 0.05	78.0 ± 2.96	4.93 ± 0.19
				Average: 4.78
Spermatophytes				
Halodule wrightii	12.8 ± 0.36	2.13 ± 0.06	78.2 ± 2.19	4.69 ± 0.13
Ruppia maritima	19.2 ± 3.08	3.08 ± 0.22	80.4 ± 5.73	5.02 ± 0.36
Acrostichum aureum	10.4 ± 0.98	1.68 ± 0.16	97.0 ± 9.15	6.00 ± 0.57
Blutaparon portulacoides	8.74 ± 0.73	1.37 ± 0.11	78.5 ± 6.56	5.01 ± 0.42
Typha domingensis	9.96 ± 1.34	1.60 ± 0.21	63.8 ± 8.56	3.97 ± 0.53
Avicennia schaueriana	8.08 ± 0.17	1.34 ± 0.03	68.9 ± 1.46	4.16 ± 0.09
Laguncularia racemosa	5.56 ± 0.34	0.92 ± 0.06	93.0 ± 5.75	5.62 ± 0.35
Spartina alterniflora	7.97 ± 0.73	1.29 ± 012	66.7 ± 6.13	4.10 ± 0.38
				Average: 4.82
Invertebrates				
Desmapsamma anchorata	19.4 ± 1.10	3.38 ± 0.19	89.2 ± 5.05	5.10 ± 0.29
Bunodosoma caissarum	52.6 ± 8.54	9.51 ± 1.55	98.0 ± 15.9	5.41 ± 0.88
Aplysia brasiliana	42.2 ± 5.61	7.30 ± 0.97	96.9 ± 12.9	5.61 ± 0.75
Eledone massyae	66.7 ± 10.4	12.7 ± 1.97	99.7 ± 15.5	5.25 ± 0.82
Echinaster brasiliensis	39.7 ± 5.75	7.49 ± 1.08	98.3 ± 14.2	5.21 ± 0.75
Echinometra lucunter	29.6 ± 2.42	5.37 ± 0.44	98.9 ± 8.08	5.46 ± 0.45
				Average: 5.34
Fishes				
Dactylopterus volitans	70.2 ± 7.82	11.9 ± 1.33	101.4 ± 11.3	5.98 ± 0.67
Genypterus brasiliensis	81.5 ± 1.70	13.9 ± 0.29	99.3 ± 2.08	5.80 ± 0.12
Mullus argentinae	66.2 ± 1.26	11.4 ± 0.22	98.5 ± 1.87	5.69 ± 0.11
Rhizoprionodon lalandii	79.3 ± 1.06	13.4 ± 0.18	92.9 ± 1.24	5.50 ± 0.07
Rhizoprionodon porosus	80.5 ± 9.69	13.6 ± 1.64	91.4 ± 11.0	5.39 ± 0.65
				Average: 5.67

Table 4. Calculation of nitrogen-to-protein conversion factors for twenty three marine organisms based on the amino acid residues to total nitrogen ratio. Values are expressed as percentage of the dry weight. Results represent the mean of three replicates ± SD (n = 3).

Nitrogen mass within total amino acid in invertebrates ranged from 3.38% (*D. anchorata*) to 12.7% (*E. massyae*). High values of protein nitrogen were estimated in invertebrates, ranged from 89.2% (*D. anchoratta*) to 99.7% (*E. massyae*). The N-Prot factors registered for invertebrates ranged from 5.10 to 5.61, with an overall average N-Prot factor of 5.34.

N mass in total amino acid of fishes ranged from 11.4% (*M. argentinae*) to 13.9% (*G. brasiliensis*). High values of protein-N were recovered, ranging from 91.4% (*R. porosus*) to 101.4% (*D. volitans*). The elasmobranch fishes (*R. porosus* and *R. lalandii*) showed the higher percentages of non proteinaceus nitrogen (NPN) (average 7.85%). The N-Prot factors ranged between 5.39 (*R. porosus*) to 5.98 (*D. volitans*). The elasmobranchs exhibited the lowest values of N-Prot factors. An overall average N-Prot factor of 5.67 was calculated for all fish species.

4. Discussion

The chemical composition of marine organisms in general may be influenced by a number of factors such as physiological characteristics, habitat and life cycle, and environmental conditions. The chemical composition of algae and plants is directly influenced by environmental conditions of the site where they live (Kamer et al., 2004; Larcher, 2000), especially the availability of inorganic nutrients and light. On the other hand, the abiotic conditions seem to be less influential on animals: the chemical composition of invertebrates and vertebrates is more influenced by biological factors, such as diet, stage of life and reproductive cycle (Ogawa & Maia, 1999).

4.1 Nitrogen and phosphorus

Wide differences in N and P tissue concentrations of seaweeds are related to taxonomic traits and species-specific capacities of taking up dissolved nutrients (Martínez-Aragón et al., 2002). Red algae tend to show higher N tissue concentrations than green and brown algae (Lourenço et al., 2002). Red algae contain phycoerithrin, a N-rich pigment that increases their nitrogen budget (Lobban & Harrison, 1994), and the two red algae tested here are fast-growing species, which account for a higher N content in comparison to other species. Conversely, *S. filipendula* has a complex thallus and low rate of growth, showing a typically low content of N (Hwang et al., 2004). A similar trend can be extended to *Chaetomorpha aerea*, an alga that dwells a sandy substrate and it is partially burred.

Low concentrations of leaf nitrogen in spermatophytes (overall average of 2.18%) can be considered as a general characteristic of plants, since several studies provided leaf N values similar to those reported in this study. According to Larcher (2000) leaves of herbaceous plants contain 2-4% in average N, while leaves of deciduous plants contain 1.5 to 3% N and leaves of sclerophyllous plants contain 1 to 2%N. Leaves of *T. domingensis* reached one of the largest concentrations of N among all plants tested. According to Lorenzen et al. (2001), *Typha* species are typically distributed in habitat rich in nutrients, which may explain the high levels of N. The mangrove plants had low leaf N concentrations (and also low concentrations of carbohydrates, protein and total phosphorus). According to Erickson et al. (2004) leaves of *Rhizophora mangle*, *Avicennia germinans* and *Laguncularia racemosa* have N concentrations of 1.2, 1.8 and 1.0% (d.w.) respectively. Ellis et al. (2006) reported N percentages ranging from 0.75 to 1.25% in sheets of *L. racemosa*. Low percentages of N in mangrove plants may be related to the high concentration of salt in the leaves, since these plants accumulate salt in the leaves and then eliminate them (Hogarth, 1999). In addition, mangrove plants also have high concentration of tannin, polyphenols of plant origin that inhibit herbivory of the leaves. Erickson et al. (2004) estimated that tannin is an average of 20% of the d.w. of leaves of mangrove plants.

The high concentration of nitrogen observed in the animals is related with the abundance of protein in the tissues, its main component, and by the presence of OTMA (oxyde of trimethylamine), a nitrogenous substance widely found in marine animals (Ogawa & Maia, 1999). The fishes and the octopus *E. massyae* showed N concentrations significantly higher than the other animals, which may be related to the part of the body used in the analysis. We analyzed only the muscles of fishes, a procedure that reflects in the high protein concentration in the samples. In octopus, the mantle was the organ considered, which is predominantly composed of muscle (Brusca & Brusca, 2003). Total N described in the muscles of fish, according to Ogawa & Maia (1999) and Puwastien et al. (1999), varies circa

15% of the d.w. The highest concentrations of N were observed in the cartilaginous fish *R. lalandii* and *R. porosus*, which may be related to higher concentration of OTMA in elasmobranchs, where the substance associated with urea in the control of osmotic pressure (Ogawa & Maia, 1999). Dantas & Attayde (2007) and Sterner & George (2000) also found high values of TN in freshwater fishes, with values fluctuating from 9.5% to 10.35% of d.w. The peak in P concentration of seaweeds was recorded in *C. clavulatum*, a fast-growing red algae. *S. filipendula* showed the lowest P concentrations in its thallus, and this trend can also be interpreted as a consequence of its low growth rates. *Chaetomorpha aerea* showed the second highest P concentration. The occurrence of high tissue concentrations of phosphorus was recorded in three seaweeds that also live partially burred in sediments (*Chaetomorpha crassa*, *Gracilaria cervicornis* and *Gracilariopsis tenuifrons*) in a seasonal study in Araruama Lagoon, a hypersaline coastal environment of Brazil (Lourenço et al., 2005).

In spermatophytes, the highest concentrations of total P were found in seagrasses. These fast-growing plants experience a greater contact with dissolved nutrients, since they take up phosphorus from both soil and water. According to Reich & Oleksyn (2004), herbs have higher concentrations of N and P than trees and shrubs. This observation may explain the higher contents of N and P in seagrasses, as well as in the salt marsh plant *S. alterniflora*.

Analysis of the phosphorus content in animals may be useful to interpret their metabolic speed. The highest concentration of P was recorded in fish and these high concentrations could be related to the characteristics of the tissues. Muscles of vertebrates store high potential phosphoryls in the form of creatinine-phosphate, which can quickly transfer its phosphate group to the ATP (Zeleznikar et al., 1995). Ogawa & Maia (1999) state that animals that move fast spend more energy and use more ATP; this requires a greater supply of phosphorus in the muscles than animals of limited movements. In invertebrates, the wider variations in P content reflect their different capacities of locomotion. The low concentration of P in the sponge *D. anchorata* may be related to a possible lower energy demand for (non-existing) locomotion, contrasting with the peak of phosphorus in the octopus *E. massayae*.

4.2 Hydrosoluble protein, carbohydrate and lipid

The availability of N is considered the most important factor influencing the levels of proteins in algae. N-poor environments tend to influence algal species to show low concentrations of protein, and the opposite is found in environments with high availability of N (Lobban & Harrison, 1994; Lourenço et al., 2005, 2006). Measurements ranging from 3% to more than 60%of the d.w. are documented in the literature. Methodological problems also contribute largely to enhance the differences among the results. Differences in cell wall composition and procedures used to extract proteins generate remarkable influence on the final results (Barbarino & Lourenço, 2005; Fleurence, 1999).

Protein content of macroalgae from tropical and subtropical coastal environments frequently show low concentrations (Kaehler & Kennish, 1996; Wong & Cheung, 2000). Our data indicate low protein concentrations in the algae, which agrees with the predominantly oligotrophic condition of the Brazilian coastal waters (Lourenço et al., 2005, 2006). In general, the protein content of brown seaweeds is low (3-15% of d.w.) compared to green and red seaweeds (10-47% of d.w.) (Fleurence, 1999; Rúperez & Saura-Calixto, 2001), despite differences in species composition and seasonal periods (Fleurence, 1999). The level of hydrosoluble protein recorded in *S. filipendula* agrees with results found for the protein content of *Sargassum* species (Lourenço et al., 2002; McDermid & Stuercke, 2003) and with

the low TN content found in its thallus (Table 2). On the other hand, the red algae showed
both protein content and TN concentrations higher than the other species tested here.
In general, plant leaves have lower concentrations of protein than macroalgae. Yeoh & Wee
(1994) analyzed the protein content of many angiosperms, and observed an average of 11%
protein in their leaves. The value of protein described for the grass *Spartina alterniflora* is in
agreement with the values reported by Yeoh & Watson (1982), about 7.5% dry weight.
Seagrasses had the highest concentrations of TN protein in the leaves (average of 16%). The
opposite can be observed for mangrove and marsh plants, in which low concentrations of
foliar N agree with low concentrations of leaf proteins.
Hydrosoluble protein in animal tissues was higher than in primary producers, indicating
the protein as the main organic compound in the heterotrophic species tested, especially in
fishes and in the octopus *E. massyae*. According to Rosa et al. (2005) the levels of soluble
protein in cephalopods range from 50 to 75% of the d.w. of muscle. Concentrations of
protein in fishes were similar to the results reported by Simões et al. (2007) and Zaboukas et
al. (2006), who found values ca. 75% of d.w. The lowest protein concentration was found in
M. argentinae, which can be related to the high lipid content of this species. Ogawa & Maia
(1999) state that there is a relationship between lipid and protein: if a large content of lipid
are accumulates in fishes, the percentage of protein tend to decrease.
Carbohydrates are the most abundant substances in most photosynthetic organisms, since
they occur in cell wall (e.g. cellulose, agar) and as storage products (ex.: starch, laminaran).
In seaweeds, brown algae tend to show lower carbohydrate concentrations than other algal
groups (Dawczynski et al., 2007) and the presence of less reactive carbohydrates may
generate underestimates of total carbohydrate (Lobban & Harrison, 1994). This might
contribute to increase differences in comparison to both green and red algae. Kumari et al.
(2010) investigated the carbohydrate contents in eighteen species of tropical seaweeds, and
found contents ranging from 15 to 43; the value reported for *Chaetomorpha* spp. (30%) was
very similar to our data for *C. aerea* (29.4%). The low concentration of carbohydrates in the
mangrove plant *L. racemosa*, less than 10%, may be related to high salt concentration in the
leaves. Larcher (2000) states that plants that inhabit saline soils have high values of ash (up
to 55% of the d.w.), and Na, Mg, Cl and S in quantities far above average.
All animal species analyzed showed low content of carbohydrate, especially the fishes and
the octopus *E. massyae*. According to Ogawa & Maia (1999) the presence of glycogen in the
fish muscles is low, varying from 1.5 to 5% of carbohydrate (d.w.). There is a remarkable
lack of data in the literature on carbohydrate in fishes, probably due the small importance of
these substances for the nutritional value of fishes. The majority of the chemical studies with
fish focus on the composition of protein and lipid (e.g. Puwastien et al., 1999; Simões et al.,
2007). Invertebrates had more carbohydrate than fishes. The high levels of carbohydrate in
Aplysia may be related to the secretion of mucopolysaccharides, characteristic of the subclass
members Opisthobranchia (Brusca & Brusca, 2003).
The fat concentration of seaweeds tended to be a little lower than in spermatophytes. The fat
content of seaweeds typically accounts for 1-6% of the d.w. (Fleurence et al., 1994; Gressler
et al., 2010). The metabolism of benthic seaweeds involves the production of large amounts
of carbohydrates as storage products (Lobban & Harrison, 1994). The production of lipid is
greater in planktonic algal species, in which they contribute for floating mechanisms.
Altogether, seaweed species were low in fat and high in carbohydrate. Both lipid and
carbohydrate contents recorded here agree with previous studies (e.g. Dawczynski et al.,
2007; McDermid & Stuercke, 2003; Rúperez & Saura-Calixto, 2001). These studies reported

crude lipid values in most seaweed predominantly lower than 5% of d.w. In our study, all seaweeds showed less than 5.5% of lipids.

Similarly to algae, spermatophytes stock carbohydrates in their tissues, especially polysaccharides. Fats and oils are important forms of carbon storage in seeds and fruits (Taiz & Zeiger, 1998), but the present study focus on leaves only. The high concentration of carbohydrates, coupled with the trend that most plants stock lipids mainly in other organs, explain the low concentrations of lipids in leaves, an overall average of 5.21%.

The concentrations of lipid were higher than the carbohydrate for all animal species, with exception of *A. brasiliana*. This trend results of the fact that animals store energy as fat content, converting the excess of sugar in fatty tissues. The highest concentration of lipids among all animals tested was found in the starfish *E. brasiliensis*. Moreover, Mathew et al. (1999), Özogul & Özogul (2007), and Zaboukas et al. (2006) point that the percentages of lipid in animals are more influenced by the environmental conditions, feeding and physiological traits than the percentages of carbohydrate and protein. Mathew et al. (1999) and Ramos Filho et al. (2008) affirm that the concentrations of lipid in fish muscles vary from 0.3 to 20% (fresh weight). In our study, the average value for lipid content for the five fishes was 8.37% (d.w.) and *M. argentinae* showed a high lipid concentration (16.1%, d.w.). The lowest concentrations of lipid were recorded in elasmobranch fishes (*R. lalandii* and *R. porosus*), fishes that store fat mainly in the liver (Ogawa & Maia, 1999). Among the invertebrates, the octopus *E. massyae* had the lowest concentration of lipids. Cephalopods stock lipids in the digestive tract and typically have low concentrations of lipids in muscles (ca. 5% d.w.) (Rosa et al., 2005). McClintock et al. (1990) also recorded high concentrations of lipids in the starfish *Ophidiaster alexandri*.

4.3 Amino acids and total protein

All organisms analyzed in this study had high concentrations of glutamic acid and low concentrations of histidine. The high concentrations of glutamic acid tend to occur because this AA is the precursor of the synthesis of all other AAs. The complexity of the biosynthesis of histidine usually leads to lower concentrations than non-essential AAs in plants, as reported by Noctor et al. (2002).

In general, algae and plants showed a similar pattern in the composition of AAs. Algae and plants showed high concentrations of aspartic acid, leucine, and glutamic acid and low levels of histidine and tyrosine. The main findings of amino acid composition of algal proteins described here are in agreement with previous studies (Dawczynski et al., 2007; Fleurence, 1999; Lourenço et al., 2002; Ramos et al., 2000; Wong & Cheung, 2000). In general, all species are rich in the acidic amino acids, glutamic and aspartic acid and poor in histidine. Aspartic and glutamic acid constituted a substantial amount of total AA of seaweeds, ranging from 19.9% (*A. taxiformis*) to 27.6% (*S. filipendula*). These two amino acids contribute to the flavour-related properties characteristic of the marine products and are responsible for the special taste of seaweeds. The concentration of these two AA was higher in the brown alga *S. filipendula* than in red algae, as previously described by Dawczynski et al. (2007). The level of glutamic and aspartic acid together can represent up 26% and 32% of the total AA of the green algae *Ulva rigida* and *U. rotundata* (Fleurence et al, 1995). Lourenço et al. (2002) showed that values for aspartic and glutamic acid together varied from 20.8 to 31.1% in 19 species of seaweeds. The highest value of lysine was observed in *C. aerea* (7.33%), in contrast to Ramos et al. (2000) who found in red algae higher values for lysine.

The AAs found in higher concentrations in the leaves (glutamic acid and aspartic acid) together account for about 20% of total amino acids. Yeoh & Watson (1982) reported that values of glutamic acid plus aspartic acid represent about 24% of the total content of AAs in leaves of grasses. The plant species analyzed had amino acid composition very similar. However, it is worth noticing the high concentration of proline recorded in leaves of *H. wrightii* and *B. portulacoides*. This may be related to the physiological adaptations to inhabit salt sites. According to Little (2000), some halophyte plants maintain their cytoplasm in the same osmotic pressure as the cell vacuole, accumulating free AAs such as proline. Martinelli et al. (2007) indicate that this feature in *Sporobolus stapfianus*, a halophilic grass, which accumulate AAs in vacuoles, especially proline, for adaptation to water stress.

The amino acid profiles of animals were more heterogeneous than that observed in algae and plants, since significant differences were observed in the concentrations of some AAs as glycine, lysine and arginine. The main findings of AA composition of fish proteins described here are in agreement with previous studies (Ogawa & Maia, 1999; Uhe et al., 1992). In general, all fish species are rich in glutamic acid and lysine and poor in histidine. The sum of the most representative AAs (lysine and glutamic acid) in fish, indicate that these two amino acids represent more than 20% of all AAs. The octopus *E. massyae* showed similarities with the chemical composition of fish. However, the analysis of AAs allowed to observe that the mollusk are very rich in arginine, totaling nearly 18% of AAs. The starfish *E. lucunter* also showed a higher content of lysine (8.31% of the total AA), while glycine achieved high concentration in the sponge *D. anchorata*. The remarkable differences in AA profiles of invertebrates possibly reflect phylogenetic traits (Dincer & Cakli, 2007; Rosa et al., 2005).

Total protein was estimated by the sum of amino acid residues, which represents the actual protein in each sample (Aitken et al., 1991; Lourenço et al., 2004; Mossé, 1990). A remarkable difference between the protein concentrations obtained with Lowry's method and the sum of AA-Res was observed for photosynthetic organisms. The protein concentration estimated with Lowry's method achieved only about 50% in red algae in comparison with protein concentration estimated with total AA-Res. The possible cause for these differences could be related to difficulties in extraction of protein from freeze-dried samples, as demonstrated by Barbarino & Lourenço (2005). Total amino acid analysis involves an acidic hydrolysis of the samples, which eliminates problems with protein extraction.

For three plant species (*H. wrightii*, *R. maritima* and *B. portulacoides*), the protein values calculated by the sum of AAs were higher than those recorded on Lowry et al. (1951). This trend could be faced as "normal", since not all protein can be extracted from the samples, especially in freeze-dried plant materials (Barbarino & Lourenço, 2005). However, for leaves of mangrove and salt marsh species, the wetland plants *A. aureum* and *T. domingensis*, and the seaweed *C. aerea* the Lowry's method yielded a higher protein content. Recording higher concentrations of protein by Lowry's method compared to the estimate of total protein by the sum of AAs is not coherent. The need of a tricky extraction of protein in Lowry's method reduces the chances of recovering 100% of the protein content in a sample. In addition, the high concentration of protein indicated by the method of Lowry et al. (1951) in these species would be unsustainable, given the low concentration of N in these organisms (Table 2). Thus, it is assumed the presence of some interference in the sample that artificially raised the quantification of proteins. According to Zaia et al. (1998) the method of Lowry is subject to many interferences, where the presence of phenolic compounds, sugars, uric acid and melamine react with Folin-Ciocalteu reagent resulting in false positive. This method also receives interference by some inorganic ions such as potassium and magnesium (Stoscheck, 1990). As mentioned above, leaves of mangrove plants have high concentrations of tannin, which may be interfering in protein analysis by the method of Lowry. Abundant literature data support the interpretation that the sum of AA-Res yields a closer estimate of the actual protein content of plants, including mangrove samples. Erickson et al. (2004) reported that *Rhizophora mangle*, *Avicennia germinans* and *Laguncularia racemosa* have low concentrations of total protein, with respective figures of 6.5%, 8.5% and 6.5%. These values were very similar to those reported in this study for total protein estimated by the sum of AAs.

Differences were also found for protein data in animals, with values protein obtained by the method of Lowry always lower (ca. 35%) than that obtained by the sum of AA-Res. This trend is coherent with the nature of animal samples, which contents of protein are easier to extract and react, compared with plant materials. In addition, animal samples typically have less chemical interferences than algal and plant sources (Zaia et al., 1998).

4.4 Nitrogen-to-protein conversion factors

There are different ways to calculate N-Prot conversion factors. Several studies calculate the N-Prot factors as the ratio between AA-Res and TN (Fujihara et al., 2008; Levey et al., 2000; Matilla et al., 2002), such as it has been done in the present study. Other studies determined conversion factors taking into account the proportion between AAs and the recovery of N from the amino acids (AA-N) (Mossé, 1990; Sosulski & Imafidon, 1990; Yeoh & Wee, 1994). The conversion factor calculated based in N-AAs are higher than the factor calculated using TN, thus the application of the conversion factor calculated only by AA-N can overestimate the actual protein content in the case of species with high NPN. This happens because the factor would multiply the TN content for calculating the percentage of protein, and TN does not distinguish between protein-N and NPN. For correct use of the conversion factor calculated with AA-N it is necessary to quantify protein-N from TN. However, this analysis is laborious and the use of this conversion factor would lose its practicability.

The total amino acid content represents not only amino acids derived from proteins, but also those AAs in free form. Free amino acids typically account for less than 7% of the total AA (Yeoh & Watson, 1982). The presence of free amino acids contributes to an overestimation of the total protein. However, according to Mossé (1990), the use of data of total amino acid, without determination of free amino acids, is a widely accepted procedure to estimate protein, since in acid hydrolysis some not-abundant amino acids are partially or totally destroyed (e.g. tryptophan, cystine, methionine and serine). The loss during acid hydrolysis might compensate for the influence of free amino acids in the quantification of protein by the sum of the total amino acid residues.

It can be seen that the conversion factors N-Prot calculated for algae and plants were similar and lower than the factors calculated for the animals. Sosulski & Imafidon (1990) and Tacon et al. (2009) compared the conversion factors in animal and vegetable products and found that the highest factors were calculated for animals. These observations can be explained mainly by the high concentration of NPN in photosynthetic organisms compared to the animals. The presence of photosynthetic pigments and the accumulation of inorganic nitrogen in the cells increase the relative importance of NNP in photosynthetic organisms. Lourenço et al. (2004) determined the concentration of NPN in 12 species of microalgae, with values ranging from 0.8% to 39% of the TN. Yeoh & Wee (1994) indicated that the NNP represents about 24% of the TN in leaves. On the other hand, the quantity of NPN in muscles can vary from 6 to 14% of the total N (Puwastein et al., 1999). NPN in animals is present in the constitution of several substances, such as OTMA, amines, guanidines, nucleotids and their degrading products such as urea and ammonia. Other non-proteinaceous substances with nitrogen that can be present are glycilbetaine, carnitine and homarine (Ogawa & Maia, 1999). The flavour of seafoods depends on the species, the fat content, and the presence and type of NPN compounds (Venugopal & Shahidi, 1996).

Despite the great scientific and practical importance of this issue, so far only Aitken et al. (1991) and Lourenço et al. (2002) established specific N-Prot conversion factors for seaweeds. The overall mean N-Prot factor calculated in this report for four seaweeds was 4.78. Lourenço et al. (2002) reported an average N-Prot factor of 4.92 for 19 seaweeds, with average specific factors for groups: 5.13 for green algae; 5.38 for brown algae and 4.59 for red algae. Aitken et al. (1991) proposed mean N-Prot factor of 5.0 obtained for two species of *Porphyra*. Red algae tend to show larger amounts of NPN (30.5% in *C. clavulatum* and 25.5 in *A. taxiformis*) than brown and green algae (22% and 24.2%, respectively). Current N-Prot factors are similar to those proposed by Lourenço et al. (2002) and Aitken et al. (1991).

Clear relationships were found between NPN and the conversion factors for spermatophytes. Plants with higher NPN had the lowest conversion factors (e.g. *T. domingensis*, *A. schaueriana* and *S. alterniflora*) and plants with less NPN had the highest conversion factors (e.g. *A. aureum* and *L. racemosa*). We observed a wide variation in the N-Prot factors calculated for plants, with an average of 4.82. Among the spermatophytes, we observed a wide variation in levels of NPN, with values ranging from 3% (*A. aureum*) to 36.2% (*T. domingensis*) of the TN, with an average of 21.7%. This variation of NPN may be related to the hypothetical presence of abundant N compounds not measured here (e.g. inorganic nitrogen, chlorophylls). Alkaloids and glycosides also contain N and take part of the complex secondary products of plants (Fowden, 1981). The N-Prot factors calculated in this study are consistent with those described in the literature for plant species. Several studies have shown that for plant tissues, the conversion factor N-Prot varies from 3.7 to 6.0 (Fujihara et al. 2008; Levey et al. 2000; Mossé, 1990; Tacon et al., 2009; Yeoh & Wee, 1994). Yeoh & Wee (1994) proposed the use of average factor of 4.43 as a reliable estimate for the determination of proteins in plant leaves.

The animals showed higher conversion factors to those reported in photosynthetic organisms, a reflex of reduced concentrations of NPN in these organisms. An overall average of 3.6% of NPN was found in invertebrate and fish samples. Conversion factors calculated in this study for animals were similar among the species, where the factors ranged from de 5.10 (*D. anchorata*) to 5.98 (*D. volitans*). The overall mean N-Prot factors calculated in this report for invertebrates and fishes were 5.34 and 5.67, respectively. Among fishes, elasmobranchs had smaller N-Prot factors, achieving 5.39 and 5.50, for *R. porosus* e *R. lalandii*, respectively. This is possibly related to the higher concentration of NPN than in the other species. In cartilaginous fish, a smaller concentration of protein-N was estimated, which may be related to the high concentration of OTMA; a high concentration of urea can be found in muscles of elasmobranchs, achieving up to 2% (Ogawa & Maia, 1999). Despite the general high concentrations of protein-N in animals, all conversion factors calculated by us were lower than the traditional factor 6.25.

For animals, there were minor differences between the use of the factor 6.25 and the sum of AA-Res for the calculation of total protein. The factor 6.25 overestimates the protein content in fish samples in ca. 9%. For elasmobranchs the overestimation is higher: 14%. For invertebrates, the 6.25 factor overestimate the protein content between 10 and 20%. Such differences are not neglectable, which means that the presence of NPN does affect the use of nitrogen-to-protein conversion factors in samples of marine animals.

Our results show clearly that the use of the traditional N-Prot factor of 6.25 overestimates the protein content of marine organisms. This statement can be illustrated by a simple hypothetical analysis, if someone estimates the content of protein of a given species of marine spermatophyte not studied here (e.g. a seagrass) from its N content. If the N percentage of this hypothetical plant is 2.0%, the use of the traditional N-Prot factor (6.25) would give an estimate of protein content of 12.5%. By the use of the specific N-Prot factors calculated by us (4.82), the estimate of the protein content would be 9.64%.

In addition, specific N-Prot factors calculated here would receive minor influences of environmental conditions, despite possible changes in internal nitrogen budget. As the N-Prot factors result from the ratio of protein-N to total N, organisms in N-rich environments tend to accumulate more nitrogen and also more protein-N. In environments with scarce availability of nitrogen species would show lower contents of both total N and protein-N (Lourenço et al., 2002, 2005). These trends point for the wide use of specific N-Prot factors, which could be applied to species in different environmental conditions without restrictions.

5. Conclusions

This study shows wide variations in the gross chemical composition of organisms from coastal subtropical environments of Brazil. Seaweeds and spermatophytes showed higher concentrations of carbohydrate, but animals tended to show higher concentrations of protein, lipid, total nitrogen and total phosphorus.

Acidic amino acids dominate the profile of seaweeds and spermatophytes. In animals, other amino acids also show high concentrations, such as lysine in echinoderms and some species of fish.

In animals, more than 89% of the nitrogen is found in protein. Photosynthetic organisms show high and variable percentages of non-protein nitrogen, which may achieve more than 35% of the total nitrogen budget.

We recommend the use of the specific nitrogen-to-protein conversion factors calculated for each species studied here. For organisms not assessed in this study, we recommend the use of overall N-Prot factors estimated for the corresponding taxonomic groups (e.g. spermatophytes, fishes). The traditional factor 6.25 should be avoided, since it overestimates the protein concentration of marine organisms. The use of the N-Prot conversion factors calculated would yield more accurate protein analysis of marine organisms, contributing for better protein analysis in marine science.

6. Acknowledgements

Authors are indebted to Brazil's National Council for Scientific and Technological Development (CNPq) and Research Support Foundation of Rio de Janeiro State (FAPERJ) for the financial support of this study. GSD thanks Coordination of Improvement of Higher Education Personnel (CAPES) for her scholarship. Authors thank Dr. Renato Crespo Pereira (UFF) for the use of laboratory facilities and to Dr. Joel C. De-Paula (UNIRIO) for confirming the identification of the seaweeds. Thanks are due to Dr. João Oiano Neto and M.Sc. Sidney Pacheco (EMBRAPA) for running the amino acid analysis.

7. References

Aitken, K.A.; Melton, L.D. & Brown, M.T. (1991). Seasonal Protein Variation in the New Zealand Seaweeds *Porphyra columbina* Mont. and *Porphyra subtumens* J. Ag. (Rhodophyceae). *Japanese Journal of Phycology*, Vol.39, No.4, (December 1991), pp. 307-317, ISSN 0038-1578.

Amsler, C.D. (ed.) (2009). *Algal Chemical Ecology*. Springer-Verlag, ISBN 978-3-540-92998-7, Berlin, Germany.

Barbarino, E. & Lourenço, S.O. (2005). An Evaluation of Methods for Extraction and Quantification of Protein from Macro- and Microalgae. *Journal of Applied Phycology*, Vol.17, No.5, (October 2005), pp. 447-460, ISSN 0921-8971.

Barbarino, E. & Lourenço. S.O. (2009). Comparison of CHN Analysis and Hach Acid Digestion to Quantify Total Nitrogen in Marine Organisms. *Limnology and Oceanography: Methods*, Vol.7, No.11, (November 2009), pp. 751-760, ISSN 1541-5856.

Bradford, M. (1976). A Rapid and Sensitive Method for the Quantitation of Microgram Quantities of Protein Utilizing the Principle of Protein Dye-Binding. *Analytical Biochemistry*, Vol.72, No.1-2, (May 1976), pp. 248-254, ISSN 0003-2697.

Brusca, R.C. & Brusca, G.J. (2003). *Invertebrates*. 2nd ed. Sinauer Associates, ISBN 0-87893-097-3, Sunderland, U.S.A.

Carvalho, H.F. & Recco-Pimentel, S.M. (eds.) (2007). *A Célula*. 2nd ed. Manole, ISBN 978-85-204-2543-5, Barueri, Brazil.

Cohen, S.A. & De Antonis, K.M. (1994). Applications of Amino Acid Derivatization with 6-Aminoquinolyl-N-Hhydroxysuccinimidyl Carbamate. Analysis of Feed Grains, Intravenous Solutions and Glycoproteins. *Journal of Chromatography A*, Vol.661, No.1-2, (February 1994), pp. 25-34, ISSN 0021-9673.

Conklin-Brittain, N.L.; Dierenfeld, E.S.; Wranghan, R.W.; Norconk, M. & Silver, S.C. (1999). Chemical Protein Analysis: a Comparasion of Kjeldhal Crude Protein and Total Ninhydrin Protein from Wild, Tropical Vegetation. *Journal of Chemical Ecology*, Vol.25, No.12, (December 1999), pp. 2601-2622, ISSN 0098-0331.

Dantas, M.C. & Attayde, J.L. (2007). Nitrogen and Phosphorus Content of Some Temperate and Tropical Freshwater Fishes. *Journal of Fish Biology*, Vol.70, No.1, (January 2007), pp. 100-108, ISSN 1095-8649.

Dawczynski, C.; Schubert, R. & Jahreis, G. (2007). Amino Acids, Fatty Acids, and Dietary Fiber in Edible Seaweed Products. *Food Chemistry*. Vol.103, No.3, (March 2007), pp. 891-899, ISSN 0308-8146.

Dincer, T. & Cakli, S. 2007. Chemical Composition and Biometrical Measurements of the Turkish Sea Urchin (*Paracentrotus lividus*, Lamarck, 1816). *Critical Reviews in Food Science and Nutrition*. Vol.47, No.1 (January 2007), pp. 21-26, ISSN 1040-8398.

Diniz, G.S.; Barbarino, E.; Oiano-Neto, J.; Pacheco, S. & Lourenço. S.O. (2011). Gross Chemical Profile and Calculation of Nitrogen-to-Protein Conversion Factors for Five Tropical Seaweeds. *American Journal of Plant Sciences*, in press, ISSN 2158-2750.

Duarte, C.M.; Augustí, S.; Berelson, W.; Gnanadesikan, A.; Regaudie-de-Gioux, S.A.; Sarmiento, J.L.; Simó, R. & Slater, R.D. (eds.) (2011). *The Role of the Marine Biota on the Functioning of the Biosphere*. Fundación BBVA, ISBN 978-84-92937-04-2. Bilbao, Spain.

Dubois, M.; Gilles, K.A.; Hamilton, J.K.; Rebers, P.A. & Smith, F. (1956). Colorimetric Method for Determination of Sugars and Related Substances. *Analytical Chemistry,* Vol.28, No.3, (March 1956), pp. 350-356, ISSN 003-2700.

Ellis, W.L.; Bowles, J.W.; Erickson, A.A.; Stafford, N.; Bell, S.S. & Thomas, M. (2006). Alteration of the Chemical Composition of Mangrove (*Laguncularia racemosa*) Leaf Litter Fall by Freeze Damage. *Estuarine, Coastal and Shelf Science,* Vol.68, No.1-2, (June 2006) pp. 363-371, ISSN 0272-7714.

Erickson, A.A.; Bella, S.S. & Dawes, C.J. (2004). Does Mangrove Leaf Chemistry Help Explain Crab Herbivory Patterns? *Biotropica,* Vol.36, No.3, (September 2004), pp. 333-343, ISSN 1744-7429.

Fleurence, J. (1999). Seaweeds Proteins: Biochemical, Nutritional Aspects and Potential Uses. *Trends Food Science Technology,* Vol.10, No.1, (January 1999), pp. 25-28, ISSN 0924-2244.

Fleurence, J.; Gutbier, G.; Mabeu, S. & Leray, C. (1994). Fatty Acids from 11 Marine Macroalgae of the French Brittany Coast. *Journal of Applied Phycology,* Vol.6, No.5-6, (December 1994), pp. 527-532, ISSN 0921-8971.

Fleurence, J.; Le Couer, C.; Mabeau, S.; Maurice, M. & Landrein, A. (1995). Comparison of Different Extractive Procedures for Proteins from the Edible Seaweeds *Ulva rigida* and *Ulva rotundata. Journal of Applied Phycology,* Vol.7, No.6, (December 1995), pp. 577-582, ISSN 0921-8971.

Folch, J.; Lees, M. & Sloanne-Stanley, G.H. (1957). A Simple Method for the Isolation and Purification of Total Lipid from Animal Tissue. *The Journal of Biological Chemistry,* Vol.226, No.1, (May 1957), pp. 497-509, ISSN 0021-9258.

Fowden, L. (1981). Non-Protein Amino Acids of Plants. *Food Chemistry,* Vol.6, No.3, (March 1981), pp. 201-211, ISSN 0308-8146.

Fujihara, S.; Sasaki, H.; Aoyagi, Y. & Sugahara, T. (2008). Nitrogen-to-Protein Conversion Factors for Some Cereal Products in Japan. *Journal of Food Science,* Vol.73, No.1, (February 2008), pp. 204-209, ISSN 0022-1147.

Gressler, V.; Yokoya, N.S.; Fujii, M.T.; Colepicolo, P.; Mancini Filho, J.; Torres R.P. & Pinto, E. (2010). Lipid, Fatty Acid, Protein, Amino Acid and Ash Contents in Four Brazilian Red Algae Species. *Food Chemistry,* Vol.120, No.2, (May 2010), pp. 585-590, ISSN 0308-8146.

Hedges, J., Peterson, M.L., & Wakeham, S.G. (2002). The Biochemical and Elemental Compositions of Marine Plankton: A NMR Perspective. *Marine Chemistry,* Vol.78, No.1, (April 2002), pp. 47-63, ISSN 0304-4203.

Hogarth, P.J. (1999). *The Biology of Mangroves.* Oxford University Press, ISBN 0-19-850222-2, Oxford, U.K.

Hwang, R.-L.; Tsai, C.-C. & Lee, T.-M. (2004). Assessment of Temperature and Nutrient Limitation on Seasonal Dynamics Among Species of *Sargassum* from a Coral Reef in Southern Taiwan. *Journal of Phycology,* Vol.40, No.3, (June 2004), pp. 463-473, ISSN 0022-3646.

Janecki, T. & Rakusa-Suszczewski, S. (2004). Chemical Composition of the Antarctic Starfish *Odontaster validus* (Koehler 1911) and Its Reactions to Glutamic and Kynurenic Acids. *Russian Journal of Marine Biology,* Vol.30, No.5, (September 2004) pp. 358–360, ISSN 1063-0740.

Jones, D.B (1931). Factors for Converting Percentages of Nitrogen in Foods and Feeds into
    Percentages of Protein. *United State Department of Agriculture Circular*, Vol. 183,
    (August 1931), pp. 1-21, ISSN 1052-5378.

Kaehler, S. & Kennish, R. (1996). Summer and Winter Comparisons in the Nutritional Value
    of Marine Macroalgae from Hong Kong. *Botanica Marina*, Vol.39, No.1, (January
    1996), pp. 11-17, ISSN 0006-8055.

Kamer, K.; Fong, P.; Kennison, R. & Schiff, K. (2004). The Relative Importance of Sediment
    and Water Column Supplies of Nutrients to the Growth and Tissue Nutrient
    Content of the Green Macroalga *Enteromorpha intestinalis* Along an Estuarine
    Resource Gradient. *Aquatic Ecology*, Vol.38, No.1, (March 2004), pp. 45-56, ISSN
    1386-2588.

Karl, D.; Michaelis, A.; Bergman, B.; Capone, D.; Carpenter, E.; Letelier, R.; Lipschultz, F.;
    Paerl, H.; Sigman, D. & Stal, L. (2002). Dinitrogen Fixation in the World's Oceans.
    Biogeochemistry, Vol.57/58, No.1, (April 2002), pp. 47-98, ISSN 1386-2588.

Kumari, P.; Kumar, M.; Gupta, V.; Reddy, C.R.K. & Jha, B. (2010). Tropical Marine
    Macroalgae as Potential Sources of Nutritionally Important PUFAs. *Food Chemistry*,
    Vol.120, No.3., (June 2010), pp. 749-757, ISSN 0308-8146.

Larcher, W. (2000). *Ecofisiologia Vegetal*. RiMa, ISBN 85-86552-03-8, São Carlos, Brazil.

Legler, G.; Müller-Platz, C.M.; Mentges-Hettkamp, M.; Pflieger, G. & Jülich, E. (1985). On the
    Chemical Basis of the Lowry Protein Determination. *Analytical Biochemistry*,
    Vol.150, No.2, (November 1985), pp. 278-287, ISSN 0003-2697.

Levey, D.J.; Bissell, H.A. & O'Keefe, S.F. (2000). Conversion of Nitrogen to Protein and
    Amino Acids in Wild Fruits. *Journal of Chemical Ecology*, Vol.26, No.7, (July 2000),
    pp. 1749-1763, ISSN 0098-0331.

Little, C. (2000). *The Biology of Shoft Shores and Estuaries*. Oxford University Press, ISBN 0-19-
    850426-8, Oxford, U.K.

Lobban, C. S. & Harrison, P.H. (1994). *Seaweed Ecology and Physiology*. Cambridge University
    Press, ISBN 0-521-40897-0, New York, USA.

Lorenzen, B.; Brix, H.; Mendelssohn, I.A.; McKee, K.L. & Miao, S.L. (2001). Growth, Biomass
    Allocation and Nutrition Use Efficiency in *Cladium jamaicense* and *Typha
    domingensis* as Affected by Phosphorus and Oxygen Availability. *Aquatic Botany*,
    Vol.70, No.2, (June 2001), pp. 117-133, ISSN 0304-3770.

Lourenço, S.O.; Barbarino, E.; De-Paula, J.C.; Pereira, L.O. da S. & Lanfer Marquez, U.M.
    (2002). Amino Acid Composition, Protein Content and Calculation of Nitrogen-to-
    Protein Conversion Factors for 19 Tropical Seaweeds. *Phycology Research*, Vol.50,
    No.3, (September 2002), pp. 233-241, ISSN 1322-0829.

Lourenço, S.O.; Barbarino, E.; Lavín, P.L.; Marquez, U.M.L. & Aidar, E. (2004). Distribution
    of Intracellular Nitrogen in Marine Microalgae. Calculation of New Nitrogen-to-
    Protein Conversion Factors. *European Journal of Phycology*, Vol.39, No.1, (February
    2004), pp. 17-32, ISSN 0967-0262.

Lourenço, S.O.; Barbarino, E.; Nascimento, A. & Paranhos, R. (2005). Seasonal Variations in
    Tissue Nitrogen and Phosphorus of Eight Macroalgae from a Tropical Hypersaline
    Coastal Environment. *Cryptogamie Algologie*, Vol.26, No.4, (November 2005), pp.
    355-371, ISSN 0181-1568.

Lourenço, S.O.; Barbarino, E.; Nascimento, A.; Freitas, J.N.P. & Diniz, G.S. (2006). Tissue Nitrogen and Phosphorus in Seaweeds in a Tropical Eutrophic Environment: What a Long-Term Study Tells Us. *Journal of Applied Phycology*, Vol.18, No.4, (October 2006), pp. 389-398, ISSN 0921-8971.

Lowry, O.H.; Rosebrough, N.J.; Farr, A.L. & Randall, R.L. (1951). Protein Measurement with the Folin Phenol Reagent. *The Journal of Biological Chemistry*, Vol.193, No.1, (May 1951), pp. 265-275, ISSN 0021-9258.

Mariotti, F.; Tomé, D. & Mirand, P.P. (2008). Converting Nitrogen into Protein-Beyond 6.25 and Jones' Factors. *Critical Reviews in Food Science and Nutrition*, Vol.48, No.2, (February 2008), pp. 177-184, ISSN 1040-8398.

Martinelli, T.; Whittaker, A.; Bochicchio, A.; Vazzana, C.; Suzuki, A. & Masclaux-Daubresse, C. (2007). Amino Acid Pattern and Glutamate Metabolism During Dehydration Stress in the "Resurrection" Plant *Sporobolus stapfianus*: A Comparison Between Desiccation-Sensitive and Desiccation-Tolerant Leaves. *Journal of Experimental Botany*, Vol.58, No.11, (August 2007), pp. 3037-3046, ISSN 0022-0957.

Martínez-Aragón, J.F.; Hernández, I.; Pérez-Looréns, J.L.; Vázquez, R. & Vergara, J.J. (2002). Biofiltering Efficiency in Removal of Dissolved Nutrients by Three Species of Estuarine Macroalgae Cultivated with Sea Bass (*Dicentrarchus labrax*) Waste Waters 1. Phosphate. *Journal of Applied Phycology*, Vol.14, No.5, (October 2002), pp. 365-374, ISSN 0921-8971.

Mathew, S.; Ammu, K.; Viswanathan Nair, P.G. & Devadasan, K. (1999). Cholesterol Content of Indian Fish and Shellfish. *Food Chemistry*, Vol.66, No.4, (September 1999), pp. 455-461, ISSN 0308-8146.

Mattila, P.; Salo-Väänänen, P.; Könkö, K.; Aro, H. & T. Jalava. (2002). Basic Composition and Amino Acid Contents of Mushrooms Cultivated in Finland. *Journal of Agricultural and Food Chemistry*, Vol.50, No.22, (October 2002), pp. 6419-6422, ISSN 0021-8561.

McClintock, J.B.; Cameron, J.L. & Young, C.M. (1990). Biochemical and Energetic Composition of Bathyal Echinoids and an Asteroid, Holothuroid and Crinoid from the Bahamas. *Marine Biology*, Vol.105, No.2, (June 1990), pp. 175-183, ISSN 0025-3162.

McDermid, K.J. & Stuercke, B. (2003). Nutritional Composition of Edible Hawaiian Seaweeds. *Journal of Applied Phycology*, Vol.15, No.6, (November 2003), pp. 513-524, ISSN 0921-8971.

Mossé, J. (1990). Nitrogen to Protein Conversion Factor for Ten Cereals and Six Legumes or Oilseeds. A Reappraisal of its Definition and Determination. Variation According to Species and to Seed Proteic content. *Journal of Agricultural and Food Chemistry*, Vol.38, No.1, (January 1990), pp. 18-24, ISSN 0021-8561.

Myklestad, S. & Haug, A. (1972). Production of Carbohydrates by the Marine Diatom *Chaetoceros affinis* var. *willei* (Gran) Hustedt. I. Effect of the Concentration of Nutrients in the Culture Medium. *Journal of Experimental Marine Biology and Ecology*, Vol.9, No.2, (August 1972), pp. 125-136, ISSN 0022-0981.

Noctor, G.; Novitskaya, L.; Lea, P.J. & Foyer, C.H. (2002). Co-Ordination of Leaf Minor Amino Acid Contents in Crop Species: Significance and Interpretation. *Journal of Experimental Botany*, Vol.53, No.370, (April 2002), pp. 939-945, ISSN 0022-0957.

Nurnadia, A.A.; Azrina, A. & Amin, I. (2011). Proximate Composition and Energetic Value of Selected Marine Fish and Shellfish from the West Coast of Peninsular Malaysia. *International Food Research Journal*, Vol.18, No.1, (January 2011), pp 137-148, ISSN 1985-4668.

Ogawa, M. & Maia, E.L. 1999. *Manual de Pesca – Ciência e Tecnologia do Pescado*. Editora Varela, ISBN 85-855-1944-4, São Paulo, Brazil.

Özogul, Y. & Özogul, F. (2007). Fatty Acid Profiles of Commercially Important Fish Species from the Mediterranean, Aegean and Black Seas. *Food Chemistry*, Vol.100, No.4, (December 2007), pp. 1634-1638, ISSN 0308-8146.

Puwastien, P.; Judprasong, K.; Kettwan, E.; Vasanachitt, K.; Nakngamanong, Y. & Bhattacharjee, L. (1999). Proximate Composition of Raw and Cooked Thai Freshwater and Marine Fish. *Journal of Food Composition and Analysis*, Vol.12, No.1, (March 1999), pp 9-16, ISSN 0889-1575.

Ramos Filho, M.M.; Ramos, M.I.L.; Hiane P.A. & Souza. E.M.T. (2008). Perfil Lipídico de Quatro Espécies de Peixes da Região Pantaneira de Mato Grosso do Sul. *Ciência e Tecnologia de Alimentos*, Vol.28, No.2, (April 2008), pp. 361-365, ISSN 0101-2061.

Ramos, M.V.; Monteiro, A.C.O.; Moreira, R.A. & Carvalho, A.F.F.U. (2000). Amino Acid Composition of Some Brazilian Seaweed Species. *Journal of Food Biochemistry*, Vol.24, No.1, (March 2000), pp. 33-39, ISSN 0145-8884.

Reich, P.B. & Oleksyn, J. (2004). Global Patterns of Plant Leaf N and P in Relation to Temperature and Latitude. *Proceedings of the National Academy of Sciences of the United States of America*, Vol.101, No.30, (June 2004), pp. 11001-11006, ISSN 0027-8424.

Rosa, R.; Pereira, J. & Nunes, M.L. (2005). Biochemical Composition of Cephalopods with Different Life Strategies, with Special Reference to Giant Squid, *Architeuthis* sp. *Marine Biology*, Vol. 146, No.4, (March 2005), pp. 739-751, ISSN 0025-3162.

Rúperez, P. & Saura-Calixto, F. (2001). Dietary Fibre and Physicochemical Properties of Edible Spanish Seaweeds. *European Food Research and Technology*, Vol.212, No.3, (February 2001), pp. 349-354, ISSN 1438-2377.

Salo-Väänänen, P.P. & Koivistoinen, P.E. (1996). Determination of Protein in Foods: Comparison of net Protein and Crude Protein (N x 6.25) Values. *Food Chemistry*, Vol.57, No.1, (September 1996), pp. 27-31, ISSN 0308-8146.

Simões, M.R.; Ribeiro, C.F.A.; Ribeiro, S.C.A.; Park K.J. & Murr, E.X. (2007). Composição Físico-Química, Microbiológica e Rendimento do Filé de Tilápia Tailandesa (*Oreochromis niloticus*). *Ciência e Tecnologia de Alimentos*, Vol.27, No.3, (July/September 2007), pp. 608-613, ISSN 0101-2061.

Sosulski, F.W. & Imafidon. G.I. (1990). Amino Acid Composition and Nitrogen-to-Protein Conversion Factors for Animal and Plant Foods. *Journal of Agricultural and Food Chemistry*, Vol.38, No.6, (June 1990), pp. 1351-1356, ISSN 0021-8561.

Sterner, R.W. & George, N.B. (2000). Carbon, Nitrogen and Phosphorus Stoichiometry of Cyprinid Fishes. *Ecology*, Vol.81, No.1, (January 2000), pp. 127-140, ISSN 0012-9658.

Stoscheck, C.M. (1990). Quantification of Protein. *Methods in Enzimology*, Vol.182, No.1, pp. 50-68, ISSN 0076-6879.

Tacon. A.G.J.; Metian, M. & Hasan, M.R. (2009). *Feed Ingredients and Fertilizers for Farmed Aquatic Animals. Sources and Composition*. FAO Fisheries and Aquaculture Technical Paper No. 540, ISBN 978-92-5-106421-4, Rome, Italy.

Taiz, L. & Zeiger, E. (1998). *Plant Physiology*. 2nd ed. Sinauer Associates, ISBN 0-87893-831-1, Sunderland , USA .

Uhe, A.M.; Collier, G.R. & O'Dea, K. 1992. A Comparison of the Effects of Beef, Chicken and Fish Protein on Satiety and Amino Acid Profiles in Lean Male Subjects. *The Journal of Nutrition*, Vol.122, No.3, (March 1992), pp. 467-472, ISSN 0022-3166.

Undeland, I.; Hall, G. & Lingnert, H. (1999). Lipid Oxidation in Fillets of Herring (*Clupea harengus*) During Ice Storage. *Journal of Agricultural and Food Chemistry*, Vol.47, No.2, (February 1999), pp. 524-532, ISSN 0021-8561.

Venugopal, V. & Shahidi. F. (1996). Structure and Composition of Fish Muscle. *Food Reviews International*, Vol.12, No.2, pp. 175-197, ISSN 8755-9129.

Wong, K.H. & Cheung, C.K. (2000). Nutritional Evaluation of Some Subtropical Red and Green Seaweeds. Part I – Proximate Composition, Amino Acid Profiles and Some Physico-Chemical Properties. *Food Chemistry*, Vol.71, No.1, (December 2000), pp. 475-482, ISSN 0308-8146.

Yeoh, H.H. & Truong, V.D. (1996). Amino Acid Composition and Nitrogen-to-Protein Conversion Factors for Sweet Potato. *Tropical Science*, Vol.36, No.4, (November 1996), pp. 243-246, ISSN 0041-3291.

Yeoh, H.H. & Watson, L. (1982). Taxonomic Variation in Total Leaf Protein Amino Acid Compositions of Grasses. *Phytochemistry*, Vol.21, No.3, (March 1982), pp. 615-626, ISSN 0031-9422.

Yeoh, H.H. & Wee, Y.C. (1994). Leaf Protein Contents and Nitrogen-to-Protein Conversion factors for 90 Plant Species. *Food Chemistry*, Vol.49, No.3, (February 1994), pp. 245-250, ISSN 0308-8146.

Zaboukas, N.; Miliou, H.; Megalofonou, P. & Moraitou-Apostolopoulou, M. (2006). Biochemical Composition of the Atlantic Bonito *Sarda sarda* from the Aegean Sea (Eastern Mediterranean Sea) in the Different Stages of Sexual Maturity. *Journal of Fish Biology*, Vol.69, No.2, (August 2006), pp. 347-362, ISSN 0022-1112.

Zaia, D.A.M.; Zaia, C.T.B.V. & Lichting, J. (1998). Determinação de Proteínas Totais Via Espectrofotometria: Vantagens e Desvantagens dos Métodos Existentes. *Química Nova*, Vol.21, No.6, (November/December 1998), pp. 787-793, ISSN 0100-4042.

Zar, J.H. (1996). *Biostatistical Analysis*. 3rd ed. Prentice Hall, ISBN 0130-84542-6, Upper Saddle River, U.S.A.

Zeleznikar, R.J.; Dzeja, P.P. & Goldberg, N.D. (1995). Adenylate Kinase-Catalyzed Phosphoryl Transfer Couples ATP Utilization with its Generation by Glycolysis in Intact Muscle. *Journal of Biological Chemistry*, Vol.270, No13, (March 1995), pp. 7311-7319, ISSN 0021-9258.

15

Fisheries and Biodiversity
in the Upper Gulf of California, Mexico

Gerardo Rodríguez-Quiroz[1], Eugenio Alberto Aragón-Noriega[2],
Miguel A. Cisneros-Mata[3] and Alfredo Ortega Rubio[4]
[1]*Centro Interdisciplinario de Investigaciones para
el Desarrollo Integral Regional, Unidad Sinaloa,*
[2]*Centro de Investigaciones Biológicas del Noroeste, Unidad Sonora,*
[3]*Instituto Nacional de Pesca, SAGARPA,*
Centro Regional de Investigaciones Pesqueras de Guaymas,
[4]*Centro de Investigaciones Biológicas del Noroeste, Unidad La Paz,*
México

1. Introduction

The Upper Gulf of California (UGC) has been recognized by its high primary productivity and abundant fishing (Aragon-Noriega & Calderon-Aguilera, 2000). Sediments and nutrients from the Colorado River, and complex hydrodynamics render this as an important site for spawning, mating and nursing for numerous species of commercial and ecological importance (Cudney & Turk, 1998; Ramirez-Rojo & Aragón-Noriega 2006). Temperature, salinity and abundance of nutrients in this region vary depending on fresh water runoff from the Colorado River (Alvarez-Borrego et al., 1975; Hernández-Ayón et al., 1993; Lavín & Sánchez, 1999).

Commercial fishing of high market value resources such as shrimp takes place in the UGC by artisanal or small scale, and industrial fishing. Artisanal fishing is done on relatively small (30 feet) fiber glass boats or artisanal boats with outboard motors, usually operated by two fishers; their primary fishing gear is drift gillnets, which they use to catch croakers, Spanish mackerel and even shrimp. This type of fishing is carried out by cooperatives and individual fishers from the three ports of the UGC: Puerto Peñasco and El Golfo de Santa Clara, in the State of Sonora, and San Felipe, in Baja California. Because marine resources in the region are migratory, fisheries are seasonal generating bursts of accumulated fishing effort over a few months depending on availability of species (see Cudney & Turk 1998). Increasing demand of economically important species has motivated a steady rise in fishing effort and use of gear and fishing practices jeopardizing critical species such as totoaba, Totoaba macdonaldi, an endemic croaker declared under risk of extinction (Cisneros-Mata et al., 1995), and the rare vaquita, Phocoena sinus. Vaquita are accidentally caught in all kinds of gillnets used in the Upper Gulf (D'Agrosa et al., 1995; Blanco 2002).

Vaquita is the world's smallest cetacean; it is endemic to the Upper Gulf of California and has the most restricted distribution range of all marine mammals (Jaramillo-Legorreta et al., 2007). In a situation of increased mortality in fishing activities, the reasons why vaquita is under risk of extinction are its historical small population size (Jaramillo-Legorreta et al., 2007) and possibly reduced habitat (Lavin et al., 1999) with decreased flow of Colorado river inflow (Hanski, 1998; Fagan et al., 2005). They occur only in the northern quarter of the Gulf of California, Mexico, mainly north of 30°45′ N and west of 114°20′ W with a highly productive core area of about 2,235 km², between San Felipe and Rocas Consag archipelago, a small upwelling spot where they have being seen feeding (Rojas-Bracho et al., 2006).

The high productivity of the upper-most portion of Gulf of California maintains a diverse number of marine species which interact with the vaquita. This porpoise is known to feed on grunt, Orthopristis reddingi, and ronco croaker, Bairdiella icistia, as well as different species of market squid (Vaquita Marina, 2007). Vaquita competes with dolphins and rays for several food items in the area; historically it was predated by sharks and killer whales (Barlow, 1986); at present shark predation and competition with ray species has lowered because of reduced abundance of these species possibly due to over fishing in the region (Rojas-Bracho et al., 2006).

The Upper Gulf of California and Colorado River Delta was declared a Biosphere Reserve (henceforth, Reserve) on June 10, 1993; it has an extension of 934,756 hectares including marine and terrestrial environments (Diario Oficial de la Federación [DOF], 1993; Fig. 1). The Reserve was implemented to protect species inhabiting that region, some of which are commercially important, endemic or under risk of extinction (Instituto Nacional de Ecología [INE], 1995; van Jaarsveld et al., 1998). A management program was designed to promote sustainable use of the biodiversity and landscape (SEMARNAT, 1995; Rojas-Bracho et al., 2006; Aragón-Noriega et al., 2010).

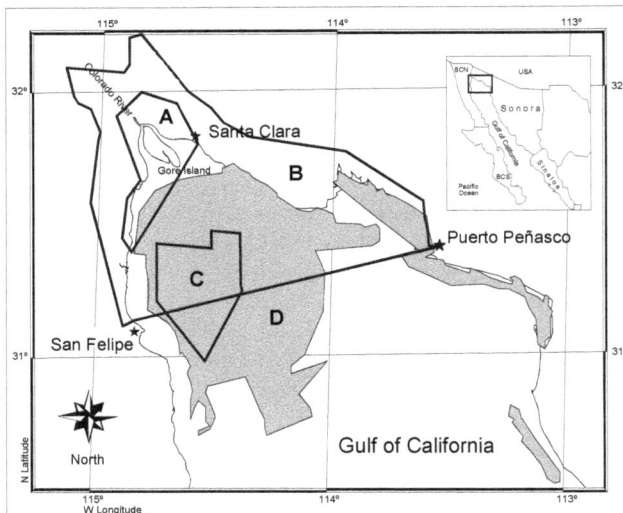

Fig. 1. The Upper Gulf of California. The thin line depicts the Biosphere Reserve declared in June 1993; the shaded area represents the vaquita refuge declared in December 2005.A) Core Zone, B) Buffer Zone in the Biosphere Reserve of the Upper Gulf of California, C) Vaquita Refuge Area, D) Shadowed are fishing grounds.

The Reserve hosts an important number of species with high commercial value. Such is the case of the gulf croaker, Cynoscion othonopterus, an endemic fish species that arrives massively into the UGC where it spawns during winter; and the blue shrimp, Litopenaeus stylirostris, which is highly priced in local and international markets (Rodriguez-Quiroz et al., 2010).

Implementation of Biosphere Reserve in June 1993 limits the use of gillnets (> 15cm mesh size) and fishing effort (up to 2,100 artisanal boats) to protect, most of all, totoaba and vaquita, considered under risk of extinction (INE, 1995; Greenberg, 2005). The most recent additional measure to protect vaquita and its habitat was a declaration in December 2005 of a Refuge to further limit fishing activities (Fig. 1). The Refuge, located in the western side of the UGC, comprises an area of 1,263.85 km^2 and is divided into two polygons: Polygon A – northern portion-, within the Reserve and with a surface of 897.09 km^2; and Polygon B – southern portion-, outside the Reserve and with a surface of 266.76 km^2. The Refuge was declared in the most likely distribution range of vaquita and includes a 65 km^2 zone where gillnets and trawl nets are prohibited (DOF, 2005).

Management of the Reserve and the Refuge imply a series of actions to achieve both protection of critical species and use of commercially important species. Consequently, fishing in the Upper Gulf becomes an economic activity with environmental implications. Conservation measures in the Reserve and the Refuge pose a challenge because they were designed to minimize negative impacts of fishing on vaquita. Several studies have been conducted in the area to determinate an average of accidental captures of vaquita in each fishing season since 1985; the most conclusive data was produced by D'Agrosa et al. (2000); who reported 39 vaquitas/year as by-catch before 1995. For those years less than 600 artisanal boats were in use, 1/3 of the boats registered in 2007 in the three fishing communities. Because of its critical condition it has been estimated that the maximum catch rate to avoid mid-term extinction is one vaquita per year (DOF, 2004). Mexican legislation recognizes that it is through participation of human communities affected by these measures that agreements can be achieved (see Palumbi et al., 2003). Therefore, solving this challenge will require a clear definition of common goals in fisheries management and conservation, expressed in a single policy (Davis, 2005).

Several management measures have been implemented both to protect vulnerable marine species and fishing resources. Amongst such measures we have: no-take zones (Mangel, 1998), subsidies (Munro & Sumaila 2002), buy-out of fishing gear and boats (Clark et al., 2005), fishing rights (Gonzalez-Laxe, 2006), and individual transferable quotas (ITQ) (Townsend et al., 2006). In the Gulf of California, some of these measures have been implemented to reduce fishing effort and protect soft bottom biological communities: buy-out of shrimp trawlers[1]; most recently, an ITQ program started aimed at rebuilding fishing stocks in the Gulf of California[2].

In this work we identify and analyze the most important artisanal fisheries of the Upper Gulf of California, which are in continuous interaction with the vaquita. We propose a scheme to reduce vaquita by-catch as a fishery management and biological conservation policy.

---

[1]http://www.conapesca.sagarpa.gob.mx/wb/cona/programa_de_retiro_voluntario_de_embarcaciones_cam
[2]http://www.conapesca.sagarpa.gob.mx/wb/cona/presentacion_del_ordenamiento_de_la_pesca_de_camar

2. Methods

Basic information used for the present analysis was generated in a series of studies conducted by World Wildlife Fund and Centro de Investigaciones Biológicas del Noroeste in the Upper Gulf of California during 2005. Additional information on fishing sites by species came from a previous report (Cudney & Turk, 1998). Artisanal fisheries data spanning since 1999 to 2007 were collected from official records in the ports of San Felipe, El Golfo de Santa Clara and Puerto Peñasco. Further information was gathered form a survey based on direct interviews to 146 artisanal fishers in those three fishing ports. Questionnaires were designed to compute direct cost structure during fishing operations, as well as fishing sites by species. A section of the survey was specifically designed to ascertain what types of activities alternative to artisanal fishing might be implemented in the Upper Gulf. Following Cochran (1989) we estimated sample size (n) for fishermen interviews:

$$n = \frac{\dfrac{Z^2 q}{E^2 p}}{1 + \dfrac{1}{N}\left[\dfrac{Z^2 q - 1}{E^2 p}\right]} \tag{1}$$

where: Z= CI=95%; p and q = Equation distribution; E= 6% Precision level; N= Fishermen community size. Following Greenberg (1993), local fishermen at each port were randomly selected.

From artisanal fishing landing records declared by fishers in local government fishery offices we obtained the following: Capture site, species, weight of landings and first-hand or "beach" economic value of landings by species.

Artisanal catch by species was processed and spatially represented in a geographic information system (GIS), identifying fishing sites within the Refuge (Fig. 1), overlapping the vaquita refuge polygon through the use of a ArcView 3.2 software using a 2002 Conica Lambert projection in maps by fishery and community in the vaquita polygon. Relative size (percentage) of the fishing activities zones in the vaquita refuge were obtained from the overall projected fishing sites.

To assess impact of fishery management measures oriented to protect the vaquita population, we arbitrarily assumed reductions in the number of artisanal boats for the three fishing communities. We establish two scenarios according to Gerrodette et al. (2011) who considered the number of vaquitas in 254 individuals. A deterministic model to describe vaquita population is defined as (Haddon, 2001):

$$P_t + 1 = P_t + rP_t(1 - \frac{P_t}{K}) - q_t f_t P_t \tag{2}$$

where for year t, P is vaquita population size, r is the per capita population growth rate, K is carrying capacity, q is catchability, and f is artisanal fishing effort in the Upper Gulf of California. Here we define catchability as the number of vaquitas incidentally caught per boat in artisanal fisheries in a given year and C = qfP is the total number of vaquitas caught in that year. Because of recent efforts to reduce vaquita incidental mortality during the first decade of 2000, we assume that these efforts have resulted in a proportional reduction in q starting in 2011. This model can thus be utilized to assess incidental mortality of vaquitas under different scenarios with respect to the number of active artisanal boats in a period of 15 years (2011-2025).

For simplicity reasons, demographic and environmental stochasticity and their impacts in the vaquita population dynamics are neglected in our model, although we are keenly aware of their potential effects. Due to chance events alone, demographic stochasticity at very low population numbers might drive populations of marine mammals such as vaquita to extinction (Burgman et al., 1993).

3. Results

3.1 Fisheries analysis

A total of 2,554 catch reports by artisanal fishers were compiled and analyzed for the three fishing communities of the Upper Gulf. Additionally, a total of 146 fishers were interviewed. Based on catch volume and beach economic value, six artisanal fisheries are the most important in the Upper Gulf: Shrimp Litopenaus stylirostris, curvina Cynoscion othonopterus, bigeye croaker Micropogonias megalops, Spanish mackerel Scomberomorus spp., rays (several species) and sharks (several species) (Table 1). Due to its high value, shrimp represents the largest gross income to artisanal fishers. The curvina is the second most economically important species for fishers of El Golfo de Santa Clara, bigeye croaker for fishers of San Felipe, and rays for those of Puerto Peñasco.

Species	Golfo de Santa Clara		Puerto Peñasco		San Felipe	
	Catch	Value	Catch	Value	Catch	Value
Shrimp	279	3'847,477	53	588,833	399	5'176,521
Curvina	1,552	806,283	43	23,843	677	376,272
Bigeye croaker	508	212,350	87	24,044	726	201,748
Spanish mackerel	888	870,345	58	54,029	95	76,769
Rays	106	91,677	121	89,713	244	180,988
Sharks	3	2,539	26	19,146	24	17,799

Table 1. Catch (Kg) and value ($US) of the main species in artisanal fisheries landed in the three ports of the Upper Gulf of California during 2007.

There are 2,100 small boats working in the Upper Gulf of California and, as discussed later in this work, fishers use different kinds of gillnets to fish for a variety of species. Table 2 shows the number of artisanal boats officially registered in each community. The number of artisanal boats is greater for El Golfo de Santa Clara, where artisanal fishing is virtually the only economic activity. Two types of fisheries concentrate the largest authorized fishing effort, shrimp with 606 artisanal boats and fishes (curvina, bigeye croaker, Spanish mackerel, sharks and rays), with 882 artisanal boats. The greatest number of authorized artisanal boats for both shrimp and fishes are registered in San Felipe and El Golfo de Santa Clara.

Species	San Felipe	El Golfo de Santa Clara	Puerto Peñasco
Clams	15	12	39
Jumbo squid			4
Shrimp	318	232	56
Snails	1		42
Fishes *	295	412	175
Swimming crab	11	39	229
Mullet	10	76	8
Octopus	2		40
Sharks	10	26	69
Total	662	797	662

Table 2. Authorized artisanal fishing vessels by group of species in the three ports of the Upper Gulf of California. Source: Federal government offices in the communities of the Upper Gulf of California. *Curvina, bigeye croaker, Spanish mackerel, rays.

Our survey data and GIS analysis showed that fishing is conducted within the Vaquita Refuge Area and in the Biosphere Reserve. Approximately 62% of the total catch in the Upper Gulf of California was caught in the marine protected areas. Approximately 77% of the marine area of the Biosphere Reserve and the entire Vaquita Refuge Area are used for fishing (Fig. 2). In the Vaquita Refuge, 97% of the total area is fished for shrimp, 94% is fished for corvine, 85% is fished for shark, 79% is fished for bigeye croaker and 69% is fished for Spanish mackerel. In the Biosphere Reserve, 56% of the total area is fished for shrimp, 55% is fished for corvine, 44% is fished for bigeye croaker, 39% is fished for shark and 30% is fished for Spanish mackerel.

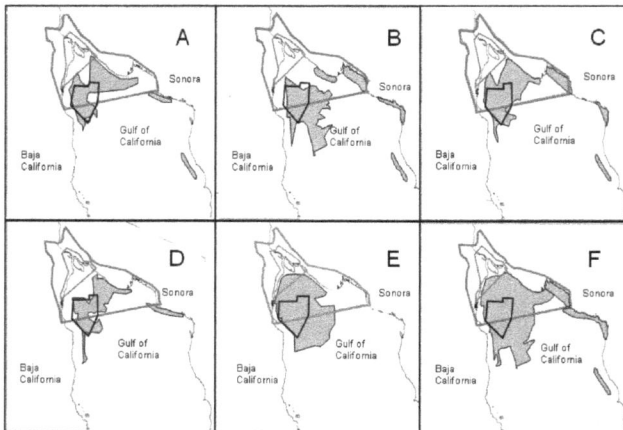

Fig. 2. Spatial distribution of artisanal fisheries as compared with the vaquita refuge declared in the Upper Gulf of California. A) bigeye croaker, B)sharks and rays , C) shrimp, D) Spanish mackerel, E) curvina, F) all fisheries.

Fishermen from Puerto Peñasco fish close to the Sonoran shoreline. 75% of the capture occurs inside the Biosphere Reserve and the fishermen fish in 20% of the northern area of the Vaquita Refuge. Fishermen from Golfo de Santa Clara carry out their fishing inside the marine protected areas and they fish in about half of the Vaquita Refuge Area. San Felipe fishermen fish near the Baja California shoreline in the UGC from the core zone to Puertecitos, which covers the entire Vaquita Refuge Area and 70% of the Biosphere Reserve (Fig. 3).

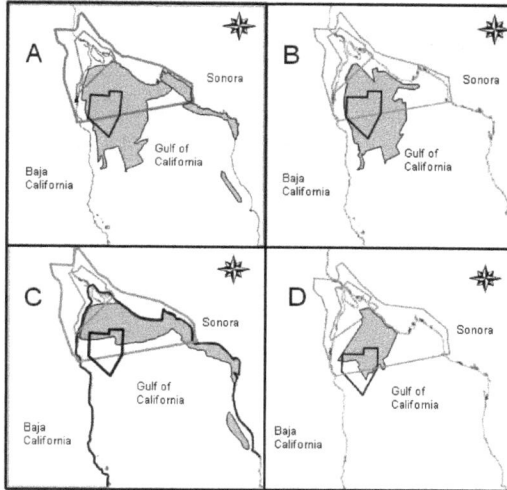

Fig. 3. Spatial distribution of the artisanal fisheries in the UGC by community. A) all communities; B) San Felipe; B) Puerto Peñasco; C) El Golfo de Santa Clara.

Within the Refuge, curvina represents the greatest annual catch with ~3,000 mt, followed by bigeye croaker with ~1,240 mt; other species amount to 2,192 mt (Table 3). In terms of economic value, shrimp represents 80% of ca. $US 1.7 million total gross incomes in the marine protected areas; sharks and rays represent the lowest gross income with only ~$US 385,000.

Indicator	Shrimp	Bigeye croaker	Curvina	Rays	Spanish mackerel	Sharks	Total
Catch (metric tons)	459	938	2,957	429	1,239	65	6,087
Value of catch *	4,791	528	2,765	357	1,812	66	10,319
Costs of catch *	3,073	165	489	29	131	9	3,897
Gross profit *	1,718	363	2,276	328	1,681	57	6,423
Return rate (%)	36	69	82	92	93	86	62

Table 3. Catch, first-hand value and operation costs by group of species in the artisanal fisheries of the Upper Gulf of California in 2007. Source: Local fishery offices in the communities. *US thousands of dollars.

Total annual catch in the marine protected areas from 1999 to 2007 were ~5,506 mt with a first-hand economic value of $US 8´563,000. The operation costs spent to obtain that total catch were $US 2´666,000 for a total gross income of $US 5´897,000, or a mean return rate of 68%. During that period we registered an increased fishing effort; gross profits provided high incomes. Therefore, estimated opportunity costs for artisanal fishers giving up their activities in the vaquita Refuge amount to ca. $US 1.7 million per year (Table 4).

	Catch	Value of catch*	Costs of catch*	Gross profit*	Return rate (%)
1995	2,510	3,444	569	2,874	83
1996	2,354	2,323	1,185	1,138	49
1997	5,466	4,327	1,640	2,688	62
1998	6,450	9,451	1,945	7,505	79
1999	7,536	9,625	2,390	7,234	75
2000	6,786	9,907	2,726	7,181	72
2001	6,050	9,181	2,994	6,186	67
2002	7,492	12,882	3,131	9,750	76
2003	5,029	8,755	3,321	5,435	62
2004	5,407	8,082	3,493	4,589	57
2005	5,888	12,660	3,609	9,051	71
2006	4,525	10,356	3,755	6,601	64
2007	6,087	10,319	3,897	6,423	62
Average	5,506	8,563	2,666	5,897	68

Table 4. Catch (metric tons), first-hand value and operation cost by group of species in the artisanal fisheries of the Upper Gulf of California inside the vaquita refuge and the Biosphere Reserve from 1995 to 2007. Source: Local fishery offices in the communities. *US thousands of dollars.

3.2 Social analysis

In our study, opinions of fishers can be interpreted as guidelines of a comprehensive strategy to achieve the purposes of the Reserve (Table 5). When asked what their activity would be if the most important fishery to them were closed, 56% of fishers responded that they would continue to fish anyway: 22% claim that they would fish on the same, and 34% other species. These responses were mostly responded by El Golfo de Santa Clara fishermen, who do not have enough employment alternatives as compared to Puerto Peñasco and San Felipe fishermen. A total of 23.8% expected an economic aid and 19.6% would ask for something else such as a credit for a new business or local employment (plumber, carpenter, construction, etc).

Option	Percentage	Frequency
Permit for another fishery	34.3	49
Economic compensation	16.1	23
Payment of permit cost	7.7	11
Continue fishing the same	22.4	32
Other	14.7	21
Nothing	4.9	7

Table 5. Response of fishers of the Upper Gulf of California to the question: If the most important fishery to you was closed, what would you ask in return?

When we posed the question: If you were asked to stop fishing, what would you do? A large number of them responded that they would switch to the tourism and trade sector (49%), 6% would like to work in aquaculture and maquilas, 25.2% in another fishery (clams, oysters, etc.) or the same, and the remaining 20.1% would seek employment in domestic duties (Table 6).

Option	Percentage	Frequency
Tourism	24.5	34
Trade	24.5	34
Work in a private sector	5.8	8
Other activity in fisheries	7.2	10
Would not stop fishing	18.0	25
Other	20.1	28

Table 6. Response of fishers of the Upper Gulf of California to the question: If you were asked to stop fishing, what would you do?

3.3 Vaquita recovery analysis

Our model was fitted to reported mean annual vaquita abundance, effort (number of artisanal boats), and incidental mortality (Fig. 4). For the four years where there is information on vaquita population size (P) and incidental catch (C), we estimated $q = C/(fN)$ and computed a weighted q (= 0.00011374) which was used for other years in the calculations. Using a fixed value for r (= 0.09531; Barlow, 1986) we fitted our model using least squares as criterion and found K = 4,640 vaquitas.

Our scenarios showed that it could take a large period of time for the vaquita population to recover to its 2010 size In scenario 1 (Fig. 4a) and according to response from fishermen (cf. table 6), starting in year 2011 we reduced the number of boats to 506, which is the quantity that would continue fishing and maintained constantly through year 2025. We observed that the number of vaquitas could recover in a relatively short time and could continue slow growth to over 349 individuals in 2025. In scenario 2 (Fig. 4b) we considered a 15% per year reduction of the artisanal fleet through year 2018 when the fleet reaches 506 boats and maintained it constant until year 2025. The model predicted a fast decrease in the vaquita population numbers until year 2015, stabilized thereafter and even showed a recovery of 6.3% per year through 2025 when the vaquita population reaches 242 individuals.

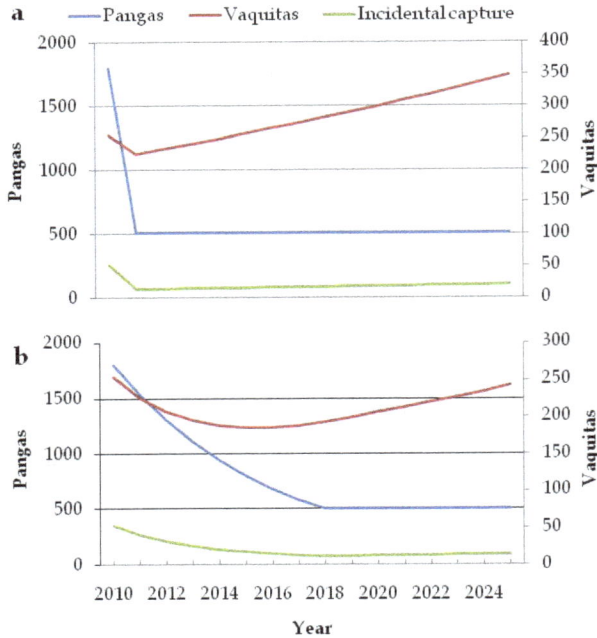

Fig. 4. Scenarios for vaquita recovery if artisanal boats are reduced in the Upper Gulf of California: a) 75% in year 2011; b) 15% yearly until 2018 and then maintained constant through year 2025.

Again, we caution that our model neglected consideration of potentially important ecological and biological aspects such as demographic and environmental stochasticity at low population numbers, as well as environmental forcing (Burgman et al., 1993). It has been well established that in theory an age-structured population will experience a sharp increase in extinction risk, even if growing geometrically, when subjected to stochastic shocks in its vital rates (Burgman et al., 1993; Caswell, 1989).

4. Discussion

Our study showed that both the Upper Gulf of California Biosphere Reserve and the recently declared vaquita Refuge are important grounds for artisanal fishing. Catch of shrimp, curvina and bigeye croaker have the largest distribution range in the Reserve and Polygon B of the Refuge. Shrimp generates the most important income for artisanal fishers. Our survey data indicates that 98% of artisanal fishers of El Golfo de Santa Clara and 100% of San Felipe fish on shrimp due to its high commercial value, gross revenues, and availability during the fishing season (September to January). This result is an important challenge to the fulfillment of goals of the Reserve and the vaquita Refuge, moreover because the number of registered artisanal boats is higher than recommended when the refuge was declared (DOF, 2005).

Operation costs determine to a great extent where fishing is conducted in the Upper Gulf of California and they depend mostly upon the distance of fishing sites to the ports and seasonal distribution of natural resources. San Felipe is the fishing port nearest to the recently declared Refuge; fishers residing in this port work in that vicinity throughout the year. Although El Golfo de Santa Clara holds the greatest number of registered permits and artisanal boats, fishers from this port do not fish near the Refuge because of high operation costs related to travel distance. Fishers of Puerto Peñasco fish near the sonoran coast to reduce operation coasts.

The high number of artisanal boats working in the Upper Gulf represents a clear threat to the vaquita (Rojas and Jaramillo 2001, Blanco 2002). The bulk of artisanal fisheries are done using gillnets to catch curvina (100%), shrimp (93%), Spanish mackerel (68%), bigeye croaker and rays (44%), and sharks (10%) (Vidal et al., 1994; D'Agrosa et al., 1995; Vidal, 1995). Gillnet mesh size varies from 5.7cm to 17.8cm; the highest vaquita mortality has been registered in gillnets with 11.43cm mesh size (Ortiz, 2002). These nets cover a great proportion of the water column where they are set and left for various hours (D'Agrosa et al., 1995). Length of these nets varies from 99 to 1,485 meters, the most common measuring between 594 and 990 meters with a mean height of 5.4 to 18 meters (Walsh et al., 2004). Because of the mesh size used, curvina and bigeye croaker fishing represent the biggest fishery-induced potential impact to the vaquita population (D'Agrosa et al., 2000; Rojas-Bracho et al., 2006).

To succeed, a strategy to protect vaquita from artisanal fishing should consider social aspects such as attachment of fishers to their activity. It is clear that 25% of the fishermen would not stop fishing because that is the only activity they feel comfortable with and have done for years. In that context, some fisheries must be assessed considering the species value and impact to the environment, and it would regulate fishing enforcement allowing a specific number of fishermen and fishing tools by species to fish within the Reserve and the vaquita Refuge.

The capture in the Biosphere Reserve and the Vaquita Refuge maintains a steady level of production with important economic incentives, which make it attractive to fishermen despite recent restrictions on their activity. However, the continued recruitment of new fishermen to the area will not enhance the welfare of the existing fishermen and there is no guarantee that the fishery will be sustained over the coming years (Ponce et al. 2006).

Our model results indicate that the vaquita population has never been too large (K < 5,000 individuals) and thus the importance of management actions to account for reductions of incidental kill in fishing activities, the only proven source of anthropogenic mortality. According to our scenarios, a fisheries management and vaquita conservation strategy can consider the reduction of the artisanal fleet by 15% every year until it reaches 506 boats in year 2018. Without compromising the vaquita population and the fishers in the region, this analysis could serve as a reference point to the buy-out program implemented since 2007, Parallel to this yearly reduction of the fleet, an integral strategy should incorporate a program to compensate and promote fishermen phase-out, including investments in equipment (sport fishing vessels with sustainable gears and refrigerated vehicles), infrastructure (storages, docks, freezers, added value of fishing products) and training in new activities. To support this program and according to table 4, ~$US 6'423,000 should be invested in the initial years to finance fishing opportunity cost to prevent a massive return of fishermen to the activity in cases where the buy-out program fails in the short term. Jobs must be accompanied by education and long-term provision for new qualifications, because many of the fishermen do not have technical business skills or experience in tourism administration or other activities. Also needed is a periodic evaluation of criteria set out in all implemented actions to increase chances of the vaquita recovery (Aragón-Noriega et al., 2010).

Through implementation since 2007 of an integral vaquita recovery program 242 artisanal boats and 340 fishing permits have been bought-out; in addition, 190 permits have been converted to alternative fishing gears. Also, a shrimp farm was rebuilt and in agreement with artisanal fishers whom will exchange their gear and permit, this could represent an additional 150 to 180 fishing boats phased-out of the region (Rojas-Bracho et al. 2010).

5. Conclusion

Our contribution is significant because we now have more information about the habitat utilized by commercially important species. We have also collected and analyzed information that can be used to elucidate which kinds of fisheries represent more risk to endangered species. Not all fisheries are necessarily a threat to biodiversity. A better understanding of the fisheries can help determine which fisheries are the most important to consider when developing a conservation strategy (Rodriguez-Quiroz et al., 2010).

Conservation success of vaquita must be based on agreements which dignify inhabitants of the Upper Gulf. Governments of all levels and conservation organizations should promote development of the region. We must strive to improve the quality of life of fishers while recovering the endangered vaquita considering socio-economic, ecological and institutional factors. Our calculations can serve as basis of a gradual compensation scheme to reduce artisanal fishing in the vaquita refuge through a buy-out scheme in the long run.

Success of most fisheries management policies to conserve species is contingent upon vulnerability of the species, size of the protected area and viable, equitable economic alternatives to fishers. Contrary to this view and given the critical situation of the vaquita, clearly enforcement of the Refuge as a no-take zone by itself will not suffice to save vaquita from extinction.

Continuation of the recently implemented integral program that includes measures for sustainable fishing, economic alternatives, and buy-out of artisanal fishing units must be guaranteed. A constant monitoring of the program must be put in place in an adaptive manner so as to ensure efficiency of interventions.

6. Acknowledgments

We thank Javier de la Cruz for aid in field work and analyzing profit information. Helen Regan and Mauricio Ramirez provided helpful comments to an earlier version of this work. Mary López processed the catch information and produced the GIS figures. WWF-México provided partial funding for field work. GRQ thanks CONACYT (112401) and COTEPABE-IPN (347) for a scholarship during his doctoral studies. EAAN thanks CONACYT Grant-48445.

7. References

Álvarez Borrego, S.; Flores Báez, B.P. & Galindo Bect. L. (1975). Hidrología del Alto Golfo de California II. Condiciones durante invierno, primavera y verano. *Ciencias Marinas*, Vol. 2, pp. 21-36

Aragón-Noriega, E.A. & Calderon-Aguilera, L.E. (2000). Does damming the Colorado River affect the nursery area of blue shrimp Litopenaeus stylirostris (Decapoda:Penaeidae) in the Upper Gulf of California?. *International Journal of Tropical Biology and Conservation*, Vol. 48, pp. 867-871

Aragón-Noriega, E.A.; Rodríguez-Quiroz, G.; Cisneros-Mata, M.A. & Ortega-Rubio, A. (2010). Managing a protected marine area for the conservation of critically endangered Vaquita Phocoena sinus (Norris, 1958) in the Upper Gulf of California. *International Journal of Sustainable Development & World Ecology*, Vol. 17, No. 5, pp. 410-416

Barlow, J. (1986). *Factors affecting the recovery of Phocoena sinus, the Vaquita or Gulf of California harbor porpoise*. U.S. National Marine Fisheries Service. Admistrative Report No. 86-37, December 1986, pp. 19

Blanco, M.L. (2002). Pobreza y explotación de los recursos pesqueros en el Alto Golfo de California, In: *Manejo de Recursos Pesqueros, Reunión Temática Nacional*, R.E. Morán Angulo, M.T. Bravo, S. Santos Guzmán & J.R. Ramírez Zavala, (Ed.), 318-338, Editorial Universidad Autónoma de Sinaloa. Culiacán, México.

Burgman, M.A.; Ferson S. & Akçakaya, H.R. (1993). *Risk assessment in conservation biology*. Chapman & Hall, N.Y.

Caswell, H. (1989.) Matrix Population Models. Sinauer Associates, Inc. Sunderland, Massachusetts. 328 pp.

Clark, C.W.; Munro, G.R. & Sumaila, U.R. (2005). Subsidies, buybacks, and sustainable fisheries. *Journal of Environmental Economics and Management*, Vol. 50, 47-58

Cisneros-Mata, M.A.; Montemayor-López, G. & Román-Rodríguez, M.J. (1995). Life history and conservation of Totoaba macdonaldi. *Conservation Biology*, Vol. 9, pp. 806-814

Cochran, GW. (1989). *Sampling Techniques*, New York, Willey and Sons

Cudney, R. & Turk, P.J. (1998). Pescando entre mareas del Alto Golfo de California. *Centro intercultural de Estudios del Desiertos y Océanos*. Puerto Peñasco, Sonora, Mexico. 166 pp.

D'agrosa, C.; Vidal O. & Gram, W.C. (1995). Mortality of the vaquita Phocoena sinus in gillnet fisheries during 1993-1994. *Reports of the International Whaling Commission* (special issue), Vol. 16, pp. 283-291

D'agrosa, C.; Lennert-Cody, C.E. & Vidal, O. (2000). Vaquita bycatch in Mexico's artisanal gillnet fisheries: driving a small population to extinction. *Conservation Biology*, Vol. 14, pp. 1110-1119

Davis, G.E. (2005). Science and society: marine reserve design for the California Channel Islands. *Conservation Biology*, Vol. 19, pp. 1745-1751

Diario Oficial de la Federación. (1993). *Decreto por el que se declara área natural protegida con el carácter de Reserva de la Biosfera, la región conocida como Alto Golfo de California y Delta del Río Colorado*. Diario Oficial de la Federación, junio de 1993.

Álvarez Borrego, S.; Flores Báez, B.P. & Galindo Bect, L. (2004). *Acuerdo mediante el cual se aprueba la actualización de la Carta Nacional Pesquera y su anexo*. Diario Oficial de la Federación, Marzo del 2004.

Aragón-Noriega, E.A. & Calderon-Aguilera, L.E. (2005). *Programa de protección de la vaquita dentro de área de Refugio ubicada en la porción occidental del Alto Golfo de California*. Diario Oficial de la Federación, Septiembre del 2005.

Fagan, W.F.; Kennedy, C.M. & Unmank, P.J. (2005). Quantifying rarity, losses and risks for native fishes of the lower Colorado River Basin: Implications for conservation listing. *Conservation Biology*, Vol. 19, pp. 1872-1882

Gerrodette, T.; Taylor, B.L.; Swift, R.; Rankin, S.; Jaramillo-Legorreta, A.M. & Rojas-Bracho, L. (2011). A combined visual and acoustic estimate of 2008 abundance, and change in abundance since 1997, for the vaquita, Phocoena sinus. *Marine Mammal Sceince*, Vol. 27, pp. 79-100

Gonzalez-Laxe, F. (2006). Transferability of fishing rights: The Spanish case. *Marine Policy*, Vol. 30, pp. 379-388

Greenberg, J.B. (1993). Local preferences for develop, In: *Marine community and Biosphere Reserve: crises and response in the Upper Gulf of California*, TR McGuire & JB Greenberg, (Ed.), p. 168. Occasional paper number 2. BARA: University of Arizona

Greenberg, J.B. (2005). Neoliberal reforms and the political ecology of fishing in the Upper Gulf of California, In: Las Dimensiones humanas en el estudio y conservación del Golfo de California, G.D. Danemann, (Ed.), 9-18, Pronatura Noroeste, Ensenada, Baja California, México

Haddon, M. (2001). *Modelling and quantitative methods in fisheries*. Chapman and Hall/CRC

Hanski, I. (1998). Metapopulation dynamics. *Nature*, Vol. 396, pp. 41-49

Hernández-Ayón, J.B.; Galindo-Bect, M.S.; Flores-Báez, B.P. & Álvarez-Borrego, S. (1993). Nutrient concentrations are high in the turbid waters of the Colorado River Delta. *Estuarine, Coastal and shelf Science*, Vol. 37, pp. 593-602

Instituto Nacional de Ecología. (1995). *Programa de Manejo: Áreas Naturales Protegidas, Reserva de la Biosfera Alto Golfo de California y Delta del Río Colorado*. pp. 94. México, D.F.

Jaramillo-Legorreta, A.M.; Rojas-Bracho, L.; Brownell, R.L.; Read, A.J.; Reeves, R.R.; Ralls, K. & Taylor, B.L. (2007). Saving the vaquita: Immediate action, not more data. *Conservation Biology*, Vol. 21, pp. 1653-1655

Lavin, M.F. & Sanchez, S. (1999). On how the Colorado River affected the hydrography of the Upper Gulf of California. *Continental Shelf Research*, Vol. 19, pp. 1545-1560

Mangel, M. (1998). No-take areas for sustainability of harvested species and a conservation invariant for marine reserves. *Ecology Letters*, Vol. 1, pp. 87-90

Munro, G. & Sumaila, R. 2002. The impact of subsidies upon fisheries management and sustainability: the case of the North Atlantic. *Fish and Fisheries*, Vol. 3, pp. 233-250

Ortiz, I. (2002). *Impacts of fishing and habitat alteration on the population dynamics of the vaquita Phocoena sinus*. Master Thesis. School of Aquatic and fishery Sciences. University of Washington, USA.

Palumbi, S.R.; Gaines, S.D.; Leslie, H. & Warner, R.R. (2003). New wave: high-tech tools to help marine reserve research. *Frontiers in ecology and the environment*, Vol. 1, pp. 73-79

Ponce, D.G.; Arreguín, F & Beltrán, L.F. (2006). Indicadores de sustentabilidad y pesca: casos en Baja California Sur, México, In: *Desarrollo sustentable: ¿Mito o realidad?*, M.L.F. Beltrán, J Urciaga & A Ortega, (Eds.), 183-221. Centro de Investigaciones Biológicas del Noroeste, S.C, México

Ramirez-Rojo, R.A. & Aragon-Noriega, E.A. (2006). Postlarval ecology of the blue shrimp Litopenaeus stylirostris and brown shrimp Farfantepenaeus californiensis in the Colorado River Estuary. *Ciencias Marinas*, Vol. 32, pp. 45-52

Rodríguez Quiroz, G.; Aragón Noriega, E.A.; Valenzuela Quinóñez, W. & Esparza Leal, H.M. (2010). Artisanal fisheries in the conservation zones of the Upper Gulf of California. *Revista de Biología Marina y Oceanografía*. Vol. 45, No. 1, pp. 89-98

Rojas-Bracho, L. & Jaramillo-Legorreta, A. (2001). Vaquita Marina, In: *Sustentabilidad y pesca responsable en México*. Instituto Nacional de la Pesca, (Ed.), 963-981 Secretaria de Agricultura, Ganadería, Desarrollo Rural, Pesca y Alimentación, México

Rojas-Bracho, L.; Reeves R.R. & Jaramillo-Legorreta, A. 2006. Conservation of the vaquita Phocoena sinus. *Mammal Rev*. Vol. 36, pp. 179-216

Rojas-Bracho, L.; Jaramillo-Legorreta, A.M.; Taylor, B.; Barlow, J.; Gerrodette, T.; Tregenza, N.; Swift, R. & Akamatsu, T. (2010). *Assessing Trends in Abundance for Vaquita using Acoustic Monitoring: Within Refuge Plan and Outside Refuge Research Needs*. Report of Vaquita Expedition 2008 and Current Conservation Actions. Paper *SC/62/SM5* presented to the IWC Scientific Committee, 11 pp.

SEMARNAT. (1995). *Programa de manejo*. Áreas Naturales Protegidas 1. Reserva de la Biosfera del Alto Golfo de California y Delta del Río Colorado. Diciembre. SEMARNAT/CONANP.

Tognelli, M. F.; Silva-Garcia, C.; Labra, F.A. & Marquet, P. A. (2005). Priority areas for the conservation of coastal marine vertebrates in Chile. *Biological Conservation*, Vol. 126, pp. 420-428

Townsend, R.E.; Mccoll, J. & Young, M.D. (2006). Design principles for individual transferable quotas. *Marine Policy*, Vol. 30, pp, 131-141

Van Jaarsveld, A.S.; Freitag, S.; Chown, S. L.; Muller, C.; Koch, S.; Hull, H.; Bellamy, C.; Kruger, M.; Endrody-Younga, S.; Mansell, M.W. & Scholtz, C. H. (1998). Biodiversity assessment and conservation strategies. *Science*, Vol. 279, pp. 2106-2108

Vaquita Marina. (2007). *Todo sobre la vaquita marina*. 17 Jun 2007, Available from: http://www.vaquitamarina.org/todo_sobre.php.

Vidal, O.; Van Waerebeek, K. & Findley, L.T. (1994). Cetaceans and gillnets fisheries in Mexico, Central America and the wider Caribbean: a preliminary Review. *Rep. Int. Whal. Comm* (special issue), Vol. 15, pp. 221-233

Vidal, O. (1995). Population biology and exploitation of the vaquita Phocoena sinus. *Rep. Int. Whal. Comm* (special issue), Vol. 16, pp. 247-272

Walsh, P.; Grant, S.M.; Winger, P.D. & Blackwood, G. (2004). An investigation of alternative harvesting methods to reduce the by-catch of vaquita porpoise in the Upper Gulf of California shrimp gillnet fishery. *Centre for Sustainable Aquatic Resources,* Fisheries and marine institute of Memorial University of Newfoundland, Canada. 32p.

Permissions

The contributors of this book come from diverse backgrounds, making this book a truly international effort. This book will bring forth new frontiers with its revolutionizing research information and detailed analysis of the nascent developments around the world.

We would like to thank Prof. Marco Marcelli, for lending his expertise to make the book truly unique. He has played a crucial role in the development of this book. Without his invaluable contribution this book wouldn't have been possible. He has made vital efforts to compile up to date information on the varied aspects of this subject to make this book a valuable addition to the collection of many professionals and students.

This book was conceptualized with the vision of imparting up-to-date information and advanced data in this field. To ensure the same, a matchless editorial board was set up. Every individual on the board went through rigorous rounds of assessment to prove their worth. After which they invested a large part of their time researching and compiling the most relevant data for our readers. Conferences and sessions were held from time to time between the editorial board and the contributing authors to present the data in the most comprehensible form. The editorial team has worked tirelessly to provide valuable and valid information to help people across the globe.

Every chapter published in this book has been scrutinized by our experts. Their significance has been extensively debated. The topics covered herein carry significant findings which will fuel the growth of the discipline. They may even be implemented as practical applications or may be referred to as a beginning point for another development. Chapters in this book were first published by InTech; hereby published with permission under the Creative Commons Attribution License or equivalent.

The editorial board has been involved in producing this book since its inception. They have spent rigorous hours researching and exploring the diverse topics which have resulted in the successful publishing of this book. They have passed on their knowledge of decades through this book. To expedite this challenging task, the publisher supported the team at every step. A small team of assistant editors was also appointed to further simplify the editing procedure and attain best results for the readers.

Our editorial team has been hand-picked from every corner of the world. Their multi-ethnicity adds dynamic inputs to the discussions which result in innovative outcomes. These outcomes are then further discussed with the researchers and contributors who give their valuable feedback and opinion regarding the same. The feedback is then collaborated with the researches and they are edited in a comprehensive manner to aid the understanding of the subject.

Apart from the editorial board, the designing team has also invested a significant amount of their time in understanding the subject and creating the most relevant covers. They scrutinized every image to scout for the most suitable representation of the subject and create an appropriate cover for the book.

The publishing team has been involved in this book since its early stages. They were actively engaged in every process, be it collecting the data, connecting with the contributors or procuring relevant information. The team has been an ardent support to the editorial, designing and production team. Their endless efforts to recruit the best for this project, has resulted in the accomplishment of this book. They are a veteran in the field of academics and their pool of knowledge is as vast as their experience in printing. Their expertise and guidance has proved useful at every step. Their uncompromising quality standards have made this book an exceptional effort. Their encouragement from time to time has been an inspiration for everyone.

The publisher and the editorial board hope that this book will prove to be a valuable piece of knowledge for researchers, students, practitioners and scholars across the globe.

List of Contributors

Marco Marcelli, Andrea Pannocchi, Viviana Piermattei and Umberto Mainardi
University of Tuscia, Laboratory of Exerimental Oceanology and Marine Ecology, Italy

Zdzisław Prokowski and Lilla Mielnik
West Pomeranian University of Technology in Szczecin, Poland

I. Puillat, N. Lanteri, J.F. Drogou, J. Blandin, L. Géli, J. Sarrazin, P.M. Sarradin, Y. Auffret, J.F. Rolin and P. Léon
IFREMER Centre de Brest, REM Department, BP 70, Plouzané, France

Sun Ling, Hu Xiuqing and Li Sanmei
National Satellite Meteorological Centre, China

Guo Maohua
National Satellite Ocean Application Service, China

Zhu Jianhua
National Ocean Technology Centre, China

Ding Lei
Shanghai Institute of Technical Physics, China

Haibin Song and Xinghui Huang
Key Laboratory of Petroleum Resources Research, Institute of Geology and Geophysics, Chinese Academy of Sciences, Beijing, China

Luis M. Pinheiro
Departamento de Geociências and CESAM, Universidade de Aveiro, Aveiro, Portugal

Barry Ruddick
Department of Oceanography, Dalhousie University, Halifax, Nova Scotia, Canada

Roman Cieśliński and Jan Drwal
University of Gdańsk, Institute of Geography, Department of Hydrology, Gdańsk, Poland

Tsukasa Hokimoto
Graduate School of Mathematical Sciences, The University of Tokyo, Japan

Zhenhua Xu and Baoshu Yin
Institute of Oceanology, Chinese Academy of Sciences, Key Laboratory of Ocean Circulation and Waves (KLOCAW), Chinese Academy of Sciences, China

Xiangming Tang, Dan Chen, Keqiang Shao and Guang Gao
State Key Laboratory of Lake Science and Environment, Nanjing Institute of Geography and Limnology, Chinese Academy of Sciences, P.R. China

Jianying Chao
Nanjing Institute of Environmental Science, Ministry of Environmental Protection, P.R. China

Warren J. de Bruyn, Catherine D. Clark and Lauren Pagel
School of Earth and Environmental Sciences, Schmid College of Science and Technology, Chapman University, Orange, California, USA

Ewa Sobecka
West Pomeranian University of Technology, Szczecin, Poland

Nicolas Le Corre and Ladd E. Johnson
Université Laval, Québec, Canada

Frédéric Guichard
McGill University, Québec, Canada

Bożena Graca, Katarzyna Łukawska-Matuszewska, Dorota Burska and Jerzy Bolałek
Institute of Oceanography, University of Gdańsk, Gdynia, Poland

Leszek Łęczyńsk
Maritime Institute in Gdańsk, Gdańsk, Poland

Graciela S. Diniz, Elisabete Barbarino and Sergio O. Lourenço
Universidade Federal Fluminense, Departamento de Biologia Marinha, Niterói, RJ, Brazil

Gerardo Rodríguez-Quiroz
Centro Interdisciplinario de Investigaciones para el Desarrollo Integral Regional, Unidad Sinaloa, México

Eugenio Alberto Aragón-Noriega
Centro de Investigaciones Biológicas del Noroeste, Unidad Sonora, México

Miguel A. Cisneros-Mata
Instituto Nacional de Pesca, SAGARPA, Centro Regional de Investigaciones Pesqueras de Guaymas, México

Alfredo Ortega Rubio
Centro de Investigaciones Biológicas del Noroeste, Unidad La Paz, México